A SECONDARY BIOLOGY SERIES

HUMAN BIOLOGY

BY

A. E. VINES, B.Sc. (Hons.)

Principal Lecturer in Biology
and Head of Science Dept.
College of S. Mark and S. John

AND

N. REES, B.Sc. (Hons.)

Formerly Senior Lecturer in Biology
College of S. Mark and S. John

Pitman Publ

First published 1966
Reprinted 1969
Second edition 1973

SIR ISAAC PITMAN AND SONS LTD
Pitman House, Parker Street, Kingsway, London, WC2B 5PB
P.O. Box 46038, Portal Street, Nairobi, Kenya

SIR ISAAC PITMAN (AUST.) PTY LTD
Pitman House, 158 Bouverie Street, Carlton, Victoria 3053, Australia

PITMAN PUBLISHING COMPANY S.A. LTD
P.O. Box 11231, Johannesburg, South Africa

PITMAN PUBLISHING CORPORATION
6 East 43rd Street, New York, N.Y. 10017, U.S.A.

SIR ISAAC PITMAN (CANADA) LTD
495 Wellington Street West, Toronto 135, Canada

THE COPP CLARK PUBLISHING COMPANY
517 Wellington Street West, Toronto 135, Canada

Cased edition ISBN: 0 273 25269 0
Paperback edition ISBN: 0 273 00130 2

Text set in 12/13 pt. Monotype Bembo, printed by letterpress, and bound in Great Britain at The Pitman Press, Bath
T.1270/1378: 77

PREFACE TO THE SECOND EDITION

SOME new material has been added to this edition; a chapter on Hygiene and Public Health, and a shorter introduction to three important topics of the moment—Population, Pollution and Conservation. Increasing recognition of the importance of the Hypothalamus has necessitated rewriting of the section on Endocrine Glands.

This country participated in the agreement to adopt an International System of Units (SI) at the Conférence Générale des Poids et Mésures in 1960. These units are already in use in some degree, and they have now been used throughout the book.

We are pleased that "Human Biology" has been favourably received and we thank those teachers and students who have offered suggestions for its improvement. In most cases, those suggestions have been adopted.

PREFACE TO THE FIRST EDITION

HUMAN biology has more intrinsic interest for the majority of people than has the study of the lower plants and animals. Nevertheless, the fact that man is intricately interrelated with, and dependent upon other living things, needs to be continually stressed. Accordingly, this book describes, in the early chapters, man's position in the living world and his probable origins. Then follows a simple treatment of human anatomy and physiology and an elementary discussion of human genetics. Throughout the text, there is constant reference to the principles of hygiene and the prevention of disease.

The book deals adequately with the various examination requirements in Human Biology at "O" level. It is suitable also as a general course for pupils in Secondary Schools, and for those preparing for a career in nursing.

Most of the text and exercises have been used in the classroom by the authors and by student teachers in training.

CONTENTS

CHAP. PAGE

Preface to the Second Edition iii

Preface to the First Edition iii

1. Living Creatures 1
2. Man's Place in the Animal Kingdom 14
3. The Races of Men 31
4. General Features: Anatomy and Skin 39
5. The Skeleton 56
6. The Muscular System 79
7. Food and Diet 97
8. The Alimentary Canal, Digestion and Utilization of Food . 127
9. The Respiratory System 157
10. The Blood Vascular System 178
11. Excretion 211
12. The Reproductive System 219
13. Development 233
14. The Nervous System 250
15. The Endocrine System 289
16. Heredity 304
17. Human Disease 325
18. Hygiene and Public Health 343
19. Population, Pollution and Conservation 394

Appendix 399

Index 403

Chapter I

LIVING CREATURES

This planet, on which we live, is inhabited by a very great variety of living creatures, ranging in size from microscopic animals and plants to the huge blue whale, which is the largest animal, and the giant Californian big-tree, which is the largest plant. Not only is there this wide range in size; there is also a vast variety in the structure of bodies and in the numbers of the various types. Apart from the living things, there is a far greater bulk of non-living material, which differs in several important respects from the living substance.

Living and Non-living Things

There are two main ways in which we can distinguish living or **animate** things from non-living or **inanimate** things. Firstly, we can observe what they do, and secondly, we can examine the substances of which they are made.

When we observe a familiar animal, such as a cat, dog, budgerigar or another human being, over a long period of time, we find that it carries out a number of different activities. For example, it feeds, breathes, grows, and passes waste materials out of its body, and, under the right conditions, it can produce more of its own kind. In addition, it can perceive changes in its environment by using its senses of sight, smell, touch, taste and hearing. Not only is it able to perceive changes; it can also respond to them.

A non-living thing, such as a stone or a piece of metal, cannot carry out any of these processes. Some non-living things, particularly complicated machines such as motor-cars or steam-engines, do appear to perform some of the living processes. A motor-car needs fuel and oxygen; it passes out waste gases through its exhaust pipe; and it may be made to move. But it must be noted carefully that unless a motor is given the correct attention by a human being, it cannot do anything except rust and finally disintegrate.

All living creatures are made of a complex substance called **protoplasm,** together with certain materials that are made by the protoplasm. This living substance can be observed with a good microscope, but the observations will tell us little about its structure. By chemical analysis, we can find out the materials of which it is made; however, that will not tell us how they are put together. We know that differences in the protoplasm are responsible for the differences between living creatures. It is the only substance that, under the correct conditions, can obtain the materials necessary for its own growth and the energy for doing its work, and that can repair and reproduce itself.

Living things eventually die, and they also disintegrate. While the protoplasm can resist this disintegration, it is alive; when it can no longer resist, it is dead. A stone or a car is not dead, for it has never been alive.

Plants and Animals

Plants carry out the same processes as animals, but it is not so easy to observe them taking place. We cannot see plants feeding, breathing, moving or excreting waste materials. Their responses to changes in the environment are, in general, very slow. But if a plant is observed over a long period of time, it is obvious that it grows and reproduces. Plants are as fully alive as animals are; their bodies are made of protoplasm and its products; they perform the same essential functions.

Plants differ from animals in the way in which they **"make a living."** To make a living, every organism must somehow obtain the necessary materials for producing more of its own substance, and it must also obtain the necessary energy for carrying out all its processes. In both these respects, plants may be described as self-supporting, whereas animals must rely on plants for their continued existence.

The protoplasm of plants has the ability to manufacture a remarkable substance called **chlorophyll.** This is green in colour and it is able to convert light energy into a form that can be used for building all the complex substances required by a plant. The building materials are simple chemicals, namely water, carbon dioxide and a variety of salts. Thus, if a green plant has sufficient sunlight, water, carbon dioxide and salts, it can manufacture all the products it needs.

Animals cannot make chlorophyll and therefore they cannot be self-supporting in the sense that plants are. They rely on green plants for complex substances, and so they must feed on plants, or on other animals that have eaten plants.

Some plants, such as the fungi and most bacteria, cannot make chlorophyll and therefore cannot be self-supporting. To obtain the complex substances that they need, they are in the same position as the animals. They must feed on green plants or on animals, alive or dead.

The Structure of Living Creatures

All living creatures are made of protoplasm, together with the non-living products of the protoplasm. Structures such as finger-nails, hair, the bark of trees and the brown, scaly covering of bulbs, are non-living, but they were made by the living protoplasm. In the great majority of living creatures, the protoplasm exists in microscopic units called **cells.** Because cells are usually very small, we use special small units to measure them. A table of micromeasurement is given below—

1000 picometres (pm) = 1 nanometre (nm)
1000 nanometres (nm) = 1 micrometre (μm)
1000 micrometres (μm) = 1 millimetre (mm)

Note that human eyes cannot normally distinguish anything smaller than 0·1 mm, i.e. 100 μm.

The processes that take place in a cell are controlled by a specialized portion of the protoplasm called the **nucleus** (plural, **nuclei**). Apart from the nucleus, the remainder of the protoplasm is called the **cytoplasm,** which is enclosed by a thin living **plasma membrane.** These three things are the essential components of all cells. In plants, the cells are separated from one another by non-living **cell-walls** (*see* Fig. 1(*a*)). Between the cell walls of adjacent cells there is a very thin layer of cementing substance. In animals, the cells are usually separated by jelly-like **intercellular substance** (*see* Fig. 1(*b*)). In many cases, the cytoplasm of a cell is in communication with that of

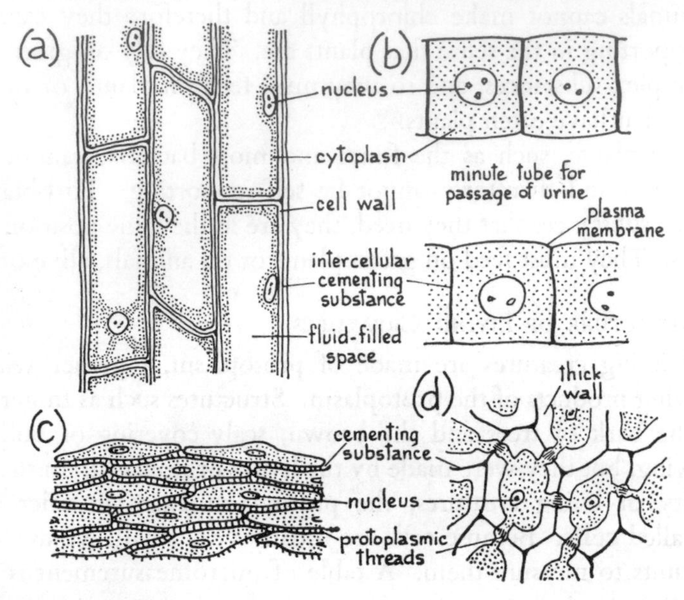

FIG. 1. Some Examples of Cells (*a*) Surface View of Cells from the Fleshy Leaf of an Onion Bulb, × 75 (*b*) Cells that form the Minute Tubes of the Kidney in Section, × 1200 (*c*) Part of a Vertical Section of the Inner Part of the Human Cheek, × 130 (*d*) Section of the Food Storage Cells in a Date Seed, × 500

adjacent cells by means of minute **protoplasmic threads,** which pass through the intervening non-living layers (*see* Figs. 1(*c*) and 1(*d*)).

Though the cellular type of structure is the usual condition, there are many examples of living things, or parts of living things, where the protoplasm is not divided into cells. Such examples are known as **coenocytes;** they are **multinucleate** masses of protoplasm (*see* Fig. 2).

The simplest living animals and plants are not made up of cells, but consist each of a small mass of protoplasm with one, or more, nuclei. There are many thousands of kinds of such simple, **non-cellular** creatures (*see* Fig. 3). Although they are usually extremely

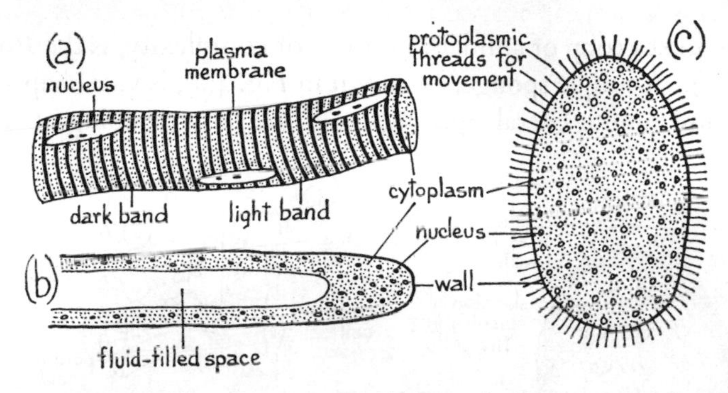

FIG. 2. Some Examples of Coenocytes (*a*) Part of a Single Muscle Fibre from the Leg of a Rat, × 200 (*b*) Section of the Tip of a Fungus Thread (hypha), × 600 (*c*) *Opalina*, a Protozoan Parasite from the Large Intestine of a Frog, × 600

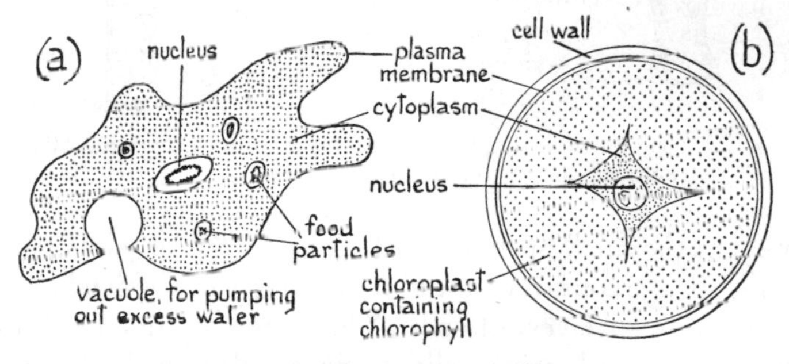

FIG. 3. (*a*) A Simple, Non-cellular Animal, a Pond Amoeba, × 300 (*b*) A Simple Plant, Forming a Green, Powdery Covering on Tree-trunks, *Pleurococcus*, × 1700

small, each is able to perform all the functions necessary to maintain itself and to produce others of its own kind.

Starting from simple animals such as the **Protozoa,** we can arrange all the higher forms in order of increasing complexity, each group being more complicated in structure than the one below it in the series. In this book only the animal kingdom is considered. A general account of both kingdoms is given in *An Introduction to Living Things*, in this series.

The next group of animals, in order of complexity, is the **Porifera** (sponges). A simple sponge, as shown in Fig. 4(*a*), is vase-shaped, with numerous minute lateral openings and one large opening at the top.

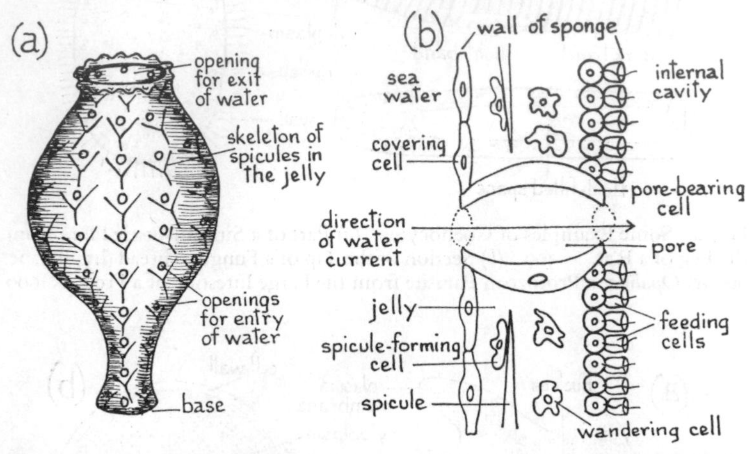

FIG. 4. (*a*) Diagram of a Simple Sponge (*b*) Section through the Wall of a Simple Sponge, to Show the Types of Cells, × 500

Water enters the small openings and leaves by the large opening. The wall of the sponge consists of a stiff, non-living jelly, which lies between the inner and outer layers of cells. There are only five kinds of cells: covering cells, pore-bearing cells, feeding cells (which also cause the water to enter and leave the body), skeleton-forming cells, and wandering cells, which act as carriers of food and waste materials and also form the reproductive structures (*see* Fig. 4(*b*)). This type of body is an advance over that found in the Protozoa, because it has different types of cells specialized to perform different functions. Sponges differ from all other multicellular animals in the fact that they have no nervous system controlling the whole body.

The next group, in order of complexity, is the **Coelenterata,** which contains the familiar jelly-fishes, sea anemones, corals and the freshwater *Hydra*. Here again the body is hollow and its wall is made up of inner and outer layers of cells, separated by jelly (*see* Fig. 5(*a*)). There are five types of cells in both layers, and apart from this, there is a great advance over the sponge condition in that some of the types of cells are present in continuous sheets, and all are controlled by a

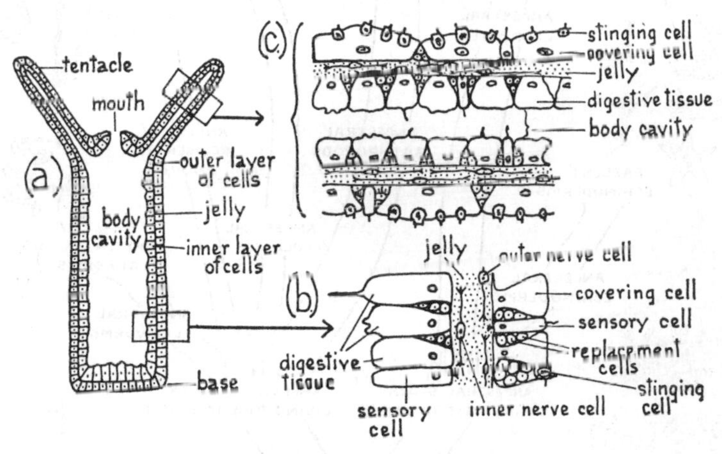

FIG. 5. (*a*) Vertical Section of a Hydra, × 6 (*b*) Section of Part of the Wall of a Hydra, × 100 (*c*) Section of a Small Portion of a Tentacle, × 100

nervous system (*see* Fig. 5(*b*)). Such a collection of cells of the same type, doing the same work and coordinated by a nervous system, forms a **tissue.** In a coelenterate, there are covering tissue, which also has a muscular function, nervous tissue and digestive tissue. In a **tentacle** of a hydra there are outer covering tissue, nervous tissue, inner digestive tissue, sting cells and sensory cells (*see* Fig. 5(*c*)). Such a structure, consisting of several types of tissue, performing special functions and co-ordinated by the nervous system, is called an **organ.** The tentacles are organs for paralysing and capturing the prey and pushing it into the mouth. Thus, a coelenterate has not only various types of cells but also tissues and even simple organs.

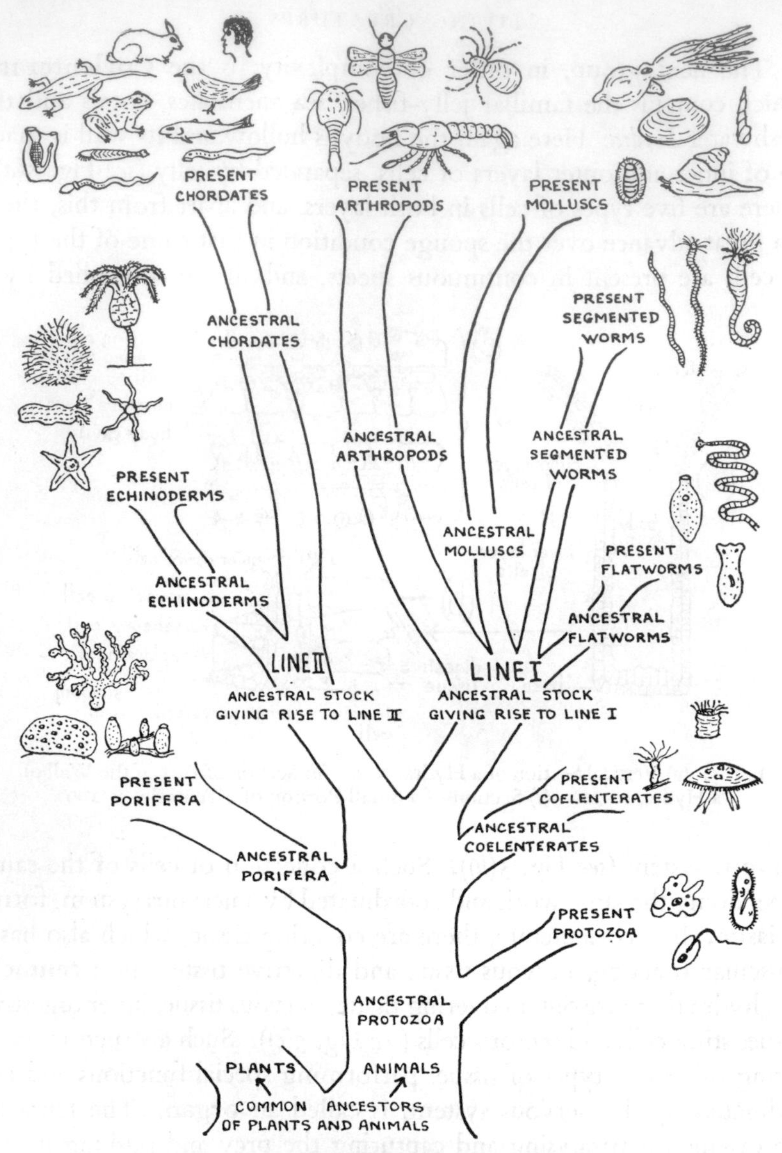

FIG. 6. Family Tree, Showing the Main Groups of Animals and the Relationships between Them

In all the higher groups of animals there are tissues and organs; the organs increase in number and in complexity. A whole group of organs may be co-ordinated for the performance of one major function. Such a group is called a **system.** Thus, in our own **alimentary** system there are the mouth, tongue, teeth, gullet, stomach, small and large intestines and anus; each of these is an organ. Similarly, there are the breathing, circulatory, excretory, nervous, skeletal, muscular and reproductive systems.

All the main groups of animals are described in *An Introduction to Living Things.* They are shown, arranged in the form of a tree, in Fig. 6.

Biological Classification

When the whole range of living things is examined, it is found that they can be arranged in order from the simplest to the most complex. This arrangement has been worked out for plants and animals, but it can never be quite complete, for two reasons. First, we do not know what the early living creatures were like, and secondly, we do not know whether any groups have perished without leaving fossil traces.

There are indications from fossils that living creatures existed on earth more than a thousand million years ago. Since that remote age, great changes have taken place in the kinds of living creatures (*see* Fig. 7). These changes constitute the process of **evolution.** As far as the evolution of living things is concerned, biologists believe that there has been no creation of successive new types, and that every creature is descended from some previously existing creature. Thus, if there was one original kind of life, then all living creatures have some relationship with one another. But, since evolution has been proceeding for hundreds of millions of years, some of the relationships are now so remote that they appear almost non-existent. Nevertheless, we can draw up a kind of **evolutionary tree** in which we try to indicate the relationships between the great groups of living things. Part of the tree, showing the main animal groups, is pictured in Fig. 6;

the complete tree for animals and plants is shown on pages 12 and 13 of *An Introduction to Living Things.*

There is such a vast number of different kinds of living creatures that it would be quite impossible for anyone to study them individually.

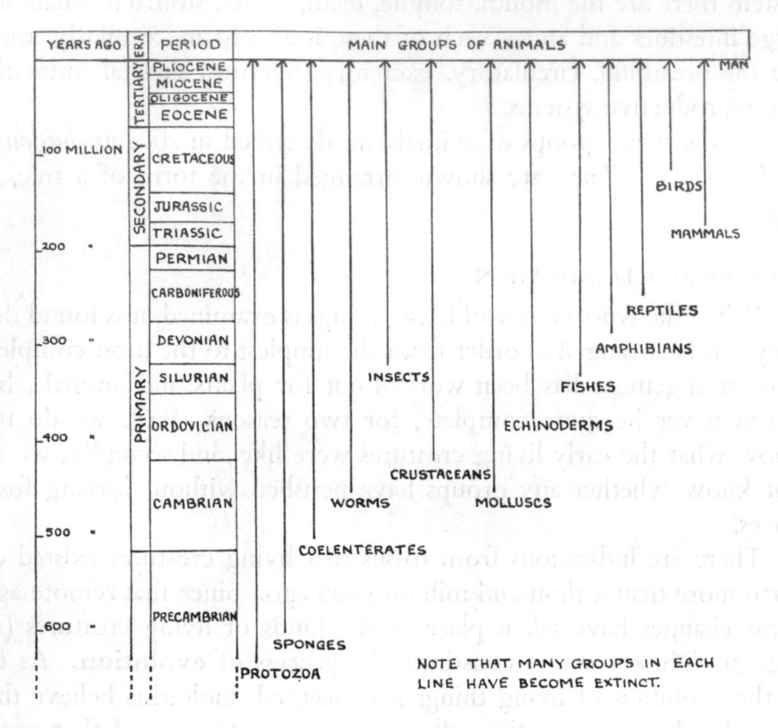

FIG. 7. Ages of the Rock Formations, with the Approximate Times of Origin of the Main Animal Groups. On this scale, the Quaternary era occupies roughly the thickness of the line above the word "man."

Therefore, we divide our subject-matter up into convenient groups. The members of each group have certain similarities, and by studying one member of a group, we begin to have some idea of the whole range.

No system of classification is of much value unless all agree to adopt the same units and to use the same methods. In biology, we attempt to draw up a system of classification that indicates natural relationships. For example, it is fairly obvious from simple examination that man is very much like a chimpanzee or a gorilla. Detailed investigation by scientific methods does in fact show that they are closely related. We think that man and the apes had common ancestors in the remote past, and therefore we place them in the same group (*see* p. 19).

There is a great variety of domestic cats, and among them, any male and female can be successfully mated to produce kittens. Such a group, all of whose members can be interbred, we call a **biological species,** and by international agreement, we give each species two Latin names. All domesticated cats belong to the species *Felis catus* (note that the first name must have a capital letter). There are, however, other well-known animals that are very like cats and so we use the same first name, but a different second name, for them. The lion is called *Felis leo* and the tiger is *Felis tigris.* Cats, lions and tigers are thus placed in the **genus** *Felis*, and that is their **generic** name. Other cat-like animals, such as jaguars, cheetahs and leopards, together with cats, lions and tigers, are placed in the cat family, the **Felidae.** All members of the family Felidae, together with members of the dog family, **Canidae** and certain other families, are placed in a larger group called an **order;** the name of this order is the **Carnivora.**

Carnivora, like human beings, mice, bats, elephants and many other animals, possess hair and suckle their young on milk. All the animals that have these two characteristics are placed together in the **class Mammalia.** Mammals, together with four other classes, the fishes, amphibians, reptiles and birds, possess backbones and well-developed heads with jaws. They are thus classified in the same **Sub-phylum, Vertebrata.** In their early stages of development, all vertebrates have a flexible rod called the **notochord** in the position of the backbone, and also they all have slits in the sides of the throat region. Some lesser-known marine animals retain the notochord throughout life. All the animals that possess, at some time in their lives, a notochord and the slits are placed in the **phylum Chordata.** There are about twenty of these phyla, which make up the **animal kingdom.** Thus, a complete classification of a common cat would be—

Kingdom:	Animalia
Phylum:	Chordata
Sub-phylum:	Vertebrata
Class:	Mammalia
Order:	Carnivora
Family:	Felidae
Genus:	*Felis*
Species:	*Felis catus*

All animals are classified in a similar manner (*see* p. 14).

SUGGESTED EXERCISES

1. Observe the movement of the protoplasm in a leaf of the common aquarium plant, Canadian pond-weed (*Elodea canadensis*). Remove a leaf gently and place it in water on a microscope slide. Cover it with a cover-slip and examine cells near the midrib with the high power of the microscope. After a few minutes, movements of the green objects (**chloroplasts**) carried by the protoplasm in the cells can be observed.

2. Observe the movement of the protoplasm in a living amoeba (obtainable from biological dealers). Place the amoeba in a drop of water on a slide. Support the cover-slip with two tiny strips of paper. The movement of the protoplasm can readily be observed with a microscope.

3. Strip a part of the outer layer (**epidermis**) from one of the fleshy leaves in an onion bulb. Mount it on a slide in water, cover with a cover-slip and examine under the microscope. Make drawings of a few cells showing the cell walls, protoplasm and nuclei.

4. Observe living animal cells in a similar manner. Scrape gently loose cells from the inside of your cheek, using a spatula or the handle of a scalpel. Make drawings of a few of the cells.

5. Examine a prepared slide of a section of liver or kidney. The cells, though dead, still show the nuclei, and the dark lines between the cells indicate the intercellular substance. Draw a few of the cells.

6. Examine drops of pond-water, and green water from an aquarium that has been near a window. You will see many examples of microscopic creatures, some swimming. Named specimens can be obtained from dealers. The following are suitable for examination—*Amoeba, Paramecium, Vorticella, Chlamydomonas, Chlorella, Euglena, Volvox.*

7. Examine and draw some simple sponges obtained from dealers. The following are suitable: *Sycon, Grantia, Leucosolenia.*

8. Examine a living hydra with the lower power of the microscope. Note the tentacles and the mouth. Study also some preserved specimens of sea anemones and jelly-fish.

9. Obtain a prepared slide of a section of a hydra. Study it under the microscope and note the inner and outer layers of cells separated by a thin line of jelly.

10. Collect pictures of any one species of animal and fasten them in your exercise book. They will show the wide range of variety found in some species. Suitable species are dogs, cats, horses, cows, budgerigars, rabbits.

11. Make lists of carnivora, herbivora and omnivora.

12. Try to make a classification of some group of objects for yourself. Suitable groups are vehicles, games, beasts of burden. First divide the group into smaller groups and then subdivide each until you come to individual units. State what problems you encounter in attempting to make a simple classification.

Chapter 2

MAN'S PLACE IN THE ANIMAL KINGDOM

HUMAN beings are animals, and like all animals they feed, breathe, grow, move, excrete and reproduce, and they can perceive and respond to changes in the environment. The following is a classification of modern man; the significance of all the names used will be explained in this chapter.

Phylum:	Chordata
Sub-phylum:	Vertebrata
Class:	Mammalia
Order:	Primates
Sub-order:	Anthropoidea
Family:	Hominidae
Genus:	*Homo*
Species:	*Homo sapiens*

Every one of us is entitled to regard himself, or herself, as belonging to the species *Homo sapiens*, which means "the wise man."

The Phylum Chordata

This is a very large group of animals, which at first sight do not appear to have much in common. Nevertheless, they all possess, at least at some stage in their lives, four important characteristics. Firstly, they all have a strong, but flexible, rod-like structure, called the **notochord,** above the gut. In most groups of chordates the notochord is clearly seen only in the very young animals. As they grow, it is almost entirely replaced by a vertebral column (backbone). Secondly, all chordates have, in the throat region, certain openings that put the gut in communication with the outside world. These openings are known as **visceral clefts.** In the fishes, these clefts form the gill slits, but in the higher groups of chordates they are well developed only in the embryonic stages. Thirdly, the **central nervous system,** i.e. the brain and spinal cord, is above the gut (dorsal) and it contains a central, fluid-filled canal. Fourthly, the chordates are the only animals that possess **true tails,** which are segmental structures lying behind the anus. These characteristics are shown in a fish-like chordate in Fig. 8. We do not know what the ancestral chordates were like,

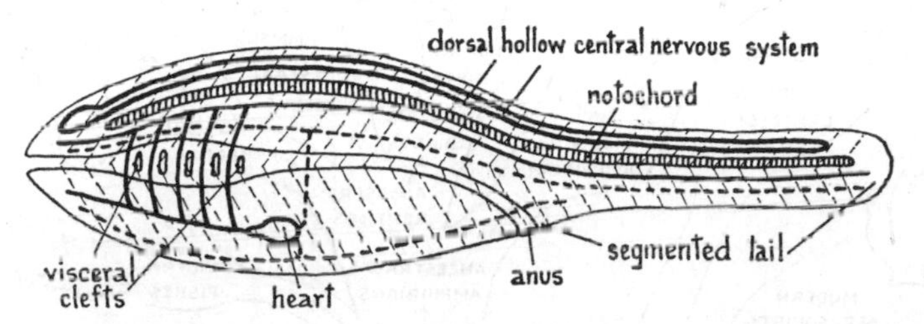

FIG. 8. Diagram Showing the Main Features of a Generalized Chordate Animal

but it is thought that they were small, soft-bodied creatures that had no hard parts. Therefore it is not likely that any fossils of them exist. The table in Fig. 9 shows the relationships between the main groups of chordate animals. **Man is a chordate.**

THE SUB-PHYLUM VERTEBRATA

Fig. 9 shows that there are two main branches of the chordates, the invertebrate and the vertebrate. (Note that the term "invertebrate" is sometimes applied incorrectly to all non-chordate animals, including worms, starfishes, snails and insects.) The invertebrate chordates are all small marine animals—the acorn worms, the lancelets and the sea squirts (*see* Fig. 9). None of them has a vertebral column and their heads are very poorly developed. The vertebrate chordates have well-developed heads and backbones made of a number of jointed **vertebrae** (*see* p. 67). In adult vertebrates, the notochord has almost entirely disappeared. The vertebrates include the slimy, eel-like lampreys and hagfishes, as well as the better-known fishes, amphibians, reptiles, birds and mammals. **Man is a vertebrate.**

THE CLASS MAMMALIA

Fig. 9 shows that an ancient group of fishes gave rise to the amphibians; they in turn gave rise to the early reptiles, some of which eventually produced the birds and mammals. The members of the class Mammalia differ from the other classes of vertebrates in many

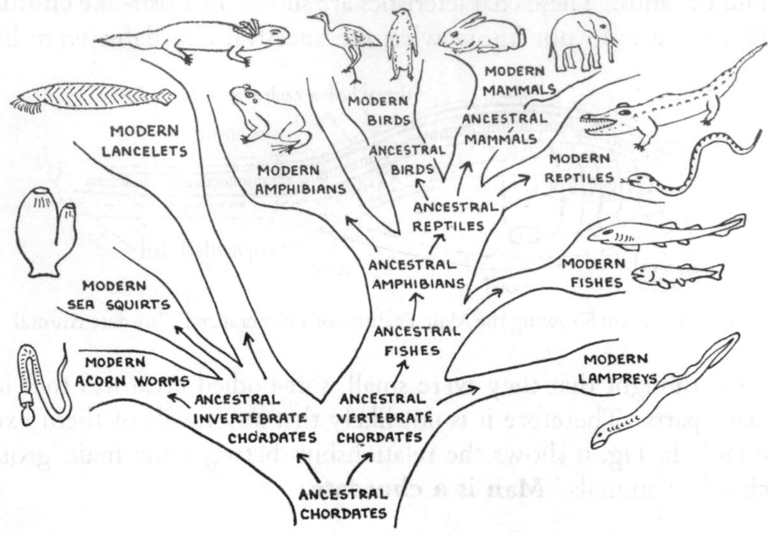

FIG. 9. Family Tree of the Phylum Chordata

respects. The most important of these are: the possession of **true hair** (*see* p. 51), **the suckling of the young on milk,** the development of the embryos in the **womb (uterus)**, the separation of the chest (thorax) from the abdomen by an internal partition called the **diaphragm,** and the possession of **sweat glands** in the skin (*see* p. 51). **Man is a mammal.**

THE ORDER PRIMATES

The class Mammalia is divided into a number of orders, the chief of which are—

1. **Monotremes.** Egg-laying mammals, e.g. the duck-billed platypus and spiny ant-eater.

2. **Marsupials.** Pouched mammals, e.g. the kangaroo, opossum and koala bear.

3. **Carnivora.** Flesh-eating mammals, e.g. the wolf, lion, weasel and seal.

4. **Ungulates.** Hoofed mammals, e.g. the cow, goat, sheep, horse, zebra, reindeer and rhinoceros.

5. **Proboscidea.** Mammals with a long trunk and tusks, e.g. the elephant.

6. **Cetaceans.** Marine mammals that breed in the water, e.g. the whale and dolphin.

7. **Edentates.** Toothless mammals, e.g. the armadillo and giant ant-eater.

8. **Rodents.** Gnawing mammals, e.g. the mouse, rat, squirrel and beaver.

9. **Chiroptera.** Flying mammals, e.g. the flying fox and long-eared bat.

10. **Insectivora.** Insect-eating mammals, e.g. the mole and hedgehog.

11. **Primates.** The most advanced mammals, mainly tree-living, e.g. the lemurs, monkeys, apes and men.

Examples of these orders are illustrated in *The Mammals*, in this series.

The order Primates includes the tree-shrews, lemurs, tarsiers, monkeys, apes and men (*see* Fig. 10). These are all separate branches of the order and there is no doubt that they were all evolved from arboreal (tree-living) insectivora, which existed more than fifty million years ago. Most modern primates live in trees, and even those that live on the ground, such as the gorilla and man, possess various adaptations for an arboreal mode of life. The main features that distinguish primates from the other orders of mammals are listed below.

1. There is considerable flexibility of the skeleton, especially in the wrist and hand and sometimes in the ankle and foot. All primates can

grasp with their hands; monkeys can grasp with both hands and feet, and New World monkeys can even use their tails to grasp branches.

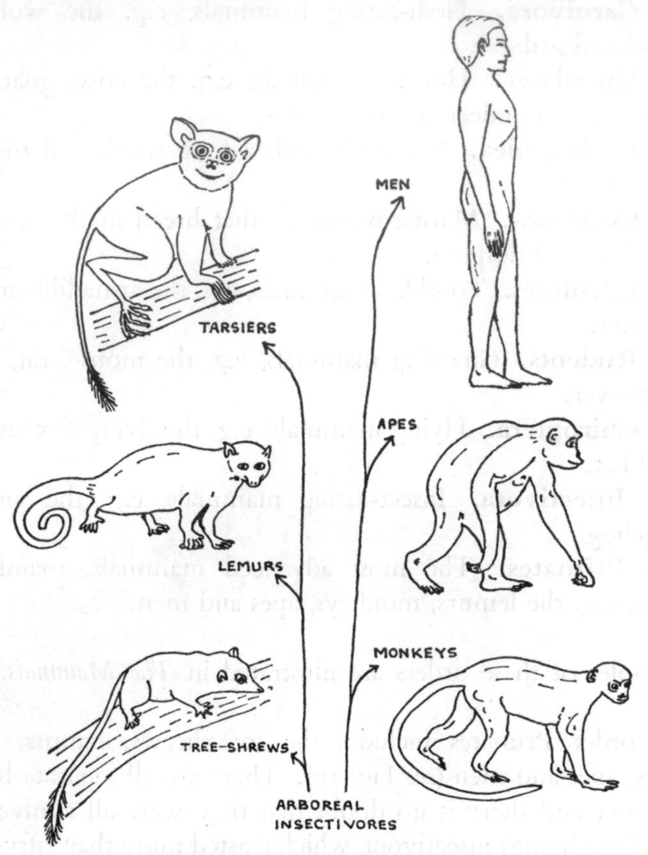

FIG. 10. Family Tree of the Primates

2. When on the ground, all primates except man normally move about on four legs, but all can squat on the hind-limbs and thus free the front-limbs for grasping things.

3. The teeth of primates are not specialized for one type of food-stuff, as are those of most other mammalian orders. This indicates an

omnivorous diet and hence the possibility of existing on almost any kind of available food.

4. The eyes face forwards, and thus both eyes form almost exactly the same image; the slight difference between the two images gives good judgment of distance. This is a great advantage to animals that jump from one bough to another.

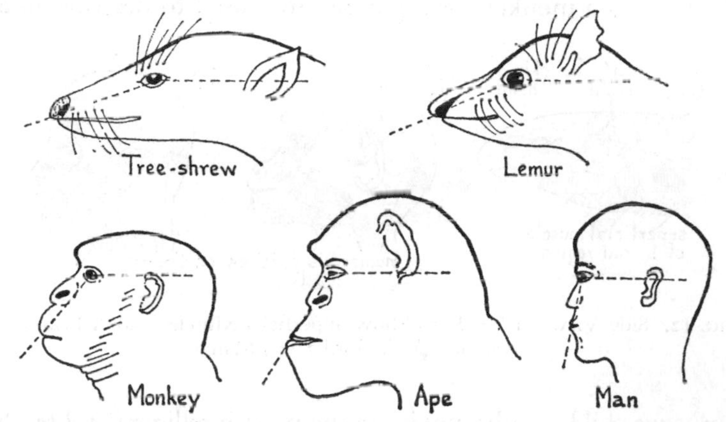

FIG. 11. Side Views of Primate Heads, Showing Gradual Increase in Cranial Arch (Thick Line) and Decrease in Facial Angle (Broken Line)

5. Although the sense of sight has become highly developed, the sense of smell has degenerated.

6. Rapid movement in trees requires great agility and a high degree of muscular co-ordination, and we find that the primate brain has evolved greatly enlarged regions of co-ordination and control.

7. There is wide variety in the shape of the head, from the long, pointed muzzle and flat cranium of the tree-shrews to the almost vertical face and rounded cranium of man (*see* Fig. 11). **Man is a primate.**

THE SUB-ORDER ANTHROPOIDEA

This is the name of the sub-order that distinguishes the monkeys, apes and man from the tree-shrews, lemurs and tarsiers. "Anthropoid" means "man-like," and certainly the monkeys and apes do look man-like when they stand erect. There are a number of characteristics that set the Anthropoidea apart from the other primates.

1. They are all alert, inquisitive and adventurous animals. They show great curiosity and examine strange objects very carefully. "As mischievous as a monkey" is a phrase often used to describe an alert,

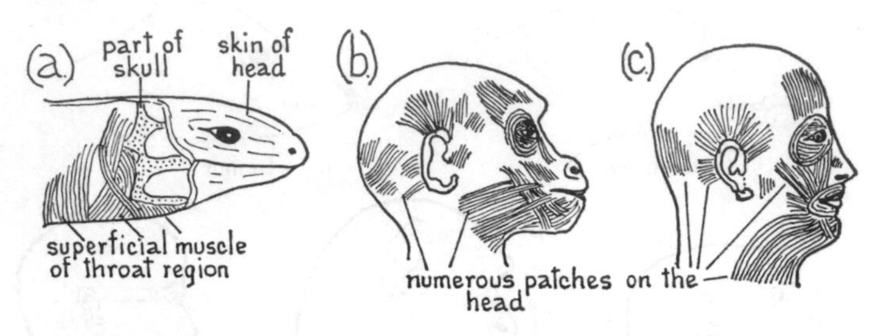

FIG. 12. Side Views of Heads, to Show Superficial Muscles (*a*) A Primitive Reptile (*b*) A Gorilla (*c*) Man

adventurous child. Anthropoids are the most intelligent and teachable of all animals.

2. The anthropoids are capable of a greater variety of facial expressions than any other animals. The so-called facial muscles, which originated in a thin, wide band round the throat of reptilian ancestors, have grown forwards in the mammals to reach the lips, eyes, nose, cheeks and ears (*see* Fig. 12). The face of a reptile is a bony mask without expression of any kind. The forward growth of these flat muscles has been greatest in the anthropoids, and facial expression is achieved by combinations of movements of the lips, cheeks, eyes, nose and ears.

3. Anthropoids are capable of a high degree of social activity. They live in troops with an acknowledged leader, and the activity of the troop is directed towards the good of all. They will help one another in trouble and show especial regard for their young. All anthropoids have developed means of communication by making

various sounds, such as howls, shrieks or grunts. This tendency has reached its climax in man with the development of speech and a large vocabulary of words.

4. The head of every anthropoid is set on the vertebral column in a more or less erect position, and the eyes, which in other mammals are at the sides, are in the front of the head. This frontal position of the eyes has been accompanied by gradual loss of the snout region, until the face appears more and more flattened (*see* Fig. 11). Throughout the anthropoids, there is also a gradual reduction in the size of the ears; these also become more flattened until they lie closely pressed to the skull.

5. There is further enlargement of the brain in the anthropoids, compared with the lower primates, and this has involved an increase in the size of the cranium until in man it is almost rounded (*see* Fig. 11).

6. All anthropoids can stand erect, though, except for man, they normally progress on all fours. Man, except during a short period of crawling in infancy, always has an upright stance and all the locomotive movements are performed by the hind-limbs.

7. A notable feature of the anthropoids is the versatility of the hand. Not only is it very flexible and capable of grasping but there is much more precision in its use. In man this versatility and precision of the hands reaches its height, with such skills as writing, drawing, painting and playing the piano. **Man is an anthropoid.**

The Family Hominidae

The sub-order Anthropoidea is divided into four families: the monkeys (Cercopithecidae), the gibbons (Hylobatidae), the apes (Pongidae), and men (Hominidae). It must be clearly understood that man did not evolve from monkeys, apes or gibbons, but that all four families had common ancestors, which lived about forty-five million years ago. Undoubtedly those ancestors were arboreal, but they did not possess the extreme specialized tendency for life in trees that is found in the modern monkeys and apes.

The evolution of the family Hominidae has taken place in the last 2 to 2½ million years, mainly in the **Pleistocene** period. During this period there were four Ice Ages, separated by Interglacial Ages. Fossils of early man-like creatures were first found in South

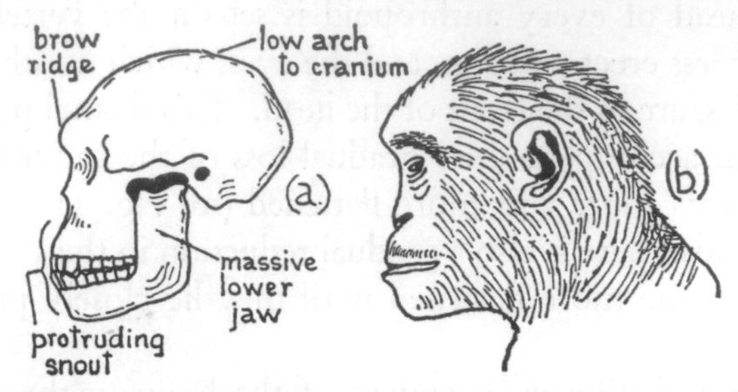

FIG. 13. (*a*) Skull of *Australopithecus*, Side View (*b*) Reconstruction of the Head

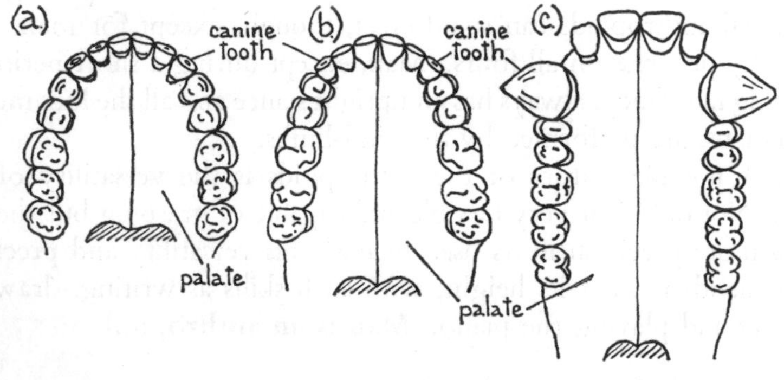

FIG. 14. Upper Jaw, Teeth and Palate of (*a*) Modern Man, (*b*) *Australopithecus*, *c*) Gorilla

Africa in 1925. These fossil specimens are placed in the genus *Australopithecus* (southern ape), and at first were considered to be man-apes, but after further discoveries and study, it is now considered that they were men of a very primitive kind. The fossil remains show that they had many ape-like features, especially in the skull, but they also had many human features (*see* Figs. 13 and 14). The structure of the limbs shows that they walked erect and were much smaller than modern man, being about 1·2 m in height and weighing about 18 to 22 kg.

The average volume of the brain was less than 600 cm^3, about half that of modern man and about the equivalent of that of the gorilla. There is no certain evidence that these early men were able to make tools, but in other regions, crude stone axes dating from a similar remote age

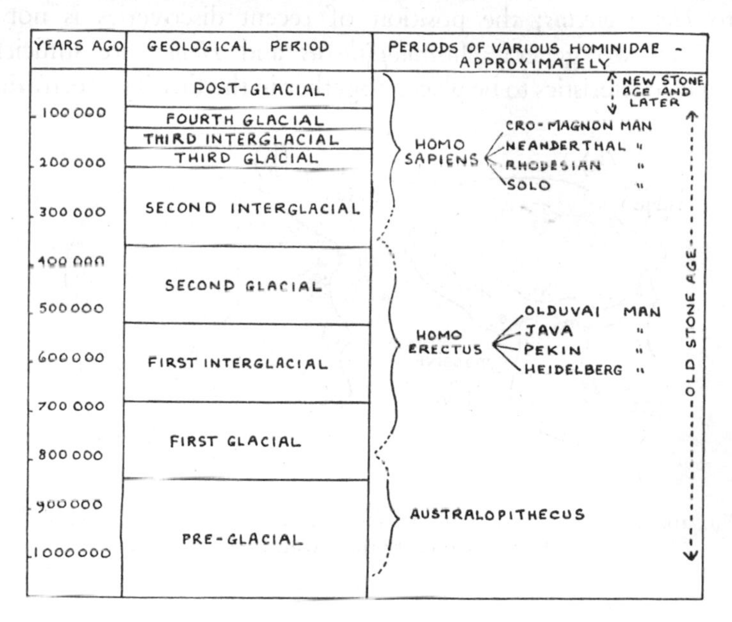

FIG. 15. Table Showing the Duration of the Ice Ages and the Corresponding Duration of the Various Types of Men

have been found. In 1959, the discovery of possibly older remains in East Africa may date man's origins to a time before the beginning of the first Ice Age (*see* Fig. 15). The latest finds, in 1970–71, seem to indicate that hominids, who used simple tools made out of large pebbles, existed more than 2 000 000 years ago.

Fossil traces of a number of later forms, dating from the second Ice Age, have been placed in the species *Homo erectus*; variants have been found in Central Europe, East and North Africa, Java and China. Men of this species were undoubtedly more advanced than *Australopithecus*.

The brain volume averaged 1000 cm^3, they made stone tools; they hunted animals for food, and had learnt the use of fire. Nevertheless, in the skull particularly, there were a number of ape-like features (*see* Fig. 16). It is considered probable that *Australopithecus* gave rise to *Homo erectus*; the position of recent discoveries is not yet clear. The two genera, *Australopithecus* and *Homo*, are sufficiently close in characteristics to be placed together in the **family Hominidae.**

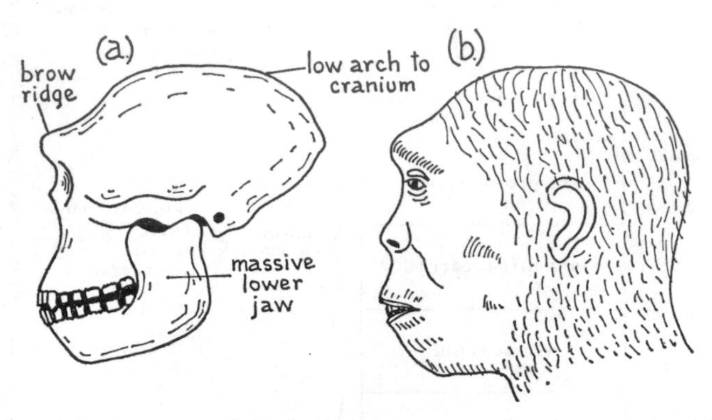

Fig. 16. (*a*) Side View of Skull of *Homo erectus* (*b*) Reconstruction of the Head (Compare these with Fig. 13.)

The Genus Homo

Before and during the fourth Ice Age there seem to have been several different types of men existing in various parts of the world. They are considered to be sufficiently close to modern man to be placed in the same species (*Homo sapiens*). They include Rhodesian man, Solo man, Steinheim man, Neanderthal man and Cro-Magnon man (*see* Figs. 17 and 18). The names refer to the places where these fossils were first found. Their culture is characterized by a wide variety of tools, somewhat crudely chipped from flint. By far the greater number of skeletal remains belong to the Neanderthal type and there is no doubt that these men were widespread over Europe, Asia and North Africa about two hundred thousand years ago.

Finally, about one hundred thousand years ago, Neanderthal men were replaced quite suddenly by men of modern type, who probably

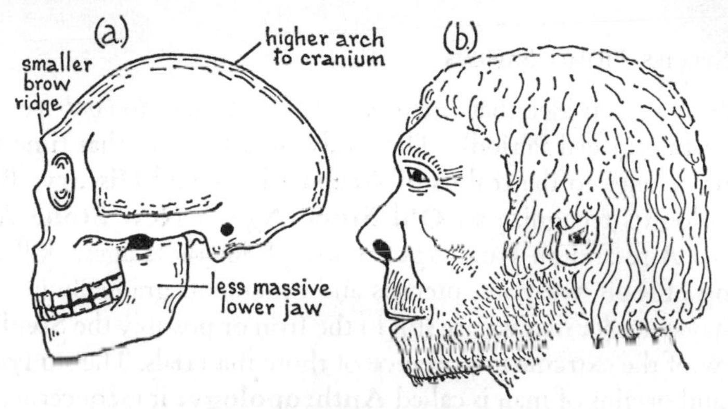

FIG. 17. (*a*) Side View of the Skull of Neanderthal Man (*b*) Reconstruction of Head (Compare these with Figs. 13 and 16.)

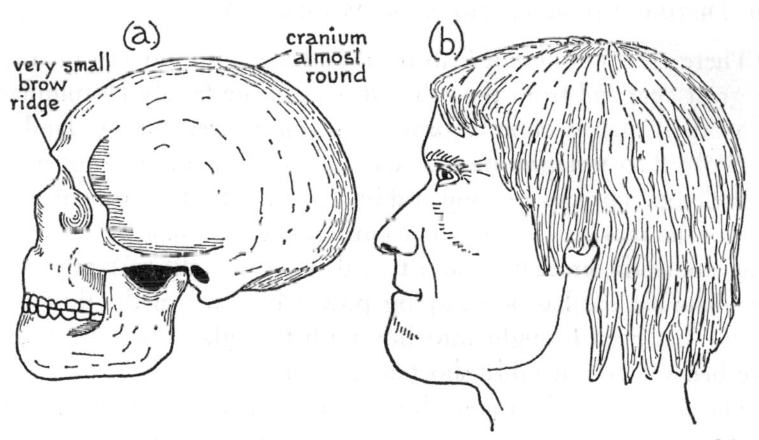

FIG. 18. (*a*) Side View of Skull of Cro-Magnon Man (*b*) Reconstruction of the Head (Compare Figs. 13, 16, and 17.)

came from Asia into Europe. These are known as **Cro-Magnon** men, after the site of the first discovery of their remains in France (*see* Fig. 18).

The species *Homo sapiens* thus includes not only modern man, but a range of primitive but now extinct sub-species.

The Species Homo Sapiens

We may consider that our own species began 300 000 years ago (second Interglacial Period). The study of man from that time to the present belongs to the realms of **Archaeology** and **History.** Briefly, archaeologists recognize an **Old Stone Age,** a **New Stone Age,** a **Bronze Age** and an **Iron Age.** Throughout these ages, skill in the making of tools, weapons, utensils and ornaments gradually increased. We may consider that we are still in the Iron or possibly the **Steel Age,** in view of the extreme importance of those materials. The study of the races and origins of man is called **Anthropology;** it is concerned with the relationships of modern men and fossil species.

The Distinguishing Features of Modern Man

There are a number of features that distinguish modern man from the great apes and from the other species in the family Hominidae.

Modern man is entirely **bipedal** in his movements on land. He has evolved an **upright stance,** which has affected numerous parts of the skeleton. His legs are long and his arms short when compared with those of the apes (*see* Fig. 19). The curvature of the spine has changed to bring about the erect back and to balance the body correctly on the feet. The big toe has no grasping power like that of the thumb and this toe has been brought into line with the other toes, all of which have become considerably shortened (*see* Fig. 20).

The skull of modern man shows many differences from that of his nearest existing relatives, the chimpanzee and the gorilla (*see* Fig. 21). The most striking difference is in the cranium, which contains the brain. In man, its capacity averages about 1500 cm^3, whereas the largest in any ape is about 650 cm^3. This enlargement has taken place mainly in the front of the cranium and has given man the high forehead, lacking the brow ridges that are so prominent in the apes. There has also been a shortening of the snout, so that apart from the projecting nose, the front of the face is almost vertical. The jaws of man are much less

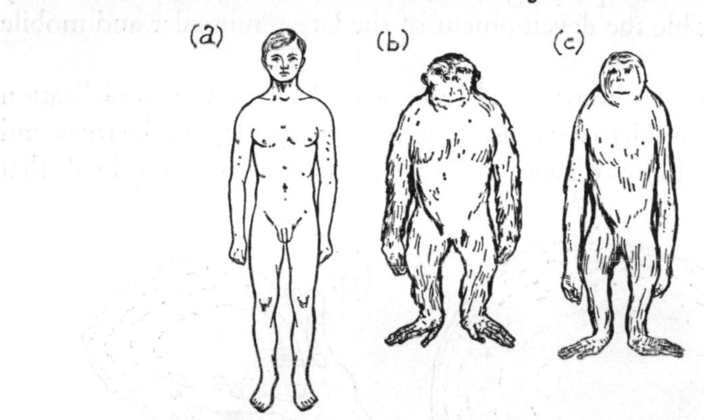

FIG. 19. (*a*) Modern Man (*b*) Gorilla (*c*) Orang-utan
All are drawn with the same trunk height to show the comparative proportions of the body. (*After A. E Schultz*)

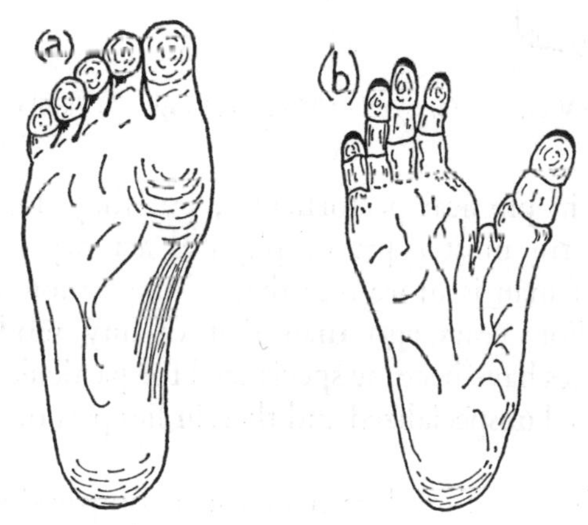

FIG. 20. Lower Surface of Right Foot of (*a*) Man (*b*) Chimpanzee

massive than those of the apes and the teeth are smaller, more even, and rather crowded in the jaws, though there are the same numbers of teeth as there are in the apes (*see* Fig. 14, p. 22). Associated with the shortening of the lower jaw, there has been increase in depth to form the chin. The accompanying increase in the size of the buccal cavity has made possible the development of the large, muscular and mobile tongue.

Altogether, the human skeleton shows lack of the specialization found in other modern animals. Man is not very adept in the trees, and he cannot run, jump or swim very well. There are many animals that

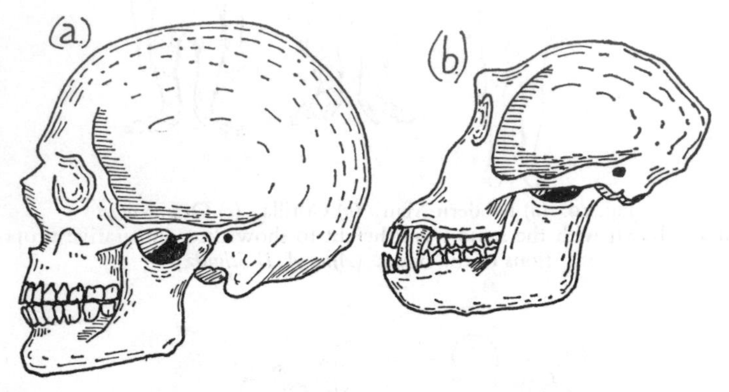

FIG. 21. Side View of (*a*) Skull of Human Being (*b*) Skull of Chimpanzee

can excel him in physical performance, but they do not possess his ability to perform most types of physical activity reasonably well. The skeleton of man is more like that of the remote anthropoids of thirty-five million years ago than that of any modern ape. The monkeys and apes have become specialized for particular modes of life; man has remained unspecialized and therein lies part of the secret of his success.

A curious feature of modern man when compared with the apes is his relative hairlessness. This may be a relic of those remote ancestors that were larger than man is now. Large animals lose relatively less heat from their body surfaces than small animals. The reason for this is that as an animal increases in volume, its surface area increases at a lower rate; that is, its area : volume ratio diminishes.

Edge of Cube	Surface Area	Volume	Ratio of Area to Volume
1 cm	6 cm^2	1 cm^3	6:1
2 cm	24 cm^2	8 cm^3	3:1
3 cm	54 cm^2	27 cm^3	2:1
6 cm	216 cm^2	216 cm^3	1:1
12 cm	864 cm^2	1728 cm^3	1:2

Thus, small mammals, with a relatively greater surface area from which to lose heat, have thick fur to prevent heat loss from the skin, while the largest mammals, e.g. the elephant, hippopotamus and rhinoceros, are almost hairless. It is interesting to note, however, that the larger mammals that colonized the colder areas of the world evolved thick fur in adaptation to the more rigorous conditions; e.g. the polar bear and the extinct woolly rhinoceros and mammoth. The seals and sea-lions living in the colder waters have thick, close fur, while the hairless whales have a thick protective layer of fat (blubber) under the skin. Primitive men inhabiting the colder regions learnt to make protective coverings from the skins of animals, and later from plant fibres. Modern man has developed various types of clothing to such an extent that, by a suitable choice of garments, he can live almost anywhere on the earth's surface; he thus has a wider distribution than any other land mammals.

The ability to make a great variety of sounds, which eventually developed into speech and music, clearly separates man from the apes. The making of sounds is widespread in the animal kingdom, but no animal has remotely approached man, with his numerous languages, enormous vocabulary and variety of voice. This faculty of communicating so much through the medium of sound may be regarded as another important factor in the success of man.

The most outstanding feature of man is the size of his brain, which is more than twice as large as that of any ape. In man, the brain is about 1/45 of the body weight; in a gorilla, it is less than 1/100; in an elephant, it is less than 1/500, and in a whale, it is about 1/8000. Coupled with the increased size of the brain has been the evolution of high intelligence, which has given man unlimited powers of learning and of invention. One of his principal attributes is his self-consciousness; he knows what he is doing and he can criticize his own efforts while he is doing anything.

Man is able to project his personality into the future. The work of people who died centuries ago still has a profound influence on our lives.

SUGGESTED EXERCISES

1. On a large sheet of cardboard, say 2 m by 1·5 m, draw ½ metre squares and fill each with pictures or drawings of the main orders of mammals.

2. Visit a Natural History Museum and examine the skeletons, and particularly the skulls, of various primates. Also, if they are available, as in the London Natural History Museum, examine the skulls of fossil types of men.

3. Most local museums have specimens of stone tools, weapons and utensils made by primitive men. If possible make a collection of drawings of these.

4. If you can pay a visit to a zoo, observe the living Primates. Look especially at the behaviour and expressions of monkeys and apes.

5. Give examples of the dexterity of the hands in monkeys and apes.

6. Make a list of people from the past who made some great contribution to our knowledge today. State what each person did that is regarded as important.

CHAPTER 3

THE RACES OF MEN

THERE have been many attempts to classify men into different races or sub-species. Throughout history, many groups of human beings have thought that their own was the master-race and that all others were inferior. It must be emphasized very strongly that all human beings belong to the one species (*Homo sapiens*) and that all members of a species are interfertile. Numerous migrations of peoples in the past have caused a great deal of mixing of different human groups. This mixing has increased very considerably in more recent times with the advent of speedy transport to all parts of the world. Nevertheless, it seems to be quite evident that there are certain clearly distinct groups, which evolved almost in isolation from one another for a considerable period of time, separated by geographical barriers such as oceans, deserts or high ranges of mountains.

The characteristics that are usually used in the classification of human races are the shape of the head, the type of hair, the colour of the skin and the shape of the nose. A number of less important characteristics are also used. For determining the shape of the head, two measurements are taken; these are the width of the head above the ears and the length from back to front. The width is then divided by the length and the answer is multiplied by one hundred. This final figure is known as the **cephalic index.** If the index is more than eighty, the type is known as broad-headed or **brachycephalic;** if between seventy-five and eighty, medium-headed or **mesocephalic;** and if below seventy-five, long-headed or **dolichocephalic.** Since the shape of the head changes during the growing period, the measurements should be taken with young adults (*see* Fig. 22).

Three principal types of hair are recognized: **woolly hair,** which is like a flattened ellipse in cross-section and curls readily;

wavy hair, which is oval in cross-section and does not curl so readily; **straight hair,** which does not wave naturally, is circular in cross-section (*see* Fig. 23(*c*)). Colour of skin ranges from black, through

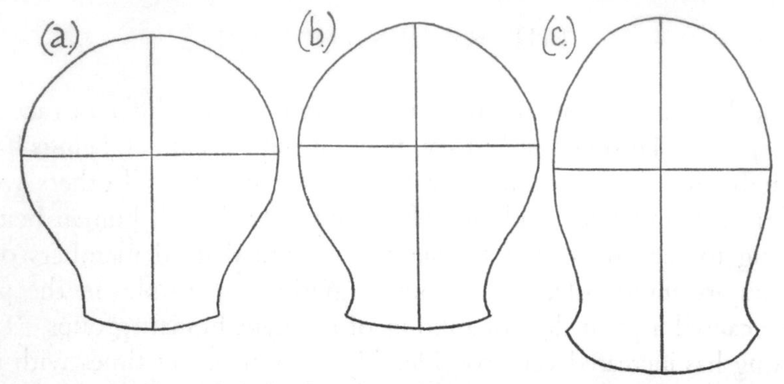

FIG. 22. Outlines of Skulls Seen from Above (*a*) Brachycephalic—round-headed (*b*) Mesocephalic—medium-headed (*c*) Dolichocephalic—long-headed.

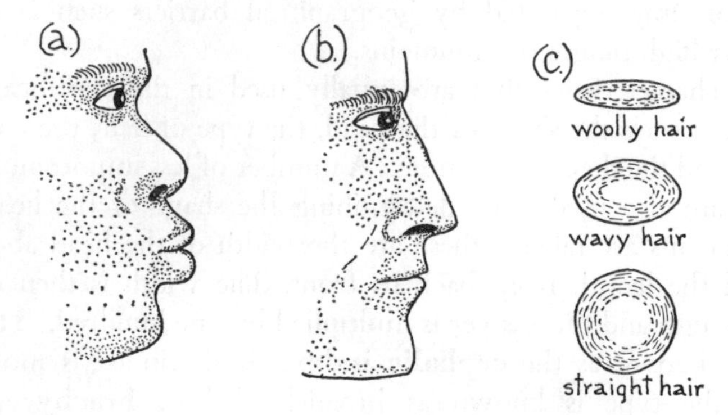

FIG. 23. *a*) Typical Negriform Profile (*b*) Typical Europiform Profile (*c*) Cross-section of the Three Principal Types of Hair

brown and yellow, to the condition of no pigment found in **albinos.** Nose-shape varies from the flat, broad type to the narrow projecting type (*see* Fig. 23(*a*) and (*b*)). For all these characteristics there are innumerable grades, ranging from one extreme to the other.

Considering these characteristics, it is generally recognized that there are four major human races. They are the **Negriform**, the **Europiform**, the **Mongoliform**, and the **Australiform.**

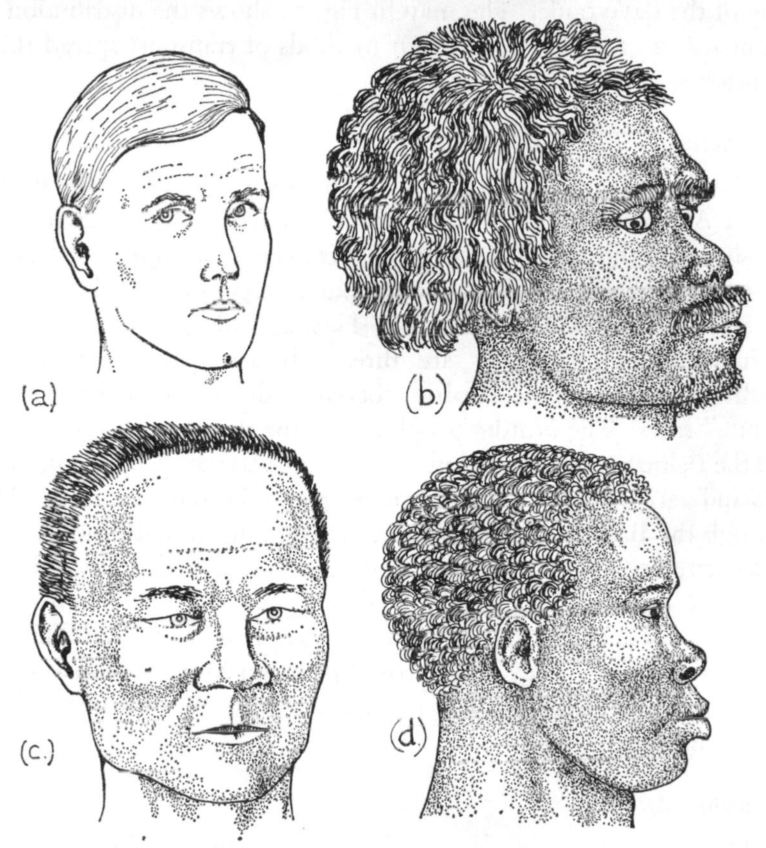

FIG. 24. The Four Races of Mankind (*a*) Europiform (*b*) Australiform (*c*) Mongoliform (*d*) Negriform

Negriforms

Typical negriforms are long-headed, with short, black, woolly hair, dark complexions and flat, broad noses (*see* Fig. 24). The most characteristic examples are found among the negroes of Africa, North

America and the West Indies. The pygmies, bushmen and Hottentots are considered to be variants of typical negriforms. The North American negroes were originally transported from Africa during the time of the slave trade. The map in Fig. 25 shows the distribution of the negriform race before modern methods of transport spread them far more widely.

Europiforms

These have a very wide distribution, including most of Europe, North Africa, Southern Asia and India. There is wide variation in the shape of the head, every grade between long and broad being present. The hair is wavy, the complexion ranges between brown and white, and the noses are narrow (*see* Figs. 24 and 25).

Included in this group are three sub-races: the **Nordic**, the **Mediterranean** and the **Alpine;** between them, they constitute the "white" races. The Nordic peoples from the shores of the North Sea and the Baltic have typically long heads, fair, wavy hair, light complexions and narrow noses. The Alpine peoples, who range from the Alps through the Balkans to Asia Minor, are typically round-headed, with dark hair, brown complexions and shorter noses than the Nordic people. The Mediterranean sub-race have spread all round the Mediterranean Sea. They have long heads, dark wavy hair, brown complexions and narrow noses. They range from Spain and Morocco eastwards to Asia Minor and down into India. All the various Arab peoples belong to this sub-race.

Mongoliforms

These have broad heads, straight hair, yellow or red complexions and noses intermediate between broad and narrow (*see* Fig. 24). Typical of this group are the Chinese, with their yellow skins, straight black hair, narrow slanting eyes and high cheek-bones. Mongoliforms range widely over Eastern Asia and were the original inhabitants of North and South America, i.e. the Red Indians and the Mayan and Inca civilizations. The Eskimos are also undoubtedly of the same

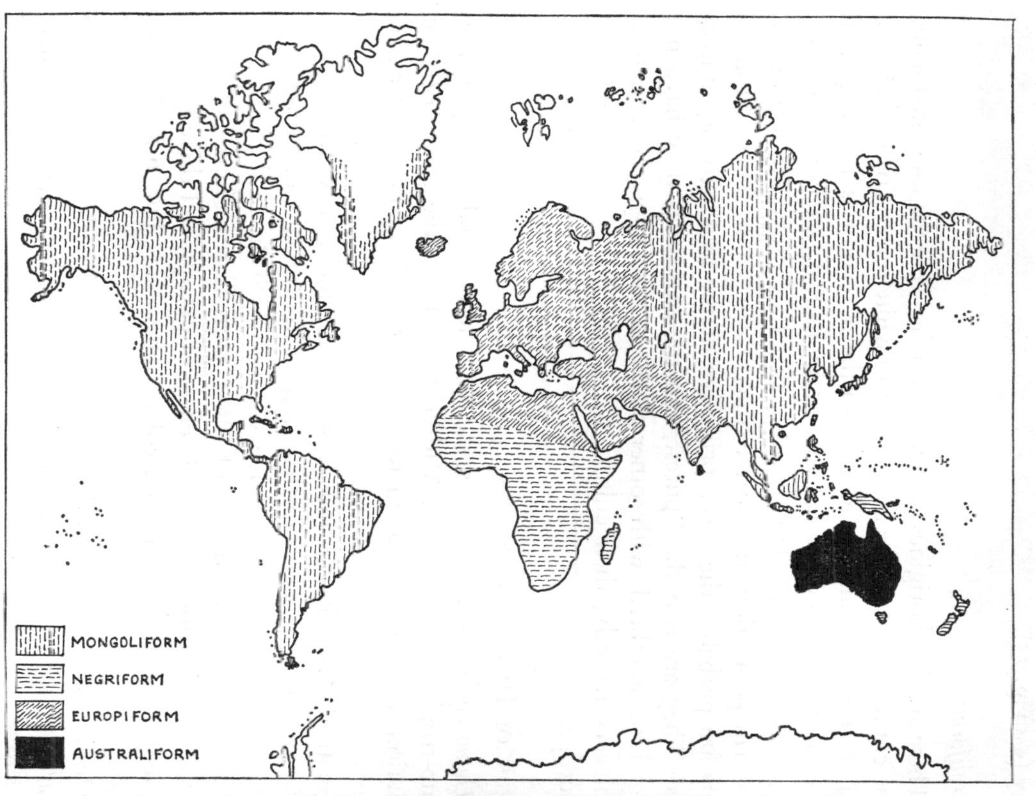

FIG. 25. Map of the World, Showing Distribution of the Human Races Before Modern Population Movements

stock. Mongoliforms are the most numerous people in the world today (*see* Fig. 25).

Australiforms

These are the aborigines of Australia, who were the sole inhabitants of the continent before the advent of the white man. Today there are about 60 000 left, mainly in the northern territories (*see* Fig. 25). Australiforms are long-headed with wavy hair, dark complexions and flat noses (*see* Fig. 24). It is possible that they drove out a previous negriform population, the now-extinct Tasmanians. These Australiforms are probably the most primitive of the four existing human races. They are also the "purest," in the sense that they have not merged or interbred with other races to any appreciable extent. Until very recently, their culture was at the stage of the Stone Age in Europe.

Origin of Human Races

Homo sapiens evolved from some previous hominid stock, probably in sub-tropical Africa or the Middle East. From the centre of their evolution, they spread out in four main directions, probably exterminating other hominids on their way: South and South-east to originate the Negriform and Australiform stock; East to originate the Mongoliform stock; North and West to originate the Europiform stock. There followed successive waves, which drove the earlier inhabitants to the extremities of their particular regions. Climatic and geological changes have isolated most groups, and there has been sufficient time for them to evolve sub-species but not different species.

Until relatively modern times, each race had within it a variety of cultures, some of which reached a high degree of variety and quality; for example, the Chinese, the Southern Indian, the Maya, Inca and Red Indian cultures. Among these, the least progress into what we call civilization has been made by the aborigines of Australia, and the bushmen and pygmies of Africa. In the last five hundred years, it is Western civilization that has made the greatest advances and that is now spreading all over the world. The great religions, Christianity, Mohammedanism, Buddhism, and great art, literature and science have all emanated from the Europiform peoples.

Social Evolution of Mankind

In their early history, human beings lived as small family groups, subsisting by hunting, fishing and gathering wild fruits, seeds and roots. For mutual protection, families came together in tribal groups. The bushmen of South Africa and some of the Australian aborigines still cling to this primitive form of existence. With the development of agriculture and the gradual domestication of animals, a more settled form of life arose, though even to this day many peoples still maintain a nomadic existence. Small settlements grew up at various points where the environment had natural advantages. With the gradual dominance of one group, states and nations came into being, the extent of each being determined by the area over which rule could be maintained. A more settled existence led to the acquisition of many skills: some became warriors, some tilled the soil, some made clothing, some built shelters, etc. The more successful and enterprising nations built up empires by the conquest and subjection of other nations.

SUGGESTED EXERCISES

1. Measure the length and width of the skulls shown in Fig. 22 and work out the cephalic index for each.

2. Make a simple, wooden measuring-gauge, as shown in Fig. 26. For the ruler,

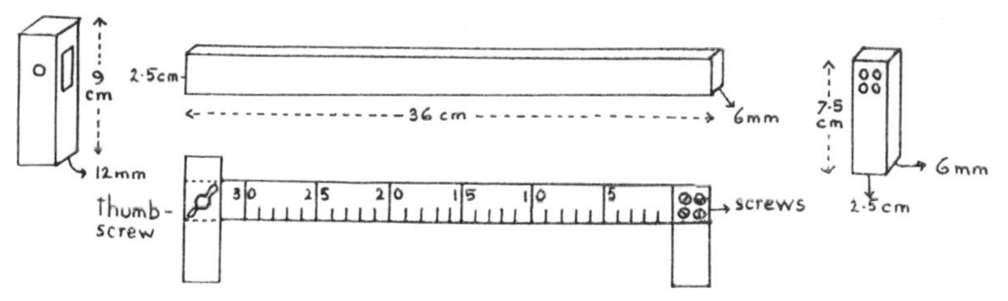

Fig. 26. Parts of a Gauge for Measuring Cephalic Indices

use a piece of wood 36 cm long, 25 mm wide and 6 mm thick. To one end fix at right-angles another piece of wood 7·5 cm long, 25 mm wide and 6 mm thick. For the other end, make a piece 9 cm long, 25 mm wide and 12 mm thick and cut out a hole so that it will slide along the measure and can be fixed by a thumb-screw. Use the slide to measure lengths and widths of heads, and so work out cephalic indices for different people.

3. Collect pictures of different human races and fasten them in your exercise book, classified as Negriform, Australiform, Mongoliform or Europiform.

4. Observe and sketch various human eye-shapes, ear-shapes and nose and lip profiles and notice how much variation there is, even among the pupils of your own class.

Chapter 4

GENERAL FEATURES: ANATOMY AND SKIN

The main characteristic features of modern man have been pointed out on pp. 26–30. In this chapter, human height, weight and longevity, and the major divisions of the body and the skin will be considered.

Height

There is considerable variation in the height of human beings. The tallest people, on average, are the Dinka negroes of the upper Nile valley, the full-grown males averaging over 180 cm. The shortest people are the pygmies of Africa, the full-grown males averaging approximately 120 cm. Men reach their maximum height by the time they are twenty-three years old and women by the time they are twenty years old. The average height for men in Great Britain is 175 cm and for women 160 cm. Height, like most of our characteristics, is affected by heredity and by environment. (*See* Chapter 16.) Environment plays a very important part in determining height; diet, sunlight and exercise all affect us and many people in the less advanced parts of the world are undersized because of their poor diet. But even the best possible environment will not bring about any marked increase in any one person's maximum height (i.e. at age twenty-three). In the more civilized countries, the height and weight of children show a gradual increase, but there is no reason to suppose that the average male height can be raised more than 6 cm or so. Careful measurements of adult skeletal remains from past ages show that average height has increased about 5 cm in the last five thousand years; all of this increase is probably due to better conditions of life.

Weight

Weight is far more affected by environment than height and hence there is far greater fluctuation even during one's own adult life.

The average weight for adult males in Western Europe is just below 70 kg. Weight is closely connected with height, but it must be remembered that some human groups are characteristically tall and slender, while others are short and broad. A Dinka negro, 180 cm tall, weighs about 57 kg, while an Eskimo of height 170 cm weighs about 77 kg. The average weights for particular heights of men and women of twenty-three years of age is shown in Fig. 27, which applies to Great

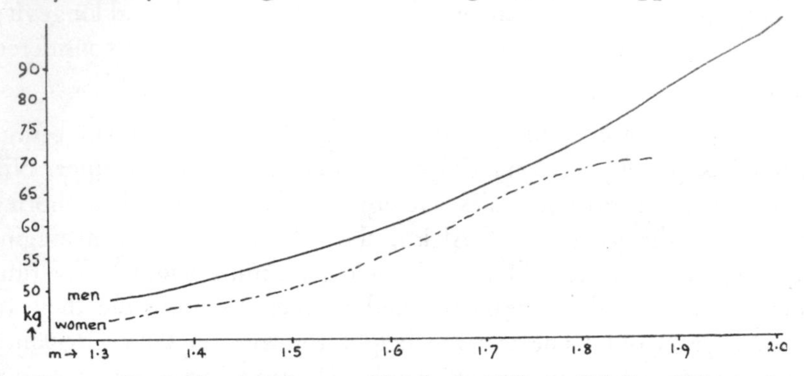

FIG. 27. Graph Showing Average Weight for Different Heights of Men and Women in Great Britain

Britain only. Factors such as diet, exercise and disease affect weight considerably.

LENGTH OF LIFE (LONGEVITY)

Modern man has a greater potential maximum length of life than any other mammal. Although length of life is undoubtedly affected by hereditary constitution, it is also considerably affected by mode of life, occupation and diseases suffered. Modern medical care and attention to diet have increased the possibility of surviving to old age. The average length of life in the Bronze Age is estimated to have been eighteen years, and even at the time of Christ it was only twenty-two (*see* Fig. 28). A great deal of the increase in average longevity is undoubtedly due to the greater knowledge about care of infants before, during and after birth. In 1850, infant mortality (i.e. before the age of one year) in Great Britain was about sixty-five per thousand live births. A century later, in 1950, the figure was forty-five per thousand.

Today, the average expectation of life in Western Europe and North America is about sixty-eight years for males and seventy-four years for females. In India, it has not yet reached thirty-five. It is a

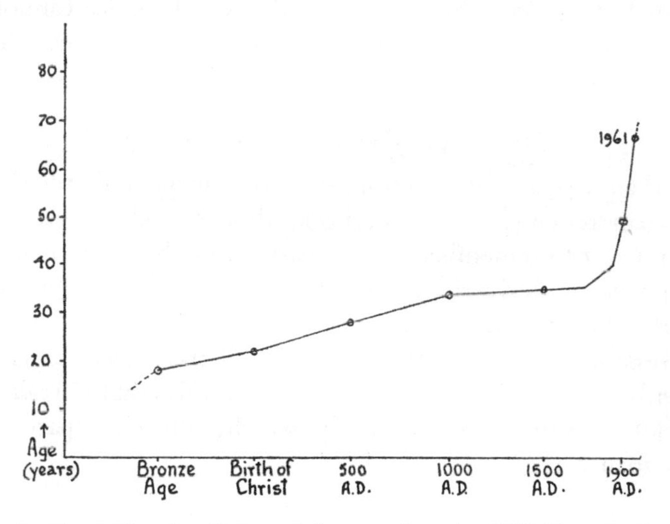

FIG. 28. Graph Showing Estimated Average Length of Life Since the Bronze Age

strange fact, as yet unexplained, that there are more boys born than girls, in the proportion of one hundred and six to one hundred. There is, however, a higher death rate among males than females, especially in infancy. By the age of twenty-three, the numbers of the sexes are approximately equal, and by middle life, i.e. fifty-five years, there are about fifteen per cent more women. In the age range seventy to eighty years there are twenty per cent more women, and over ninety years, twice as many women as men survive. These sex differences in average longevity are explained by the facts that males have less resistance to disease and more occupational risks, and in general lead more adventurous lives. There is no doubt that average longevity is increasing and by the middle of the next century may approach one hundred.

It must be emphasized strongly that these figures for height, weight and length of life are average figures, calculated from large populations; they do not apply to any particular case. The three characteristics mentioned are affected by heredity and sex, which we cannot alter, and by mode of life, medical care and occupational dangers, which we are continually striving to influence for the better.

The Main Divisions of the Body

For the purpose of description, it is convenient to divide the body into certain regions; the head, neck, thorax, abdomen and limbs. It is important to remember that all parts of the body are made up of cells, together with the substances they secrete. All the regions of the body are covered by the skin; all contain certain parts of the skeleton, blood-vessels, nerves and muscles; and all have their parts bound together by **connective tissue.** It is useful at this point to realize what the major constituents of the body weigh; the chief parts, for an average man of 70 kg, are listed below.

Part of the Body	Per cent Weight	Actual Weight in Kilograms
Muscles and tendons	42	29·5
Fat	19	13·2
Skeleton	16	10·9
Skin	7	5·0
Contents of abdomen	7	5·0
Blood	5	3·6
Brain	2	1·4
Contents of thorax	2	1·4

The following description applies to the erect position of the body, so that front and back, above and below have precise meanings.

THE HEAD

Apart from the general structures mentioned above, the head contains the **brain,** the opening for the intake of food (the **mouth**) and the first part of the food passage (the **buccal cavity**), containing

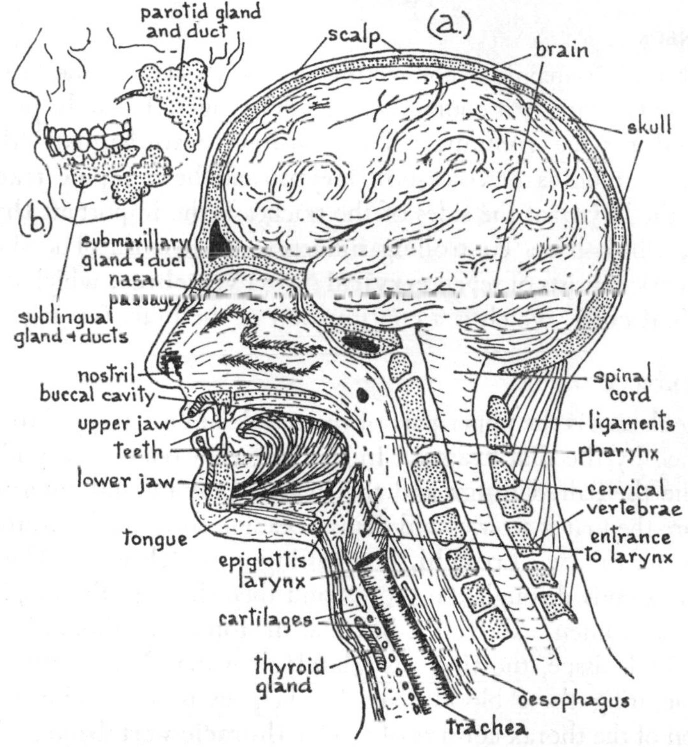

FIG. 29. (*a*) Median Section of Head and Neck (*b*) Positions of the Salivary Glands and Ducts Seen from One Side

the **tongue** and **teeth.** Above the mouth are the openings for the intake of air (the **nostrils**) and the first part of the breathing passage (the **nasal passage**). These two passages meet in the throat (**pharynx**). Below and at the sides of the buccal cavity are the **salivary glands,** with small tubes (**ducts**) leading the saliva into the cavity. These structures are shown in Fig. 29. Also in the head are the major sense organs, the **eyes,** the **ears,** the part of the **nose** concerned with smell, and the parts of the **tongue** concerned with taste. All these sense organs are more fully described in Chapter 14. The skeleton of the head is the **skull.**

The Neck

The neck is mainly a connecting structure between the head and thorax and essentially it contains a series of passages. From the pharynx, the food passage continues as the **gullet** (**oesophagus**); the air passage continues as the voice-box (**larynx**) and the windpipe (**trachea**). Below the larynx at the sides of the trachea is the important **thyroid gland,** which has a controlling influence on growth. The skeleton of the neck consists of seven **cervical** (neck) **vertebrae,** which contain the **spinal cord,** which is a continuation of the brain (*see* Fig. 29).

The Thorax

The thorax is the upper part of the **trunk** and is separated from the abdomen by the **diaphragm.** In the thoracic cavity, surrounded by fluid, lie the paired **lungs** and the **heart.** Leaving and entering the heart are the major arteries and veins. Leading from the bottom of the trachea are the two **bronchi,** one passing into each lung. The oesophagus extends through the thorax and then through the diaphragm into the abdomen. Above the heart, at the top of the thorax is a mass of soft pink tissue, the **thymus gland;** it is mainly concerned with manufacturing those blood cells that help us to resist disease. The skeleton of the thorax consists of twelve **thoracic vertebrae** enclosing the spinal cord, the twelve pairs of **ribs,** the **breastbone (sternum),** and the **pectoral girdle.** All the contents of a fluid-filled cavity such as the thorax are called **viscera.** Thus we say that the heart and lungs are the **thoracic viscera** (*see* Fig. 30).

The Abdomen

This is the lower part of the trunk, below the diaphragm. It has another fluid-filled cavity, in which the viscera lie (*see* Fig. 31).

Below the diaphragm, the oesophagus joins the **stomach;** the food passage continues as the small and large **intestines** and the **rectum,** and ends at the **anus.** The intestines occupy most of the abdominal cavity. In front of the stomach is the **liver,** the largest gland in the body, and behind the stomach is another gland, the **spleen.** A third gland, the **pancreas,** lies in the first loop of the small intestine. On the back (dorsal) wall of the abdomen are the two **kidneys,** which are concerned with extracting **urine** from the blood. From the kidneys,

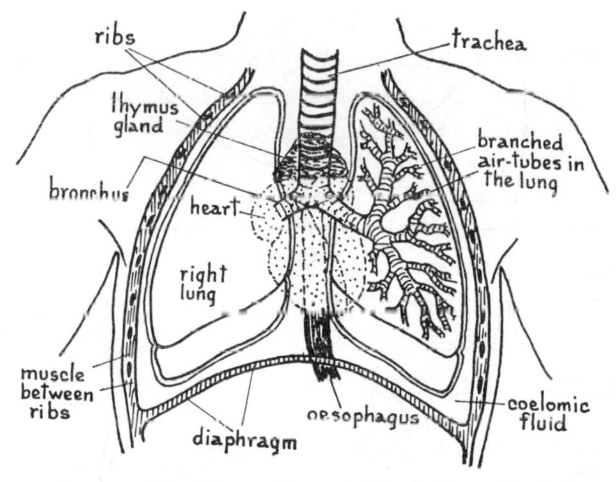

FIG. 30. Organs of the Thorax Shown with the Front Walls Removed

the urine passes down two fine tubes, the **ureters,** into the **bladder.** From the bladder the urine is expelled from the body along the **urethra,** the external opening of which is the **penis** in the male, and the **vulva** in the female. The male and female reproductive organs, **testes** and **ovaries** respectively, together with their ducts, lie in the lower part of the abdominal cavity. The skeleton of the abdomen consists of five **lumbar** and five **sacral** vertebrae (containing the end of the spinal cord), and the **pelvic** or hip **girdle.** The vertebral column ends in a small tail of four fused vertebrae, the **coccyx.**

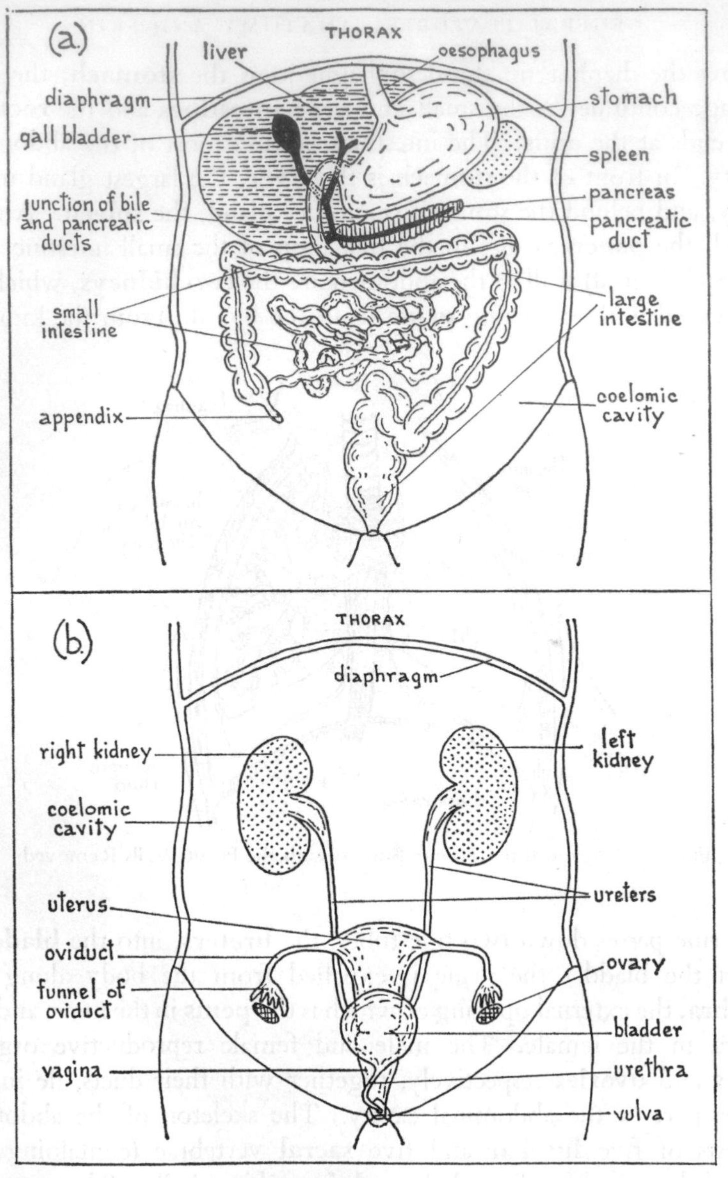

FIG. 31. General Internal Anatomy of the Abdomen (*a*) The Alimentary Canal and its Glands (Liver Turned over to the Right) (*b*) The Urinary and Reproductive Systems of the Female (For male organs, see page 227)

THE LIMBS

The limbs are essentially organs of locomotion, though in man, the front-limbs are not normally used for that purpose. They contain no large fluid cavities, and hence no viscera, and consist of little but the skeleton, the muscles, skin, blood-vessels, nerves and the all-important connective tissue.

GROWTH OF THE BODY

The human being grows for almost nine months in the mother's uterus before it is born. The fertilized egg is only 0·1 mm in diameter;

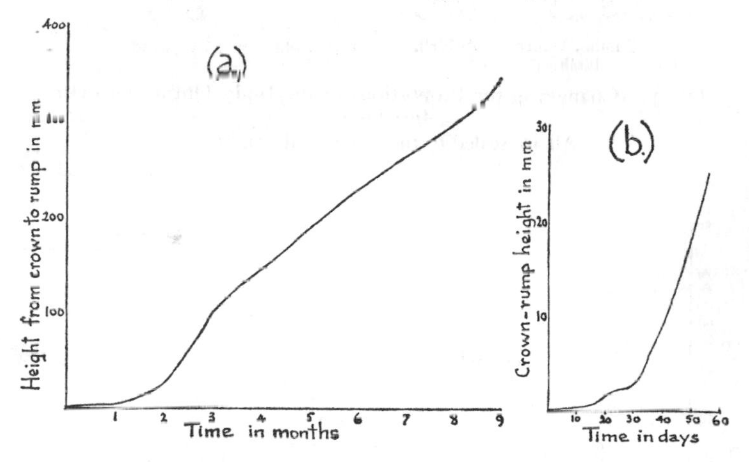

FIG. 32. Graphs Showing Growth of the Human Embyro (*a*) For the First Nine Months (*b*) For the First Eight Weeks

at birth, babies average 520 mm in total length. For the first month of this incubation period, growth is slow and fairly regular, for the second month it is rapid, and then fairly regular until birth. The graph in Fig. 32 illustrates the growth of an embryo up to the time of birth. Fig. 33 illustrates changes in the proportions of the body.

After birth, there is a short period of slow growth and then, as the infant settles down to its milky diet, it grows more rapidly. The birth weight is approximately doubled in five months and trebled in one year. Then there is slower growth for some years until the adolescent period begins. During this time, growth speeds up

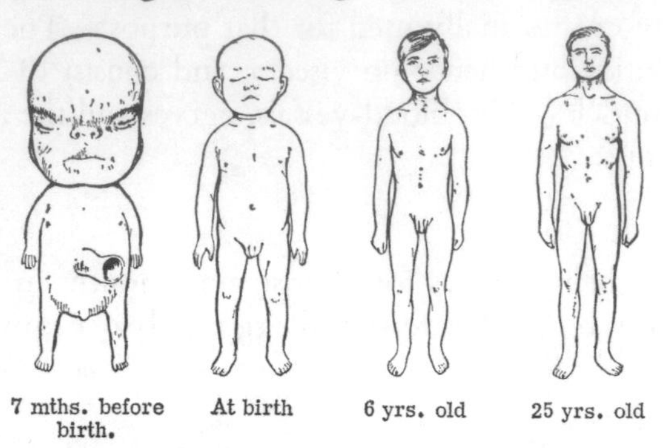

FIG. 33. Changes in the Proportions of the Body During Growth (*After Stratz*)
All are scaled to the same total height.

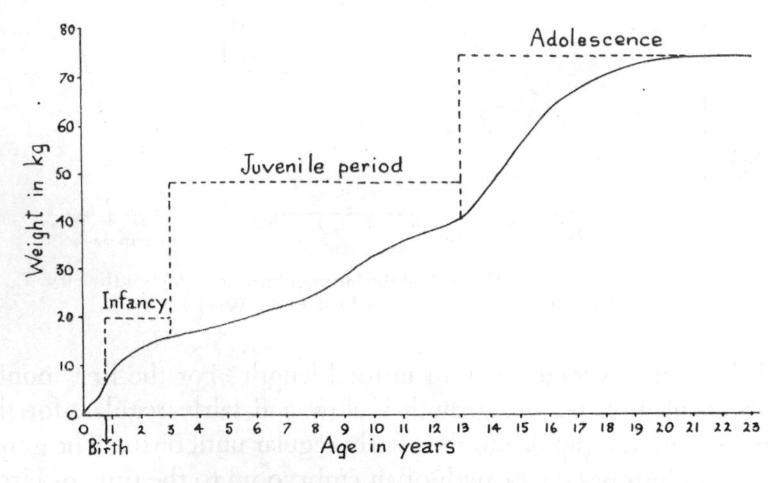

FIG. 34. Graph Showing Average Rate of Growth in Human Beings
For details of growth before birth, see Figs. 32 and 33.

considerably and then slows down again at the age of nineteen or so. Fig. 34 shows the average growth-rate for a male in Great Britain.

The Skin

The skin covers the whole external surface and is also, at an early stage of development, turned in to form the lining of the buccal cavity and of the rectum. It is a very important structure, performing a number of functions. Loss of more than three-quarters of the skin by accident, such as burning or scalding, means that death is inevitable.

Structure of the Skin

There are two distinct layers in the skin, the outer **epidermis,** which varies from 0·3 to 1·0 mm in thickness, and the inner **dermis,**

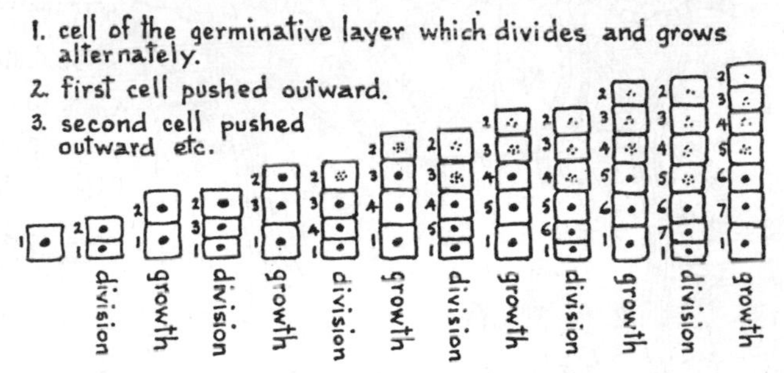

Fig. 35. Diagram Showing How a Cell of the Germinative Layers Continues to Produce More Cells, and How the Thickness is Built up and Maintained

which ranges from 1·2 to 4·0 mm in thickness. The skin is thickest on the palms of the hands and soles of the feet and thinnest on the lips.

The epidermis consists of four different layers of cells, the lowest of which, the **germinative layer,** produces all the other cells by dividing so that the outer cells are gradually pushed towards the surface (*see* Fig. 35). No blood-vessels penetrate the epidermis, and these cells die of lack of food and oxygen as they are pushed further away from the germinative layer. The outer cells are thin, like small flakes, and they are constantly removed by rubbing and washing the surface. Above the germinative layer are several layers of **granular cells,** which contain minute granules of the protein **keratin;** this substance ultimately makes the cells hard and dry. Above the granular layers are several **clear layers** of cells, so called because they are difficult to stain for microscopic work. Finally there are several layers of **horny cells,** extending to the outer surface (*see* Fig. 36).

Among the lower layers of the epidermis there are branched **pigment cells,** which produce the brown pigment, **melanin.** This

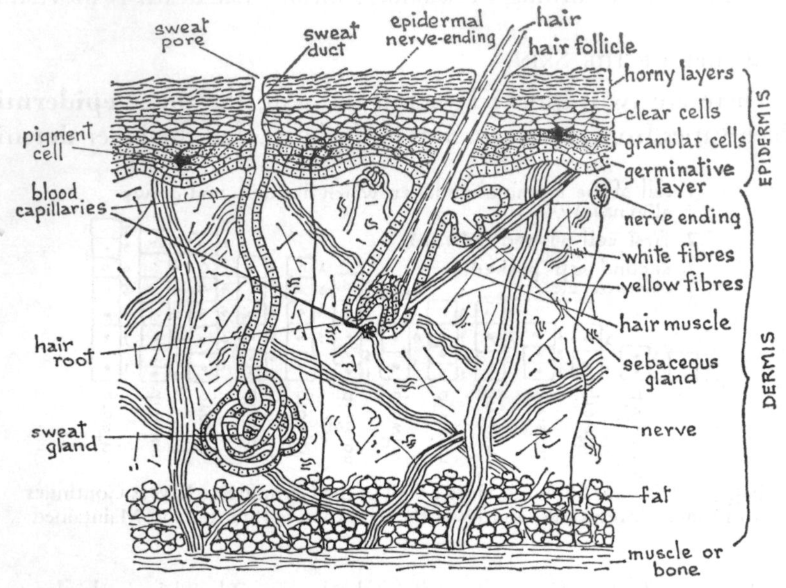

FIG. 36. Vertical Section of Human Skin, × 30

pigment is passed into the other epidermal cells, thus giving the skin a characteristic colour, ranging from white to dark brown or black. (*Note:* It is only in albinos that there is no dark pigment in the cells.) In the darker-coloured races, the pigment cells are very active; in the lighter races they are less active, though the numbers of pigment cells are about the same. Where the skin is pale, the colour of the blood also helps to make up the final shade.

The germinative layer grows down into the dermis to form two types of structure, **sweat-glands** and **hair follicles,** or pits, which are both therefore epidermal in origin (*see* Fig. 37). Each sweat-gland becomes coiled into a small knot at its base and is supplied by a network of **blood capillaries** in the dermis. The germinative layer of a hair follicle produces the hair; the cells are living at the root but they become impregnated with the protein keratin, and then die. The shaft of the hair is hollow; it contains fluid while it is below the surface and air above the surface. Every hair is moulted about once in three years and replaced by new growth from the root. At the side of

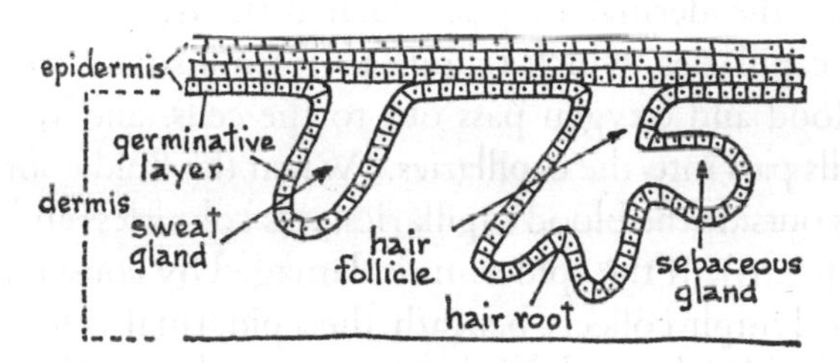

FIG. 37. Diagrammatic Vertical Section of Human Embryo's Skin, to Show the Origin of Sweat Glands and Hairs, × 60

each follicle, the germinative layer produces several flask-shaped **sebaceous glands.** These secrete an oily substance, which passes into the follicle, up to the surface and spreads over the skin and hair. It is said to make the hair supple and less easily broken; it also tends to prevent evaporation of water from the skin surface.

THE DERMIS

The dermis consists of numerous different structures and is a much more flexible layer than the epidermis. The main structure present is the **connective tissue,** consisting mainly of the **white fibres,** which are grouped into wavy bundles, and the **yellow elastic fibres,** which are fine and straight and which form networks. The white fibres give strength of attachment and impose a limit on the stretching of the skin; it can be stretched without injury only until the waves are straightened out. The yellow fibres give the skin the elasticity to recover after stretching or distortion.

There are numerous nerve-endings in the dermis; some of these penetrate for a short distance into the epidermis. Different types of nerve-endings are sensitive to touch, pain, temperature and pressure. On most parts of the body, the dermis pushes the epidermis into a series of bulges; these are very prominent on the hands and feet and give rise to the small grooves and lines visible on the palms and fingers. These grooves give the hands and feet greater gripping power.

There are networks of blood-vessels supplying the sweat-glands, the hairs and the dermal bulges. Each little artery branches into a network of capillaries, which reunite to form a small vein. From the capillaries, food and oxygen pass out to the cells, and waste-products from the cells pass into the capillaries. When the fluid containing these substances is outside the blood capillaries, it is colourless and is known as **lymph** (*see* p. 203). If the epidermis is damaged by constant rubbing or scalding, the lymph collects beneath the epidermal cells and forms a **blister;** if the blood-vessels of the dermis are damaged as well, blood is mixed with the lymph, forming a **blood blister.**

Fat is stored in the deeper parts of the dermis in the form of liquid drops in rounded cells; fat is always liquid in the living body; it congeals after death. The fat is thickest on the buttocks, the abdomen and the small of the back (**lumbar** region); it is thinnest in the skin of the eyelids and of the lips.

Each hair follicle has a small **erector** muscle attached near its base. In most mammals, these muscles are used to erect the hair, either in cold weather or when the mammal is attacked or frightened by other animals. The erection of the hair allows a thicker layer of air to be trapped and this acts as an insulating layer in cold weather. Erection of the hair also allows the animal to present a more frightening appearance to an enemy. This hair-raising can be seen in a cat, for example, when confronted by a dog. Human beings cannot erect their hair, but the contraction of the muscles causes **goose-pimples** on the skin.

Beneath the skin lies bone in some parts of the body, such as the scalp, the knuckles and the knee-caps; or muscle, as on the abdomen, the calves and the thighs. In either case, the skin is firmly bound to the underlying structure by the white and yellow fibres of the connective tissue.

FUNCTIONS OF THE SKIN

The skin serves as an external surface layer so that the muscles and bones and other structures are not exposed. It provides **protection,** gives **sensitivity** to various stimuli, and is an organ of **excretion** and **temperature-regulation.**

Protection from cuts and scratches is afforded by the horny layers, and by the hair in some parts, such as the head. These same dead cells,

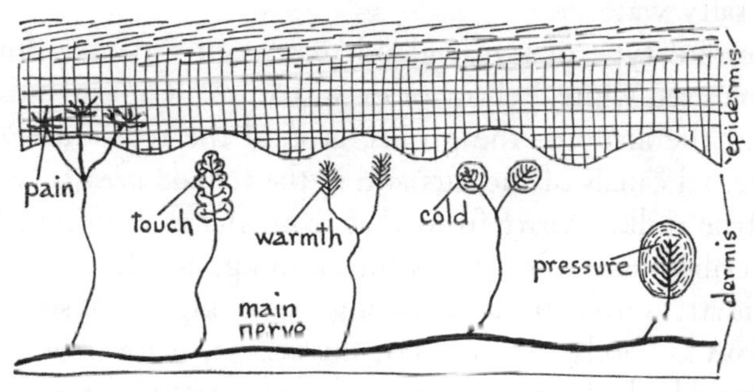

FIG. 38. Simplified Vertical Section of Skin, to Show the Different Types of Sense Organ

together with the oily secretion, protect the body from excessive evaporation of water and from penetration by germs. Man's relative hairlessness is a disadvantage and he has learnt how to make clothing, using at first the hair of other animals, in the form of fur garments, and later making garments from wool. Later still, man learnt to use vegetable fibres, such as cotton and flax. In modern times we have synthetic fibres, such as nylon and acrilan.

Sensitivity is provided by the various types of nerve-endings, each sensitive to one particular type of stimulus (*see* Fig. 38). Each nerve-ending, when stimulated, initiates an impulse, which passes along a nerve to the brain, where it is interpreted as pain, heat, pressure, etc. We can also distinguish grades of roughness and smoothness, which we call a sense of **texture.** Coiled around the base of each hair follicle is a small nerve-ending sensitive to touch.

The skin is an excretory organ, passing out sweat of two kinds from two different types of sweat-glands. Those which pass out a watery solution of salt and a little urea are present all over the body, except on the lips. In a full-grown man, there are over two million of these glands; the openings can be seen with a hand-lens on the ridges of the finger-tips. Continued loss of the sodium chloride in sweat leads to muscular cramp, such as is often experienced by workers in deep mines and in foundries. This condition is easily remedied by drinking a glass of salty water.

The second type of sweat-gland passes out a denser fluid, which may be coloured white, yellow or even reddish. These glands are most plentiful in the armpits, the groins, around the reproductive organs, in the external canals of the ears and in the female breasts, where they produce true milk. Apart from that from the mammary glands, the sweat described here often stains the clothing, and if left on the body surface it is attacked by bacteria, giving rise to unpleasant smells, known in these days as "body odour." For this reason, if for no other, human beings should bathe frequently to wash the secretion from their bodies, and clothing, particularly underclothing, should be frequently washed.

Through the medium of sweat, the skin helps to regulate body temperature. Heat is produced in all cells of the body, particularly in organs that are constantly active, such as the liver. During exercise, a great deal of heat is produced by the muscles. The heat is circulated by the blood, which acts as a heating system, and thus the temperature is made fairly even throughout the body, i.e. 36·9°C. But we cannot continue to produce heat without losing some of it, or the body would reach a temperature too high for survival. We lose heat from every aperture in the body: in the exhaled air, in the urine and in the faeces; but most heat is lost by evaporation of the sweat, the heat necessary for evaporation being taken from the body. To boil water and thus evaporate it, heat must be supplied from a fire, from burning gas, or from a hot electric element. To evaporate sweat from the skin, heat must be supplied; it comes from the hot body. The more we sweat, the more heat we lose. The process is regulated from a **temperature centre** in the brain. From this centre, nerves of two kinds pass to the blood-vessels of the skin: **constrictor nerves,** which cause the vessels to contract, and **dilator nerves,** which cause the vessels to expand. Thus the supply of blood to the skin and hence to the sweat-glands is regulated automatically. More blood passing to the sweat-glands means that more sweat is passed to the surface, there is more evaporation and more heat is lost. In very cold weather, we cannot afford to lose much heat and so the blood supply is restricted and the skin becomes pale and cold. Sweating is also affected by emotion and by fright. When one is frightened, the blood supply to the skin is reduced suddenly, giving rise to the sensation known as a "cold sweat."

SUGGESTED EXERCISES

1. Compare the rate of return of skin pulled up at the back of the hand on yourself and on a much older person.

2. Examine a prepared slide of a vertical section of human skin, with a microscope or microprojector.

3. Examine a prepared injected section of mammalian skin. This will show the blood-vessels.

4. Test your texture-sense on a number of grades of sandpaper. With eyes closed, try to grade them from smoothest to roughest.

5. Obtain a piece of the braille used by blind people for reading. With eyes closed, try to feel numbers and positions of the dots in a word.

6. With a hand lens, examine the ridges on your finger tips and note the small sweat-pores.

7. Cut a preserved rat's head vertically into two, keeping the cut slightly to one side of the mid-line. For soft structures use a sharp knife or scalpel, and for bone, a fine-toothed hacksaw. Examine the cut surface and try to identify the major structures in the head. Make a large, labelled drawing of the specimen.

8. Instructions for dissection of the neck, thorax and abdomen of the rat are given in *The Mammals*, in this series, on pp. 91, 101, 118, 124, 172.

CHAPTER 5

THE SKELETON

THE human skeleton consists largely of **bone,** but there are a number of places where **cartilage** is present. In the young embryo, much of the skeleton is first formed in cartilage; after the second month, the

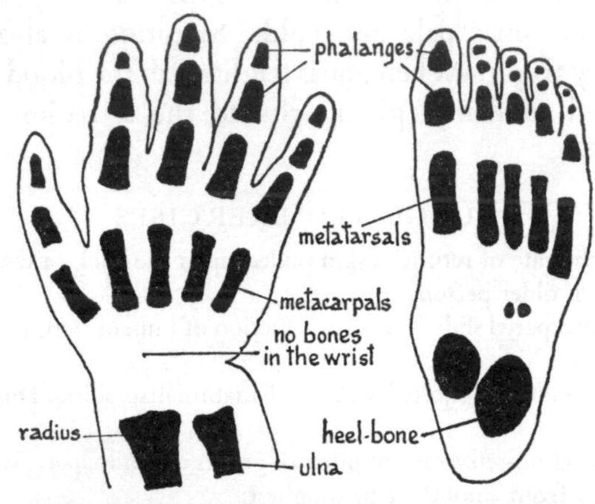

FIG. 39. The Stage of Ossification in the Hand and Foot at Birth (*drawn from a radiograph*)

process of **ossification,** i.e. formation of bone, begins and most of the cartilage is ossified by the time of birth. Fig. 39 shows a **radiograph,** or X-ray photograph, showing ossification in a hand and a foot of a new-born baby. Some parts of the skeleton, especially the skull, are not pre-formed in cartilage, but are developed in connective tissue such as the dermis of the skin.

Bones are not dead structures, but living organs that grow and change shape. Growth in the size of the bones is completed by the age of about nineteen in females and twenty-one in males.

STRUCTURE OF CARTILAGE AND BONE

In the living body, clear cartilage is a bluish-white, translucent material, consisting of individual cells or small groups of cells lying in a stiff gelatinous substance called the **matrix,** or **ground substance.** Numerous very fine fibres, difficult to demonstrate, traverse this matrix (*see* Fig. 40). Each cartilage is encased in a tough membrane called the **perichondrium,** in which blood-vessels and nerves are

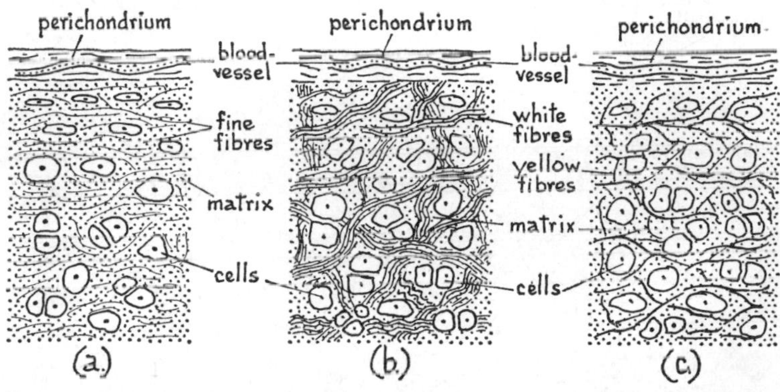

FIG. 40. Sections of the Three Kinds of Cartilage (*a*) Hyaline (*b*) Fibro-cartilage (*c*) Elastic Cartilage: All × 150

present. These structures do not penetrate into the matrix of the cartilage. This type of cartilage is known as **hyaline,** or **clear cartilage.** It will bend fairly easily and recover its shape readily; we say that it is flexible and elastic. This type of skeleton is adapted for early embryonic life, when a rigid skeleton would be a hazard to survival. In the adult skeleton, hyaline cartilage is present as incomplete rings in the larynx, the trachea and the bronchi, at the ventral ends of the ribs and at the ends of bones.

There are two other types of cartilage in the body; in **fibro-cartilage** there are numerous wavy bundles of strong, white fibres in the matrix (*see* Fig. 40(*b*)). These fibres give added strength but less flexibility and elasticity. Fibro-cartilage is present in the discs between the vertebrae (*see* p. 62) and in the junction between the two pubic bones at the lower end of the abdomen. In **elastic cartilage** there are single, fine, yellow fibres in the matrix, giving somewhat more elasticity (*see* Fig. 40(*c*)). Elastic cartilage forms the skeleton of the outer ear, the **epiglottis** and the **Eustachian tube** (*see* pp. 136 and 159).

Bone is very hard, owing to secretion of calcium salts by the cells. Two thirds of the weight of a bone consists of salts in the following

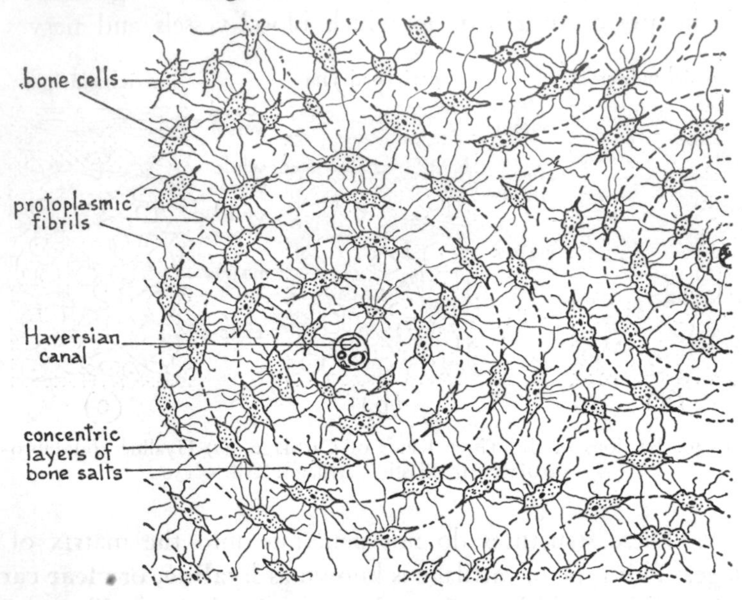

FIG. 41. Transverse Section of Compact Bone, × 100

proportions: eighty-five per cent calcium phosphate, ten per cent calcium carbonate, four per cent magnesium chloride and one per cent calcium fluoride. Among this matrix of salts, there are fine fibres forming a delicate meshwork, **bone-cells** trapped in small spaces, and a system of **Haversian canals** (*see* Fig. 41) containing blood-vessels, nerves and soft, connective tissue. Around each canal, the salts are deposited in concentric layers, the cells lying in cavities between the layers. Each cell has many fine, **protoplasmic fibrils,** which interconnect to form a delicate network of living substance. The salts of a bone may be dissolved in a dilute acid, and then a pliable fibrous network remains. Thin sections of bone are prepared by grinding; the cavities are then all filled with bone dust.

Structure of a Long Bone

Bones are described according to their shape, as long, short, flat or irregular. Long bones are found in the arms, legs and ribs; short bones in the wrists, ankles, fingers and toes; flat bones in the skull,

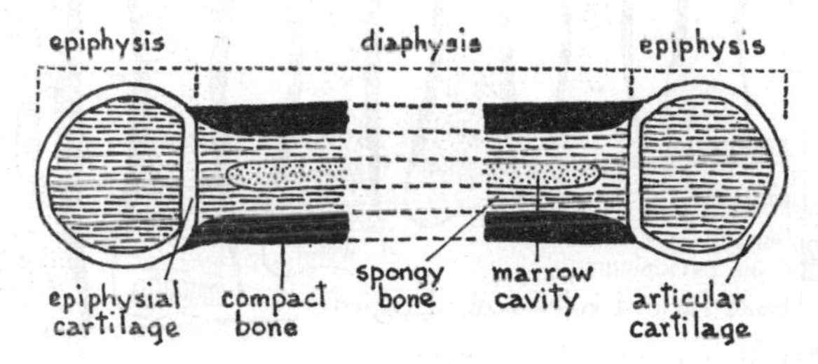

Fig. 42. Longitudinal Section of a Long Bone (Diagrammatic)

and in the pectoral and pelvic girdles; irregular bones in the vertebral column.

Each bone is encased in a tough, fibrous membrane called the **periosteum,** beneath which is a layer of very hard, **compact bone** with Haversian canals running longitudinally and occasionally transversely. Then there is a thicker layer of **spongy bone** with many large spaces. In the centre of the bone, there is a **marrow cavity** containing numerous cells. It is here that the red cells of the blood, and many of the white cells, are made. The shaft of the bone is called the **diaphysis** and the two rounded ends are called **epiphyses** (*see* Fig. 42). Between each epiphysis and the diaphysis there is a thin layer of cartilage, and another layer of cartilage surrounds the ends.

Ossification

There are two types of ossification, depending on whether the bone is formed from cartilage or from connective-tissue membranes;

thus, according to their origin, bones are known as **cartilage-bones** or **membrane-bones.** Most of the bones of the skull are membrane-bones, which are characteristically flat. With the exception of the **clavicle** (collar-bone), all the rest of the skeleton consists of cartilage-bones. A cartilage is ossified in the following sequence (*see* Fig. 43).

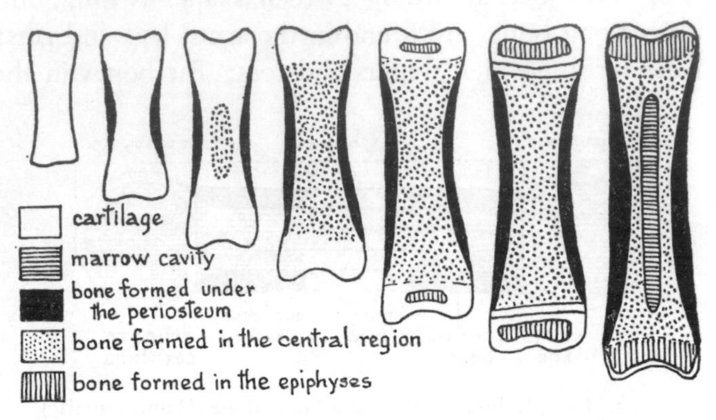

FIG. 43. Stages in the Ossification of a Long Bone

1. **Bone-forming** cells in the periosteum secrete (give out) calcium salts, which are deposited in the form of a cylinder round the middle of the shaft.
2. The cartilage cells in the central region swell and die. Then the matrix is invaded by the bone-forming cells, which deposit more salts.
3. Ossification by degeneration of the cartilage cells, and penetration by the bone-forming cells, continue from the centre to the ends of the shaft.
4. Other cells from the periosteum, **bone-destroying cells,** migrate into the calcified matrix, eating away channels, which become the Haversian canals. Blood-vessels grow from the periosteum into these canals. In the central region, almost all the calcified matrix is eaten away to form the marrow cavity surrounded by the spongy bone.
5. Many of the bone-forming cells become surrounded by the

salts they secrete and thus become trapped, communicating with one another by means of their protoplasmic fibrils.

6. Ossification next takes place in the two epiphyses, in the same sequence as above.

At birth, the shafts are ossified but the ends are still cartilaginous. Soon the epiphyses ossify, but a thin layer of cartilage separates them from the shaft. Growth in length takes place by increase and then ossification of this layer of cartilage. Growth in girth occurs by deposition of bone under the periosteum. When growth is complete, these epiphysial cartilages ossify and leave a characteristic, ring-like groove.

The thin layers of cartilage at the ends of the bone, called articular cartilages, remain throughout life; they help to absorb shock and to make the action of the joint smooth. The bones change in shape as they grow, owing to the various pulls exerted by the muscles and also to the weights to which they are subjected. This remodelling of the skeleton is constantly taking place, by absorption of bone in some regions and by further ossification in others. It is for this reason that children are encouraged to perform a variety of exercises and to adopt good posture in sitting or standing. These things affect the final shape and carriage of the body. In the skull and vertebrae, containing the brain and spinal cord respectively, special conditions apply. As the brain and spinal cord grow, bone is removed from the inner surface by bone-destroying cells, and then further bone is deposited on the outer surface by bone-forming cells. Thus the skull keeps pace with the growing brain.

How Bones Are Joined Together

One bone does not normally make contact with another; between the two adjacent surfaces there is a **joint.** According to the amount of movement they make possible, joints are classified as immovable, slightly movable and freely movable.

Immovable joints are present between the bones of the cranium (brain-case) and between the pelvic girdle and sacrum. The bone surfaces are serrated, and adjacent surfaces interlock, but between them there is always a thin layer of connective tissue (*see* Fig. 44). Such a serrated junction is called a **suture.** Often, in older people, the sutures become ossified and almost obliterated.

Slightly movable joints have pads of fibro-cartilage between the bones; they are present particularly as the **intervertebral discs**

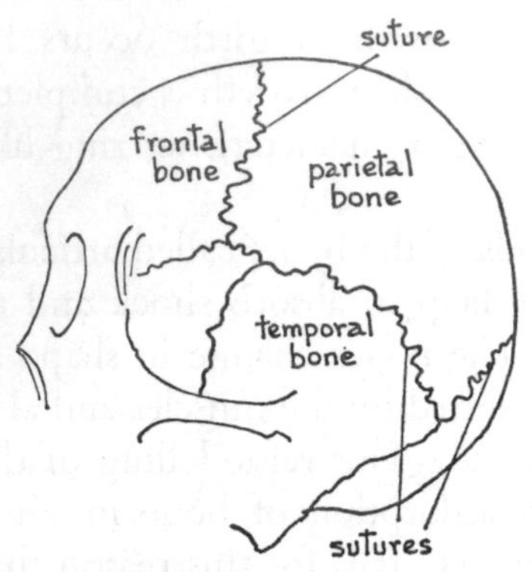

FIG. 44. Side View of Skull of Young Adult, Showing Sutures

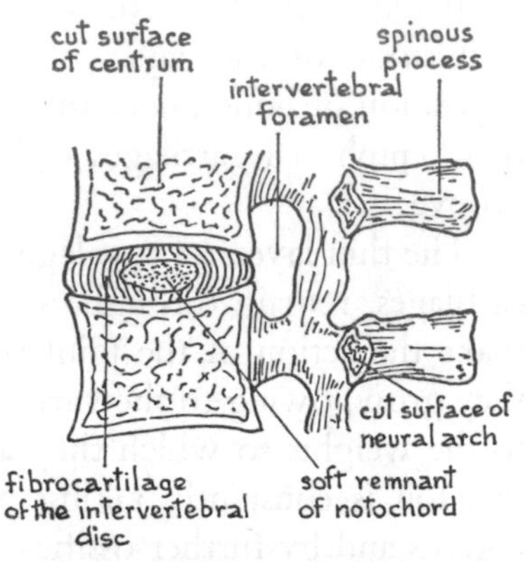

FIG. 45. Median Vertical Section of Two Lumbar Vertebrae, Showing Inter-vertebral Disc

between the vertebrae (*see* Fig. 45). These discs act as shock-absorbers and reduce jarring of the spine. In the centre of each disc, there is a soft, jelly-like mass, the remains of the embryonic notochord (*see* p. 14). Sometimes a sudden or awkward movement displaces this mass from its central position, thus distorting the shape of the disc. The disc may then press against one of the spinal nerves, giving pain in some region of the trunk or limbs; if the pain is in the legs, it is known as sciatica, if in the small of the back, as lumbago. The whole condition is known as a **slipped disc.** Another slightly movable joint is present between the two pubic bones. In males, this joint is almost immovable, but in females the cartilage contains a good deal of elastic tissue, which will part to widen the pelvic girdle during the birth process.

In freely movable joints, both bone-ends are encased in a sleeve-like ligament made almost entirely of yellow elastic fibres. Within the cavity enclosed by this **capsular ligament,** there is a connective-tissue bag called the **synovial capsule,** containing **synovial fluid.** In some of the larger joints, such as the knee, the capsule is divided into chambers by cartilages (*see* Fig. 46). Apart from the capsular ligament, smaller ligaments, somewhat like strong rubber bands, join bone to bone.

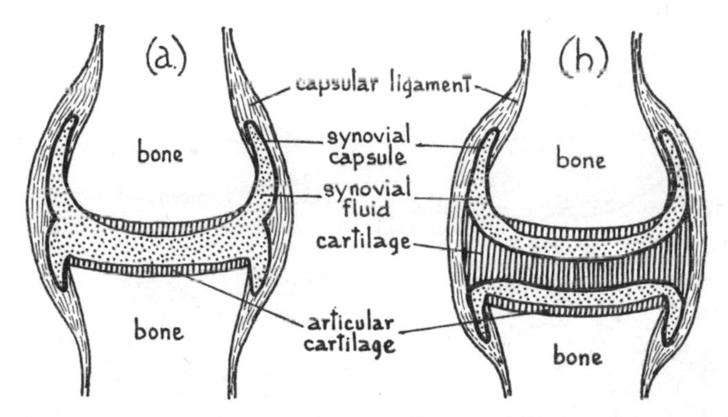

FIG. 46. *a*) Typical Synovial Joint (*b*) Synovial Joint with Cartilage

Thus, in a freely movable joint there is strength and flexibility, due to the yellow elastic ligaments, and also reduction of friction and shock, due to the articular cartilages and synovial capsules.

CLASSIFICATION OF SYNOVIAL JOINTS

Synovial joints are classified according to the shapes of the two adjacent bones and the amount of movement possible between them. There are four types—

1. In **ball-and-socket joints,** the end of one bone is rounded, while the other is hollowed out to form a shallow cup. Such a joint allows freedom of movement in any plane; in the case of the shoulder joint, complete circular rotation of the arm is possible. The hip joint is somewhat more limited in movement (*see* Fig. 47(*a*) and (*b*)).

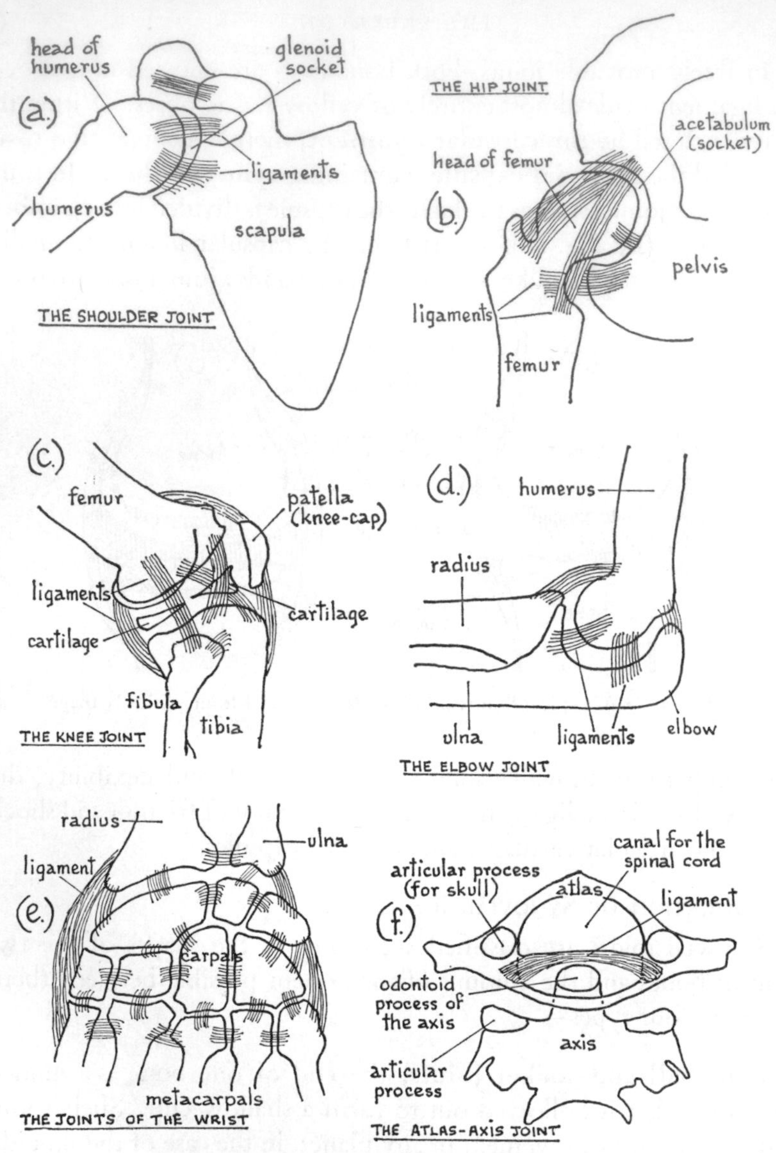

FIG. 47. Types of Synovial Joints
(*a*) and (*b*) Ball-and-socket (*c*) and (*d*) Hinge (*e*) Sliding Joints (*f*) Pivot Joint
In each case the synovial ligament and synovial capsule are omitted.

2. With **hinge joints,** there is freedom of movement in one plane only. Good examples are present in the knee and the elbow (*see* Fig. 47(*c*) and (*d*)).

3. **Plane joints** occur between the small bones of the wrist and ankle (*see* Fig. 47(*e*)). The two articulating surfaces are almost flat and slide over one another easily, giving flexibility but limited strength. Hence wrists and ankles are easily sprained.

4. There is one **pivot joint** in the body, that between the first vertebra, the **atlas,** and the second vertebra, the **axis** (*see* Fig. 37(*f*)). The axis has a small, peg-like projection, the **odontoid process,** which fits into the atlas and allows a certain amount of rotation. By this rotation, the head and atlas turn on the axis. The head alone cannot be turned. The nodding of the head up and down is due to the movement of the skull on the atlas, where there is a hinge joint.

General Plan of the Skeleton

There are two hundred and twelve named bones in the adult body; the unnamed ones are tiny bones that are not always present; they are called **sesamoid bones** and are developed in ligaments. The knee-cap (**patella**) is the largest of these sesamoid bones and others are often found in the hands and feet.

The skeleton of all vertebrates is based on a common plan. There is a main axis, or **axial skeleton,** and the remaining parts articulate with this axis and constitute the **appendicular skeleton.** The axial skeleton consists of the **skull,** the **vertebral column,** the **sternum** and the **ribs.** The appendicular skeleton consists of the **pectoral girdle,** the **pelvic girdle** and the front- and hind-limb skeletons. The whole skeleton is shown in Fig. 48, with the major parts named.

The Skull

The skull is the skeleton of the head and, strictly, it includes the **cranium,** the bony parts of the **auditory** and **nasal capsules,** the **jaws,** the **hyoid body** in the floor of the mouth, and the cartilages that support the **larynx, trachea** and **bronchi.** In our remote

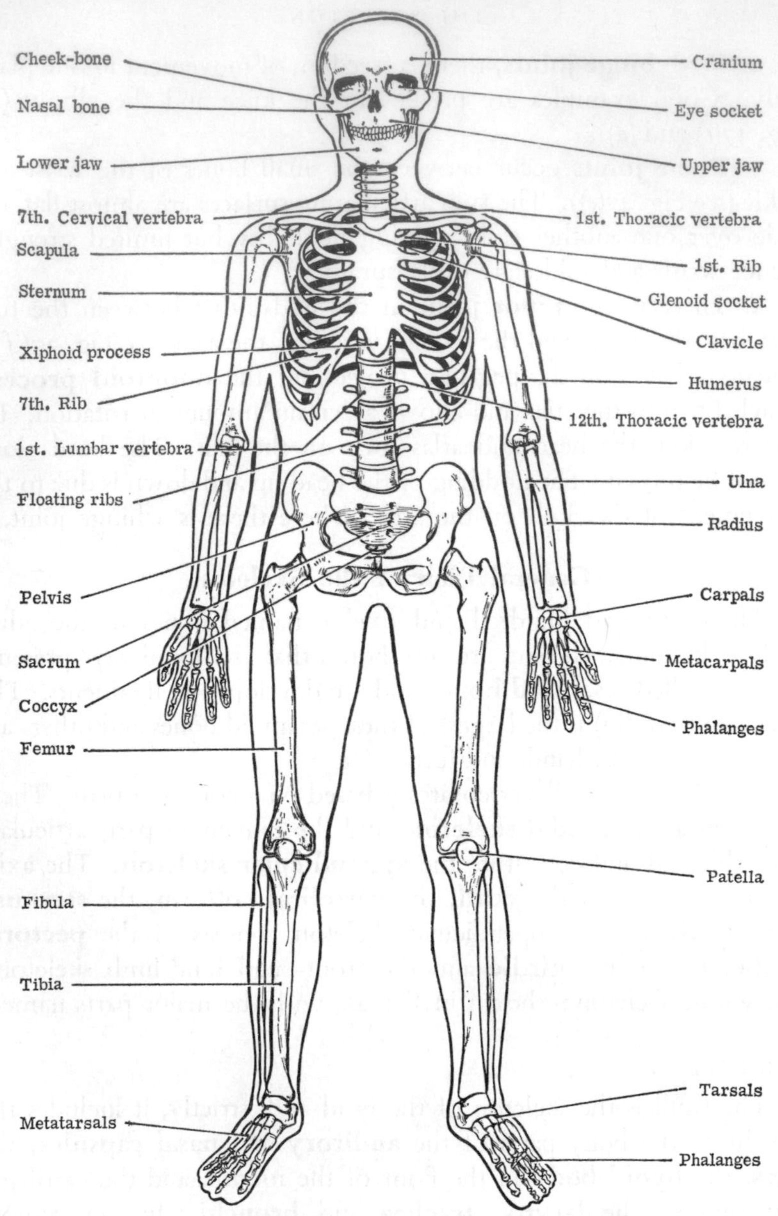

FIG. 48. Human Skeleton Viewed from the Front

ancestors among the fishes of the Primary era (*see* p. 10), the skull was made entirely of cartilage and is shown in Fig. 49. Throughout the long course of evolution, many changes have taken place in the skull. New parts have been added, old parts have changed their position and function, while some parts have disappeared. The whole skull is shown in side view in Fig. 50 and the main part of it in front view in Fig. 51.

Functions of the Skull

The cranium encloses and protects the brain and is perforated by a number of holes, or **foramina** (singular, **foramen**), through which nerves or blood-vessels pass. The largest of these holes is the **foramen magnum,** where the brain and spinal cord merge. The cranium articulates with the atlas by the two **occipital condyles,** and thus rocking movement of the head is possible. The inner and middle parts of the ear (*see* p. 262) are protected by the **temporal bone,** while the inner part of the nose is protected by the **nasal bones** above and the **vomer bones** beneath.

The upper jaw is immovable, the lower jaw articulating with it by a joint that allows vertical, and a small amount of sideways, movement. The jaws contain the teeth for biting and chewing (*see* p. 129). The cartilages of the larynx, trachea and bronchi keep the air passages permanently open (*see* p. 161).

The Vertebral Column

This is commonly called the spine, or backbone, and it consists of twenty-six vertebrae in the adult. There are thirty-three developed in the embryo but some fuse together later in life. The vertebrae of the neck are called **cervical,** and there are seven; those of the thorax are called **thoracic,** and there are twelve; those of the loin or small of the back are called **lumbar** vertebrae and are five in number. The **sacral,** or hip, vertebrae are five in number, and are fused to form a single bone, the **sacrum.** Finally man has the remnants of the skeleton

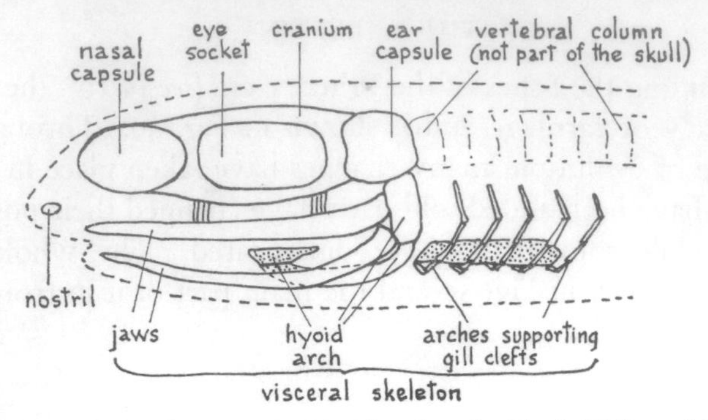

FIG. 49. Skull of Ancestral Fish, Showing the Typical Visceral Skeleton

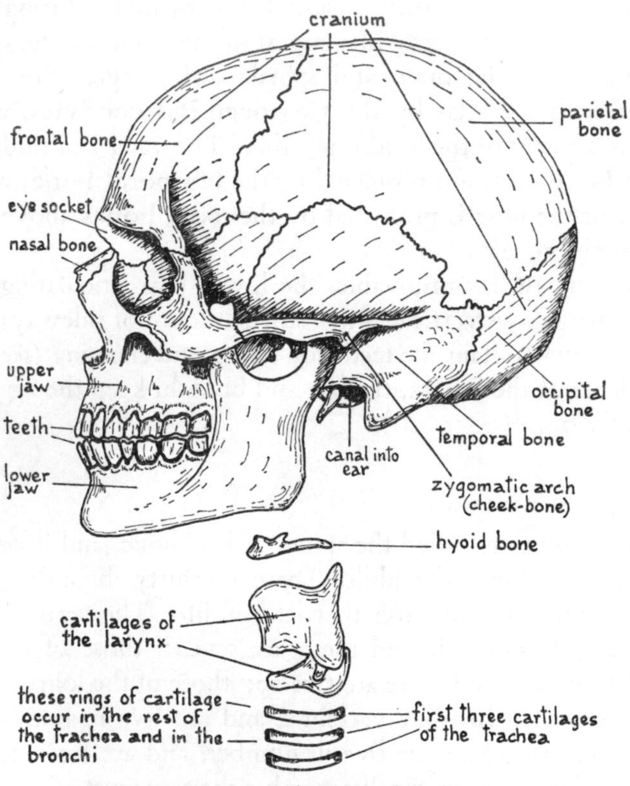

FIG. 50. Side View of Skull

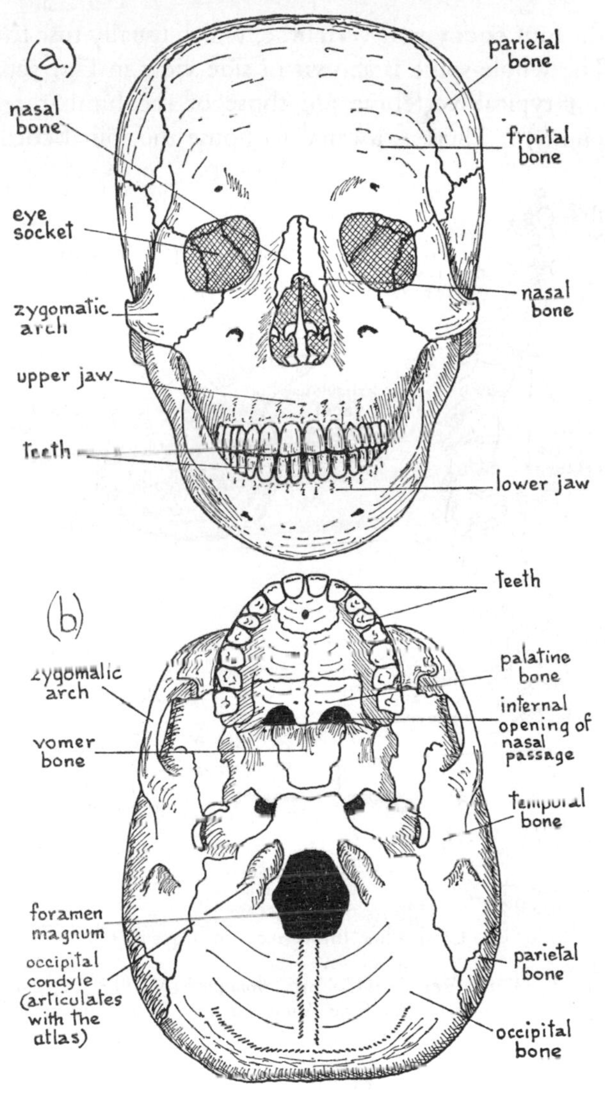

FIG. 51. (*a*) Front View of Skull (*b*) View from Beneath

of a tail in the four **coccygeal** vertebrae, which usually fuse to form the **coccyx.** The whole spine is shown in side view in Fig. 52(*a*).

The most typical vertebrae are those of the lumbar region (*see* Fig. 52(*b*) and (*c*)). There is a canal to house the spinal cord with an

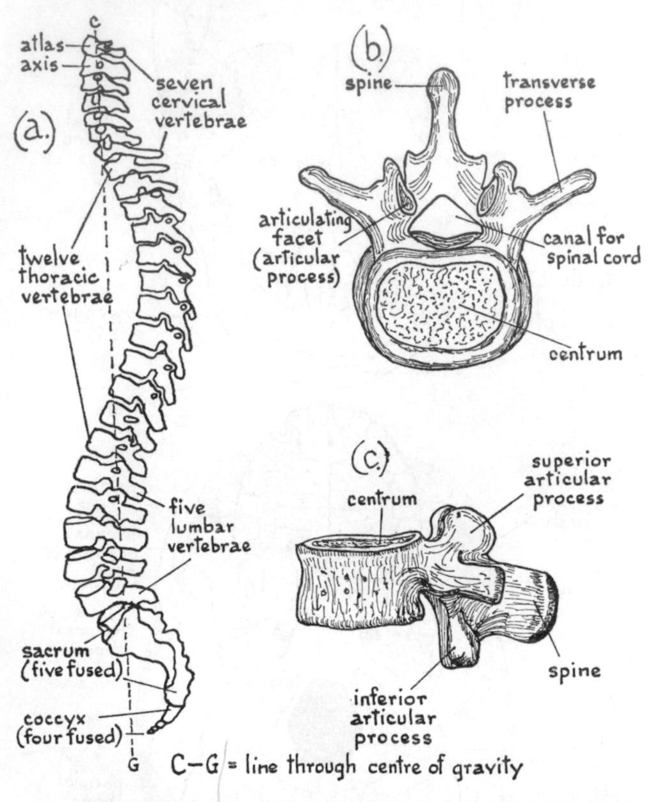

FIG. 52. (*a*) The Entire Vertebral Column (*from the left*) (*b*) A Typical Lumbar Vertebra, Anterior View (*c*) Side View of the Same Lumbar Vertebra

arch of bone over the canal. From this **neural arch** a number of processes project; they provide attachment for muscles and ligaments. Each vertebra has two **articular processes,** anterior and posterior; these form synovial joints with adjacent vertebrae. The **centrum** is the solid portion of a vertebra. Adjacent vertebrae are separated by intervertebral discs, which are located between the centra. All the vertebrae are strongly bound together by ligaments, which have a

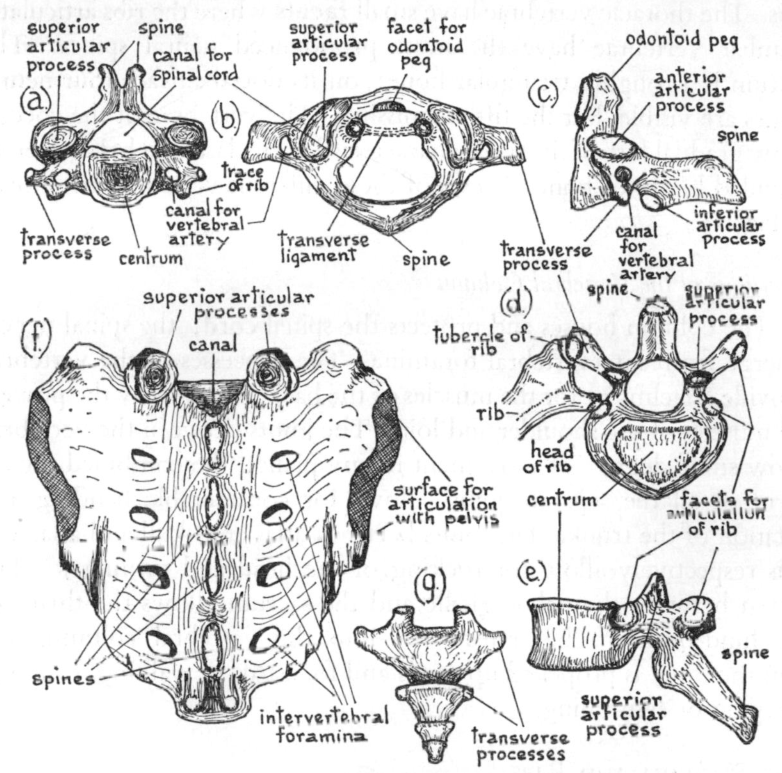

FIG. 53. (*a*) Anterior View of a Typical Cervical Vertebra (*b*) Atlas, Anterior View (*c*) Axis, Side View (*d*) Thoracic Vertebra, Anterior View, Showing Rib Articulation (*e*) Thoracic Vertebra, Side View (*f*) Sacrum, View from the Back (*g*) Coccyx, View from the Back

very complex arrangement. A very strong ligament binds the neck vertebrae to the back of the skull.

The different types of vertebrae can easily be identified. All cervical vertebrae have lateral canals in which the vertebral artery is housed during life (*see* Fig. 53(*a*)). The atlas has broad, wing-like transverse processes (*see* Fig. 47(*f*), p. 64), and no centrum. A narrow ligament divides its canal into two portions, one for the spinal cord and one to accommodate the odontoid process, which identifies the axis. The thoracic vertebrae have small **facets** where the ribs articulate. Lumbar vertebrae have the most pronounced neural spines. The sacrum is a roughly triangular bone; on its dorsal surface four neural spines are visible, but the fifth is missing. The vertebrae of the coccyx show gradual loss of the usual characteristics, the last two being merely rounded knobs of bone. Views of these different vertebrae are shown in Fig. 53.

Functions of the Vertebral Column

The column houses and protects the spinal cord; the spinal nerves emerge from intervertebral foramina. The processes of the vertebrae provide attachment for the muscles of the back, particularly the powerful muscles of the shoulder and loin. The joints between the vertebrae allow small degrees of movement in any plane; the combined movement of all the separate joints allows for considerable bending and rotation of the trunk. The joints between atlas and skull and atlas and axis respectively allow for rocking of the head and turning it. The fusion between the pelvic girdle and the sacrum allows the thrust of the hind-limbs to be transmitted into the vertebral column, and thus the body is propelled upwards and forwards in walking, running, jumping or swimming.

The Sternum and Ribs

The sternum, or breastbone, arises in the embryo as six separate cartilage bones, forming a series of **sternebrae** (singular, **sternebra**) corresponding to the vertebrae. Later in life, only three bones are evident, the **manubrium,** the **body of the sternum** and the **xiphoid process.** Even these may fuse in old age. The **clavicles** (collar-bones) articulate with the top of the manubrium and seven of the rib cartilages also join the sternum; one cartilage joins the manubrium and six join the body of the sternum (*see* Fig. 54(*a*)).

The ribs arch downwards from the vertebrae and upwards towards the sternum. Each rib consists of a bony portion and a cartilaginous portion (*see* Fig. 54(*b*)). The bony portion articulates with the thoracic vertebrae, while the cartilaginous portion joins the sternum, in the

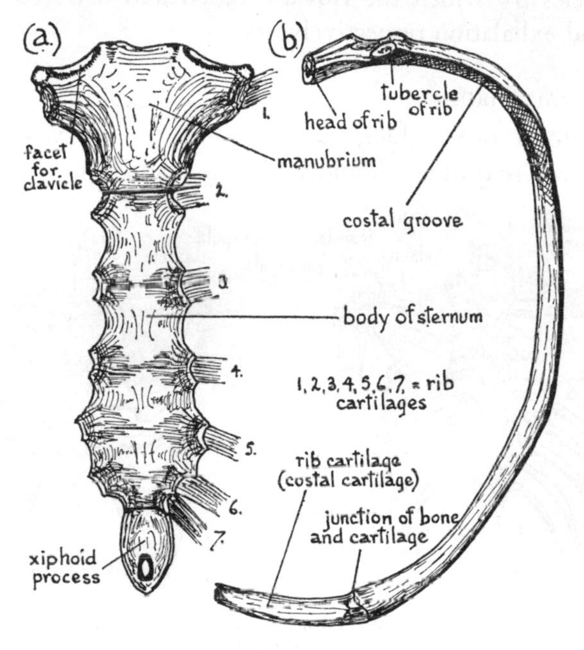

FIG. 54. (*a*) The Sternum, from the Front (*b*) Sixth Rib, from Below

case of the first seven pairs of ribs. The cartilages of the eighth and ninth pairs join the cartilage of the seventh. The tenth, eleventh and twelfth ribs are free in front and are thus called the **floating** or **short ribs** (*see* Figs. 55(*a*) and (*b*)).

Functions of the Sternum and Ribs

The bony cage formed by the sternum, the ribs and the thoracic part of the vertebral column houses and protects the thoracic viscera.

The cartilaginous articulation of the ribs with the sternum and the synovial joints of the ribs with the vertebral column provides the flexibility necessary for breathing. Finally, the sternum and ribs exhibit surfaces for attachment of muscles, particularly for the intercostal muscles by which the ribs are raised and lowered during inhalation and exhalation respectively.

THE PECTORAL GIRDLE

The pectoral, or shoulder, girdle is composed of two **clavicles** in front and two **scapulae** (shoulder-blades) behind (*see* Fig. 55). It is

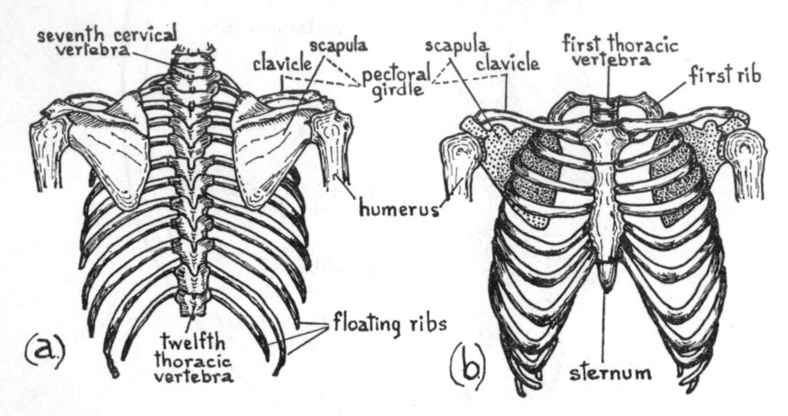

FIG. 55. (*a*) Pectoral Girdle and Ribs, from the Back (*b*) Pectoral Girdle and Ribs, from the Front

not a complete girdle, since the clavicles are separated by the sternum and the scapulae are attached to the trunk only by muscles. The clavicles are slender, somewhat curved bones, which bind the arm-sockets to the sternum. They are better developed in man than in other mammals, giving a square appearance to the shoulders. They are the only membrane-bones in the body, apart from the skull. The scapulae present broad, flattened surfaces, each of which is subdivided by a raised spine. At the outer end of each scapula is a **glenoid cavity,** to which the arm skeleton articulates.

Functions of the Pectoral Girdle

The clavicles and scapulae provide additional protection for the thoracic viscera and additional surface, especially on the scapula, for the attachment of muscles. The scapula provides the socket for the front-limb, and ligaments and muscles for moving the limb. Because of the lack of rigid attachment of the girdle to the spine, there is great freedom of movement of the shoulders and arms, much more so than of the hips and legs.

THE PELVIC GIRDLE

The pelvic, or hip, girdle consists of two bones in the adult, the left and right **pelvis.** Each pelvis develops in the embryo as three

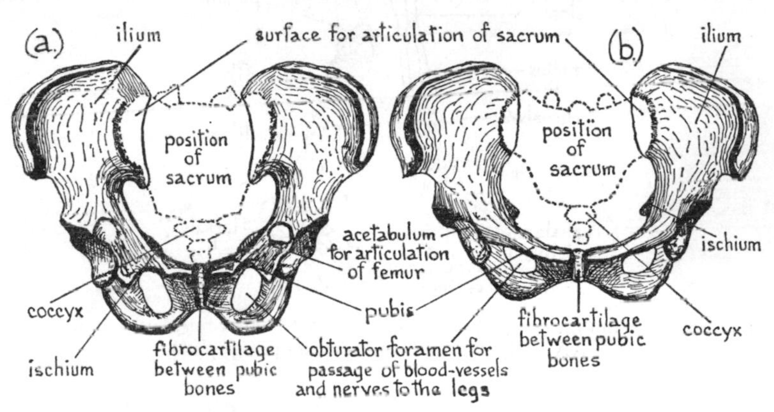

FIG. 56. The Pelvis from the Front
(*a*) Male (*b*) Female

separate bones, the **ilium, ischium** and **pubis,** which are shown in Fig. 56. Fusion to the sacrum dorsally makes the girdle complete and rigid.

Functions of the Pelvic Girdle

The girdle, with its broad, almost cup-like, shape, supports and protects the viscera of the lower abdomen. In females it is broader and shallower and bears the weight of the embryo in the uterus. The powerful muscles of the loin are attached to the girdle, as also are the muscles and ligaments to the upper parts of the leg. The upper bone of each leg is jointed to an **acetabulum,** a deep, cup-shaped socket on the outer side of each pelvis. The fusion of the pelvic girdle with the

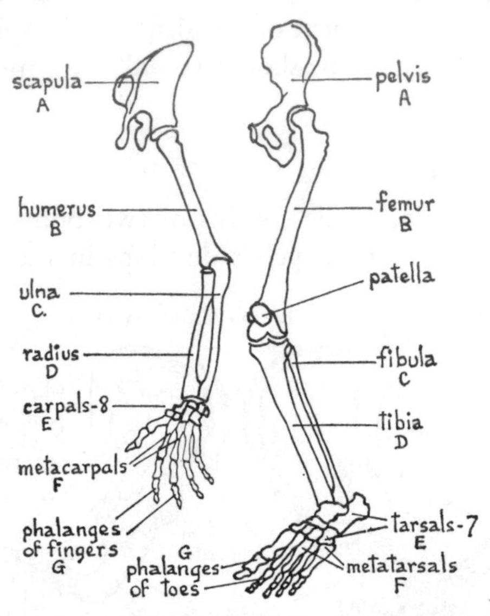

FIG. 57. (*a*) Front Limb (*b*) Hind Limb, Drawn to Show Similarity of Structure
AA, BB, etc., indicate corresponding parts.

sacrum provides for transmission of the thrust of the hind-limbs during movement.

THE LIMBS

The front- and hind-limbs (*see* Fig. 57) are based on a common pattern, known as the **pentadactyl limb.** This evolved originally in the early Amphibia and is so named from the fact that, typically, it ended in five fingers or toes (penta = five; dactyl = finger). Man still has very typical pentadactyl limbs, though in many other mammals they have been considerably altered, for example, in the horse, the bat, the seal, and in some they have almost completely disappeared, as in the whale (hind-limb). Since man has evolved his upright stance, the front-limbs are not normally used for locomotion, which is almost entirely restricted to the hind-limbs, leaving the front for a variety of movements which, during the course of evolution and by training, have come to possess a high degree of skill and dexterity in many kinds of manipulation. An outstanding feature of man's arms is their flexibility, particularly in the forearm and the wrist. Man can rotate the **radius** round the **ulna,** to present the palm or back of the hand either upwards or downwards. His ability (shared with other Primates) to rotate the thumb to a position opposite the fingers has given him the grasping and manipulative powers that have been so important in his evolution.

General Functions of the Skeleton

The functions of particular parts of the skeleton have been described on previous pages. Here, the general functions are summarized.

1. The skeleton forms a strong framework, which supports the softer tissues.

2. Together with the muscles and the joints, the skeleton provides a system of levers, with movement at the joints due to pull of the muscles. Thus movement of the whole body, or any part of it, is possible.

3. Many parts of the skeleton enclose and protect important organs, e.g. the cranium (brain), the vertebral column (spinal cord), the ribs and sternum (thoracic viscera).

4. In many of the long bones, most of the blood-cells (corpuscles) are manufactured in the marrow cavities.

Posture and Exercise

The shapes of the bones, and hence the carriage of the body, are considerably affected by the strains exerted on them, particularly during the growing period. Hence it is important that children and adolescents should be advised about correct posture in standing, sitting or, indeed, in any other activity. In standing still, for example, the weight should be evenly balanced on both feet, and the ear, shoulder, hip, knee and ankle should all lie in the same vertical line. In sitting still, as for example when writing, the back should be straight, not bowed in the thoracic region. Bending, to pick up weights, should take place at the knees and hips and not be done by arching the back. Furniture for children, e.g. school desks and benches, etc., should be of a size suitable for the child.

During exercise, the muscles exert pulls on the bones. Not only do the muscles themselves develop by usage, they also affect the final shape of the bones. Hence, exercise of the widest possible variety should be encouraged, particularly during the growing period. Infants should not be encouraged to sit, stand or walk before they are ready for these activities (*see* pp. 344–8).

SUGGESTED EXERCISES

1. Prepared sections of the following tissues should be examined: (*a*) hyaline cartilage, (*b*) fibro-cartilage, (*c*) elastic cartilage, (*d*) bone.

2. Permanent stained preparations of sections of cartilage may be made by students by the methods described in the Appendix, p. 399.

3. If an articulated human skeleton is available, the whole structure, general plan, types of joints, etc., should be examined. Plastic half-size skeletons are sold by most of the biological dealers. If neither of these is available, the skeleton of a rabbit or rat may be used to give the general plan.

4. Exercises on the structure of a bone are given in *The Mammals* in this series, p. 51, Exs. 1, 2 and 3.

Chapter 6

THE MUSCULAR SYSTEM

Muscular tissue constitutes by far the greatest proportion of the weight of the body (*see* p. 42). Its characteristic property is its power of contraction, and movements of the whole body, or of parts of it, are due to the work done by the muscular tissue. On average, muscular tissue can contract to about one half its stretched length, though some muscle fibres can contract to as little as one third of their stretched length.

There are three distinct kinds of muscular tissue, known as—

1. **Striated, striped, voluntary**, or **skeletal** muscle. 2. **Unstriated, unstriped, smooth, involuntary** or **visceral** muscle. 3. **Cardiac muscle.**

Striated Muscle

This may be known by any of the names given in (1) above. A whole muscle is composed of **fibres,** each of which is coenocytic (*see* p. 4). None of the fibres extends the whole length of the muscle; there are many more fibres in the middle, and hence most muscles have a characteristically bulged appearance, especially when contracted (*see* Fig. 58). All the fibres are separated by very delicate connective tissue, and all are enclosed in a strong membrane of connective tissue, the **epimysium** (*see* Fig. 58(*b*)). This type of muscle is so called, because of its characteristic, cross-striated appearance, which is visible in a stained preparation even under the low power of a microscope (*see* Fig. 58(*c*)). Careful examination with the high power of a microscope will show that each fibre contains longitudinal **fibrils** (*see* Fig. 58(*d*)). By the use of the electron microscope, it has been shown that each fibril is made up of many extremely slender, longitudinal **filaments,** lying in soft protoplasm. Thus, apart from connective tissue, blood-vessels and nerves, a whole muscle consists of the following, in descending order of size—muscle→fibres→fibrils→ filaments. (*See* Fig. 58(*f*).) The filaments consist of two kinds of proteins (*see* p. 106), **myosin** and **actin.** The shorter, thicker filaments

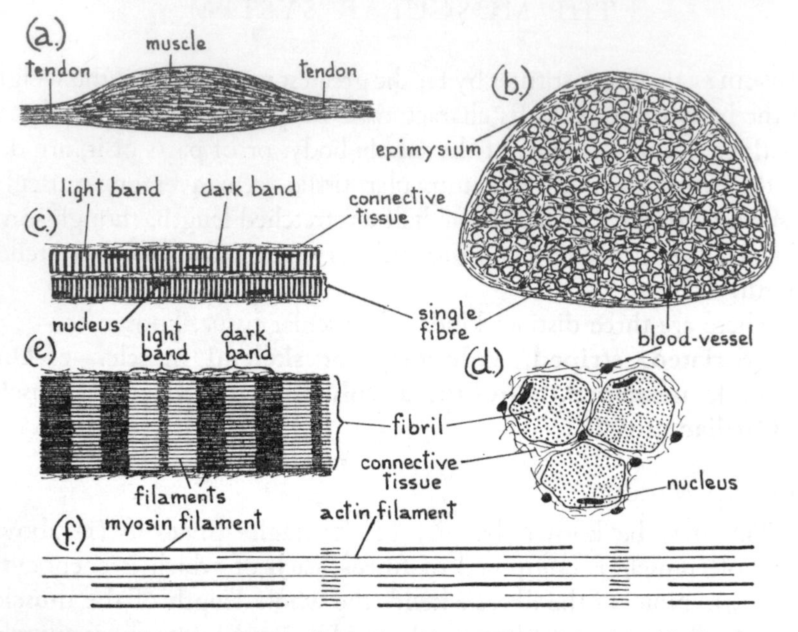

FIG. 58. Structure of Striated Muscle (*a*) A Whole Muscle with Tendons (*b*) Transverse Section of a Muscle Showing Fibres (*c*) Two Muscle Fibres, × 200 (*d*) Transverse Section of Three Muscle Fibres, Showing Fibrils, × 750 (*e*) A Single Fibril, Showing Filaments, × 10,000 (*f*) Diagram Showing the Arrangement of the Actin and Myosin Filaments

consist of myosin and the longer filaments of actin; they are partly interlocked, and when contraction occurs, the actin filaments touch one another and the myosin filaments almost touch. Note that the actual filaments do not contract, but they interlock more closely (*see* Fig. 59). It is not yet known how the actual interlocking takes place, but two things are essential for contraction. These are a supply of energy and a stimulus from the nervous system (*see* p. 84).

Striated muscle is also known as voluntary muscle, because the nerves that supply it travel direct from the brain and the contraction can be controlled by the will. It is also called skeletal muscle because it is always attached to bones or cartilages of the skeleton (*see* p. 85).

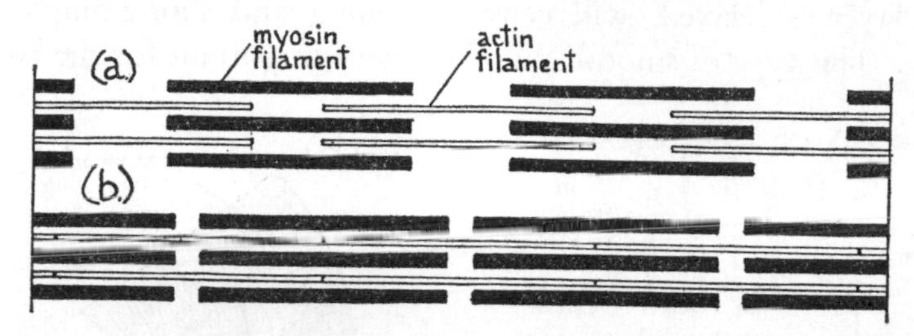

FIG. 59. (*a*) Positions of the Actin and Myosin Filaments in Relaxed Muscle
(*b*) Positions When Contracted

Contraction causes bending of the bones about the joints, and thus movement is effected.

UNSTRIATED OR SMOOTH MUSCLE

This type of muscle lacks the cross striations, though longitudinal fibrils are visible. It consists of elongated, spindle-shaped cells, which are united by protoplasmic fibrils to form a continuous network (*see* Fig. 60). There are two types of smooth muscle, which differ in the

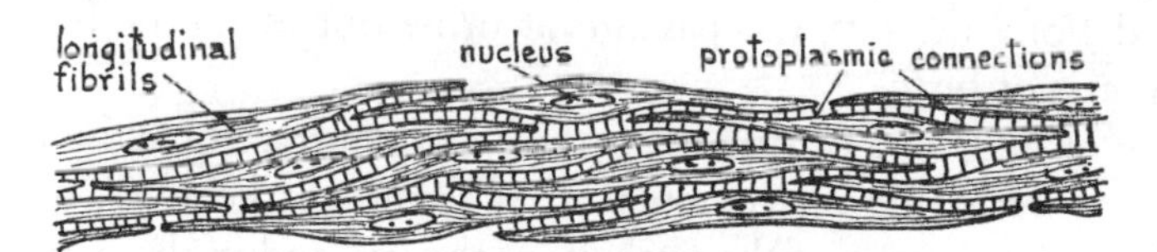

FIG. 60. Unstriated Muscle Tissue, × 300

way they contract. The more common type contracts slowly; waves of contraction, known as **peristaltic waves,** spread through the tissue, which exists in two layers, longitudinal and circular, surrounding hollow cavities. Examples of this type of muscle are found in the walls of the oesophagus, stomach and intestine, and in the walls of the bladder, uterus and ureters (*see* Fig. 61(*a*)). Contraction of the circular layer will cause narrowing of the cavity, and thus the contents are squeezed along. Contraction of the longitudinal layer, while the circular layer is relaxed, will cause widening and shortening of the cavity. This type of smooth muscle is very important for the passage

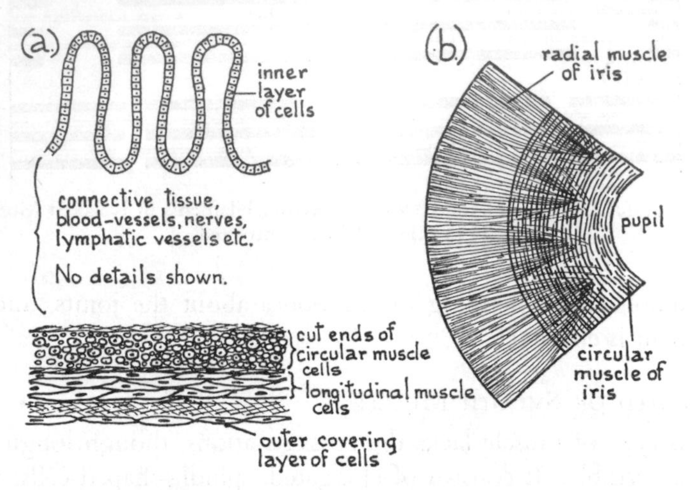

FIG. 61. (*a*) Longitudinal Section of Wall of Small Intestine, × 50 (*b*) Front View of Part of the Iris of the Eye, × 6

of food along the gut, the passage of urine out of the body, and for the process of birth.

The second type of smooth muscle can contract quickly and does not show peristaltic movements. It is found in the walls of blood-vessels, in the muscles of the eyeball, of the iris, and in the hair muscles (*see* Fig. 61(*b*)). This type of smooth muscle is very important for the blood circulation, for movements of the eyeball, for regulating the amount of light that enters the eye, and, in many mammals, for the sudden erection of the hair.

Smooth muscle is not under conscious control and the nerves that supply it do not pass directly from the brain; they are nerves of the

autonomic system (*see* p. 282). Hence smooth muscle is sometimes known as involuntary muscle. In the great majority of cases it is not attached to the skeleton.

Cardiac Muscle

This type is found only in the heart; it consists of cylindrical fibres united by diagonal cross-connections, the whole system again forming a continuous network (*see* Fig. 62). Both longitudinal and

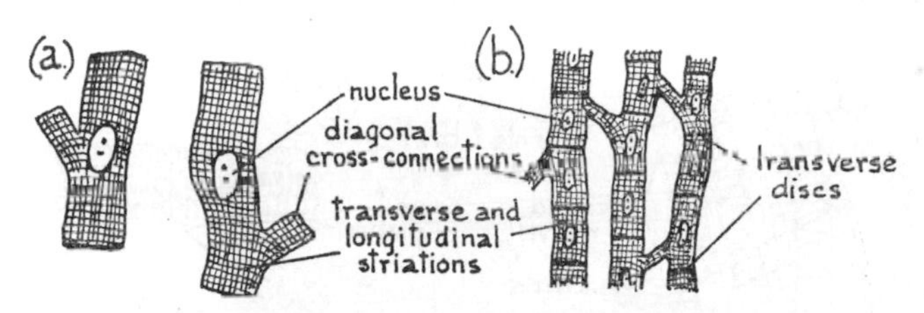

Fig. 62. Cardiac Muscle (*a*) Two Isolated Cells, × 250 (*b*) Longitudinal Section, Teased Apart, × 50

transverse striations are visible; the nuclei are central and the fibres appear to be divided into short cells by transverse discs.

Cardiac muscle contracts and expands alternately, without stimulation from the nervous system. It has its own **pacemaker,** a small area of special tissue, and a network of conducting tissue that spreads throughout the muscle (*see* p. 202). The heart has also a nervous supply of autonomic fibres, some of which cause acceleration of the beat, and others that slow it down. Control of the heart is described on p. 201.

"All or None" Law

Until fairly recently, it was considered that during life all skeletal muscular tissue was kept in a state of slight contraction called tone or tonus, even when the body is in a state of complete rest, as in deep sleep. It is now known that this is not the case. A single muscle fibre, if it is sufficiently stimulated, will contract to its limit; if insufficiently stimulated, it will not contract at all. This is the **"all or none" law.** A person can stand erect for a long period because, in each muscle concerned, groups of fibres act in succession, some contracted and some relaxed. Maximal contraction of a whole muscle cannot be maintained for more than a few minutes.

Nerve Supply of Muscles

All muscles, skeletal, cardiac and smooth, have nerve-endings and nerves of two kinds. First there are **sensory nerve-endings,** which send impulses along sensory nerves to the central nervous system.

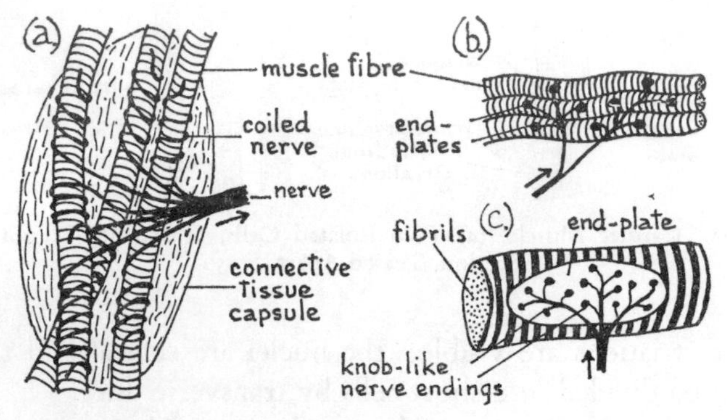

Fig. 63. Nerve-endings in Striated Muscle (*a*) Sensory Nerve-ending, a Muscle Spindle, × 100 (*b*) Motor Nerve-endings, × 75 (*c*) A Single End-plate, × 300

Then also, from the central nervous system, motor nerves lead to the muscle and end in **motor nerve-endings.** The former, which are sensitive to contraction and relaxation, send impulses to the central nervous system, which is thus made aware of these changes. Motor nerves carry impulses out from the central nervous system to the muscle, where the motor nerve-endings stimulate the process of contraction of the fibres. These types of nerve-endings are shown in Fig. 63.

Smooth muscle is supplied by two kinds of motor nerve-fibres, both belonging to the autonomic system. One of these kinds is called **sympathetic** and the other **parasympathetic.** They cause opposite effects when they stimulate the muscle; if one causes contraction, the other causes relaxation and vice versa. A nerve supply of this kind is known as **reciprocal innervation.** The table below shows some of the effects of these two kinds of nerves, in some regions of the body—

Region Affected	Sympathetic Stimulation	Parasympathetic Stimulation
Gut smooth muscle	Causes relaxation	Causes contraction
Gut sphincters	Causes contraction	Causes relaxation
Heart and arteries	Causes contraction	Causes relaxation
Skin muscles	Causes contraction	Causes relaxation
Iris radial muscle	Causes contraction	Causes relaxation
Iris circular muscle	Causes relaxation	Causes contraction

The skeletal muscles do not have reciprocal innervation. The motor nerves carry impulses that cause contraction; cessation of these causes slackening to the normal relaxed state. The majority of skeletal muscles work in pairs, and therefore there is **reciprocal action** between two members of a pair. A muscle that causes bending at a joint is called a **flexor**; one that causes straightening of the joint is called an **extensor.**

Action of Skeletal Muscles

The two ends of a skeletal muscle are respectively known as the **origin** and **insertion.** The origin is the end that is attached to a bone that is stationary, or nearly so, while the insertion is attached to a freely movable bone. For example (*see* Fig. 64), the **biceps muscle** in the arm has its origin in the scapula, while its insertion is in the radius. The **triceps muscle** of the arm has its origins in the scapula and the humerus and its insertion in the ulna.

Most of the muscles are attached to bone by a **tendon** at each end. The tendons are made mainly of longitudinal bundles of white fibres and are extremely strong. The fibres at one end of a tendon pass into the periosteum and become fused with it, while at the broader end, the fibres of the tendon pass among the muscle-fibres and fuse with the

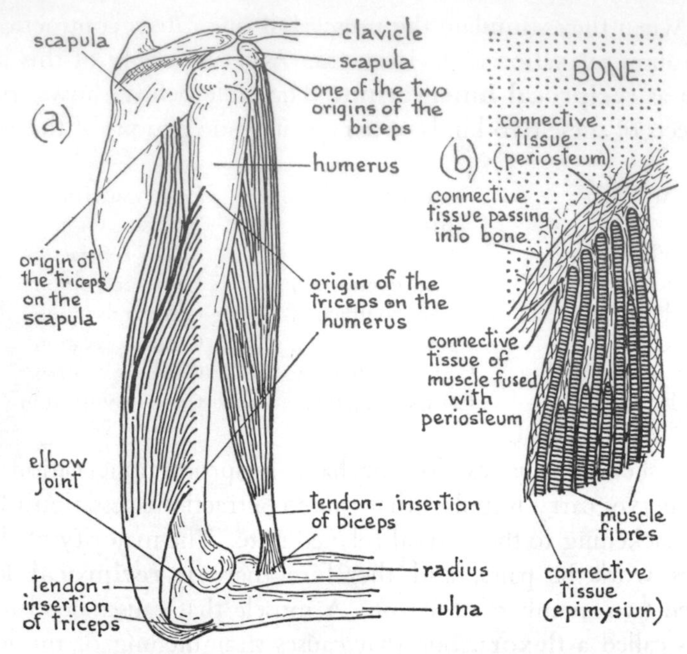

FIG. 64. (*a*) Location of Two Antagonistic Muscles, the Biceps and Triceps (*b*) Diagram showing Origin of a Muscle Without a Tendon

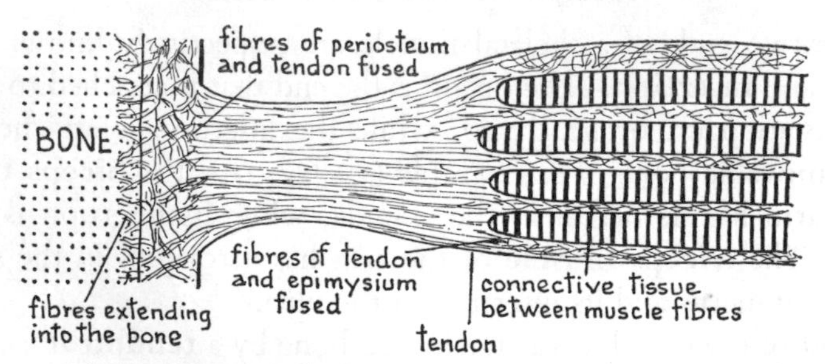

FIG. 65. Diagram Showing the Junctions between Muscle and Tendon, and between Tendon and Bone

connective tissue there (*see* Fig. 65). In some cases, the muscle-fibres themselves appear to pass into the periosteum of the bone, but actually there is fusion between the connective tissue surrounding the muscle and its fibres and that of the bone, the periosteum. The biceps and triceps have this kind of origin in the scapula (*see* Fig. 64).

Group Action of Muscles

When any given movement is performed, a number of muscles act together. There is no case in the body of a skeletal muscle working

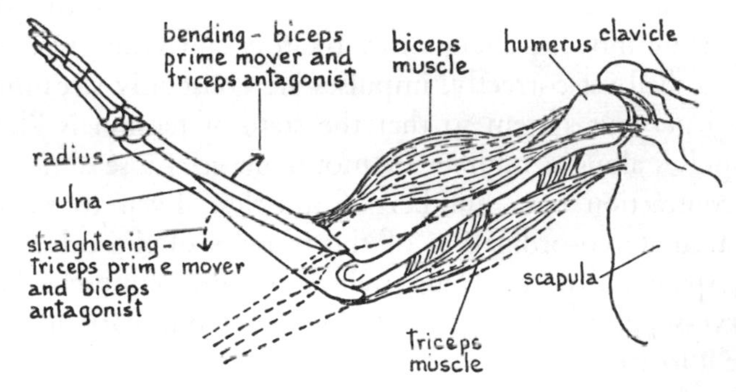

Fig. 66. Diagram Showing Positions of Biceps and Triceps as Prime Movers and as Antagonists

entirely on its own. The muscles that carry out the actual movement are the **prime movers**; for example, when the arm is bent at the elbow, the prime mover is the biceps muscle. If the prime mover is to carry out its work efficiently, the muscle that causes the opposite action, in this case the triceps, which straightens the elbow joint, must be relaxed. These muscles that have the opposite action from the prime movers are called **antagonists.** In the case quoted, at the bending of the elbow the biceps is the prime mover and the triceps is the antagonist. But in straightening the elbow, the triceps is the prime mover and the biceps is the antagonist (*see* Fig. 66). In addition, **fixation muscles** are brought into action to hold the bone that gives origin to the muscle in a stationary condition. Again, in the case given above, muscles about the shoulder-joint hold the scapula and clavicle steady. Finally, to enable the prime mover to carry out its work most efficiently, other muscles steady neighbouring joints at the same time; they are called **synergists.** In the case quoted, synergists act on the upper ribs and on the cervical and upper thoracic vertebrae. Thus, an apparently simple movement, like bending the arm at the elbow, involves a great many muscles and a number of joints. It is this co-ordination of muscles that constitutes **group action.**

Clenching the hand involves all the muscles and joints of the wrist and of the fingers. Even the simplest movement of the skeleton involves prime movers, antagonists, fixators and synergists. So that the muscles shall act correctly, impulses are constantly streaming into the central nervous system so that the state of tension is "known." Then impulses along a number of motor nerves cause almost simultaneous contraction of a number of muscles. Even to stand erect requires accurate co-ordination of the muscles of the calves, thighs, buttocks, spine and neck. Hence whole groups of muscles have the same nerve-supply and we cannot single out one particular muscle and cause it to act.

The Leverage Mechanism

Any rod that turns about a point along its length is called a **lever.** The point about which it turns is called the **axis** or **fulcrum.** For example, a bent crowbar may be used to lift a heavy block of stone (*see* Fig. 67). In the case shown we have—$2 \times x = 200 \times 0{\cdot}5 = 100$ N m, therefore $x = 50$ newton. Thus, to keep the stone balanced in position (*b*) of Fig. 67, an effort of 50 newton must be exerted. In other types of levers the load may be between effort and fulcrum or the effort may lie between the other two.

Any joint moved in one plane works on the lever principle. For example, in lifting the forearm, we have the effort exerted by the biceps muscle acting at the point of insertion on the radius; the fulcrum is at the elbow-joint and the load lifted is the weight of the forearm acting at its centre of gravity (*see* Fig. 68). In practice, it would be very complicated to measure the effort exerted by the biceps alone, since the fixators and synergists would have to be taken into account.

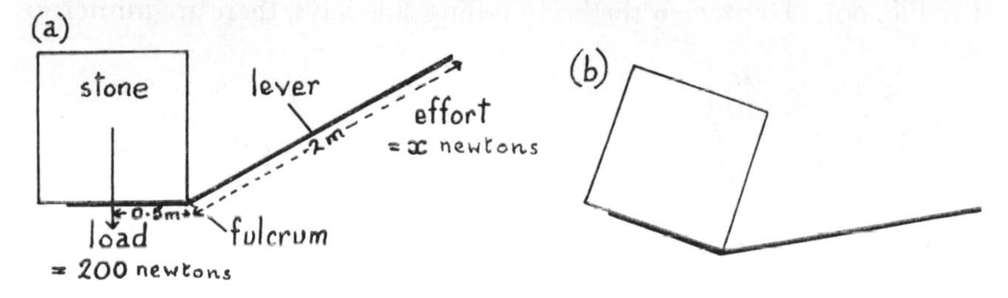

FIG. 67. The Principle of a Simple Lever; Lifting a Mass Clear of the Ground with a Crowbar

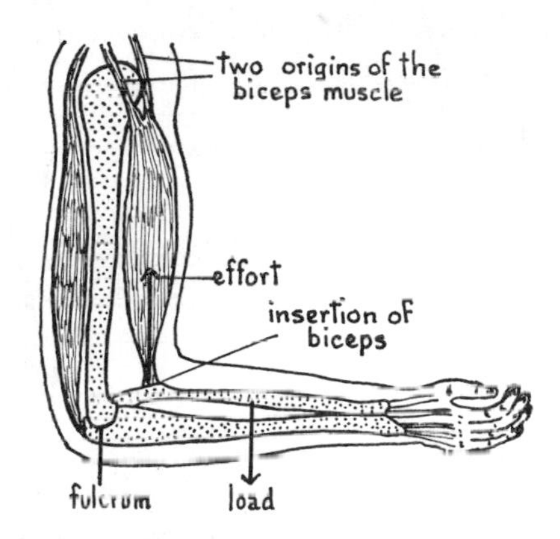

FIG. 68. Using the Forearm as a Lever with the Fulcrum at the Elbow

Some Group Actions

POSTURE

By posture we mean balancing the body, more or less stationary, in any position. It refers normally to standing and sitting, though at times other more demanding postures are necessary, such as balancing on one leg, crouching or standing on the hands. In standing erect, a number of muscles have to be kept in a state of slight contraction.

Good posture in standing erect is considered to be achieved if the ear, shoulder, hip, knee and ankle are all in the same vertical straight line (*see* Fig. 69). To prevent the body falling sideways, there are numerous

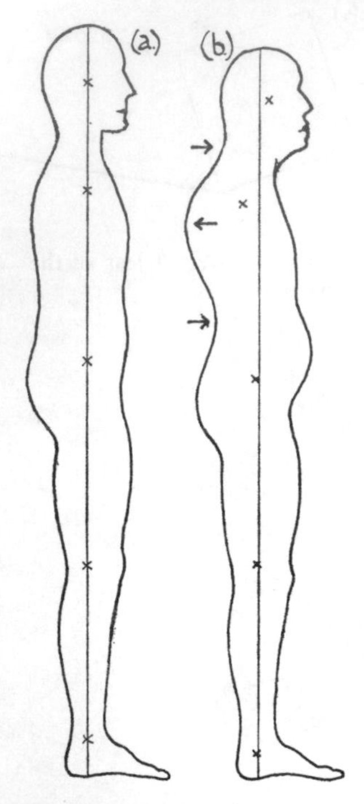

FIG. 69. (*a*) Good Posture
(*b*) A Common Type of Bad Posture

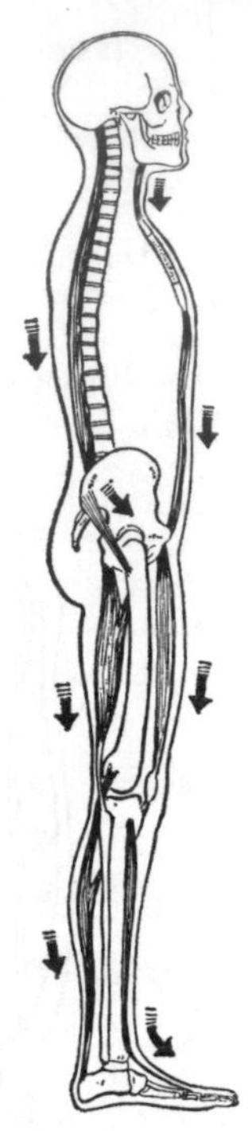

FIG. 70. Main Groups of Muscles that Keep the Body in an Erect Posture
Muscles are in black; arrows show the direction of pull.

paired lateral muscles, all of which are slightly contracted. To prevent the body falling forwards, the muscles at the back of the spine, the back of the thigh and the back of the calf exert a downward pull (*see* Fig. 70). Finally, to prevent the body falling backwards, antagonistic muscles in front of the body also exert a downward pull. The muscles at the back of the body tending to be more powerful than the muscles at the front, the whole system is balanced by the somewhat diagonal pull of the muscles of the buttocks.

WALKING

Walking is an extremely complex operation, at least as far as the correct adjustment of numerous muscles is concerned. It is outlined

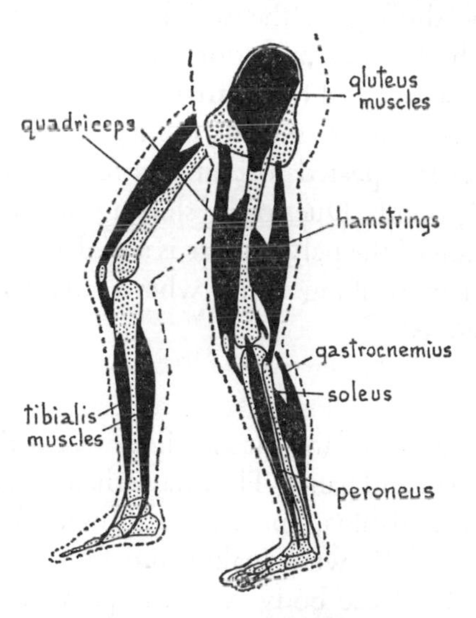

FIG. 71. Side View, Showing the Major Muscles Used in Walking
Muscles are in black, bones stippled.

here in the form of a sequence, but it is to be noted that this explanation is a simplified one.

1. The weight of the body is transferred to the left leg by tilting the pelvis to the left. This is achieved by contraction of a pair of **gluteus** muscles at the side of the buttocks, which pass from the pelvis to the femur (*see* Fig. 71). The right foot is then lifted clear of the ground and placed forward (*see* below).

2. The left foot then thrusts against the ground, moving the centre

of gravity forwards. The weight moves forwards from the heel to the ball of the foot; the heel is raised from the ground by contraction of the calf muscles, the **soleus** and **gastrocnemius.** To develop maximum thrust, the foot and the knee are kept rigid by the **peroneus** and **quadriceps** muscles.

3. The left foot leaves the ground, partly owing to its own thrust and partly to the shifting of the weight to the right. The bending upwards at the thigh, knee and foot, due to the **tibialis** muscles on the inner side of the calf and the **hamstring** muscles at the back of the thigh, completes the thrust and lifts the leg.

4. The process is repeated so that one foot swings past the other, which is on the ground. Due to the shifting of the weight from side to side by the tilting of the pelvis, there is a slight lateral rolling motion. This can be seen in walking races, where lateral movements of the pelvis are exaggerated.

Grasping

There are a number of muscles in the forearm that are flexors and extensors of the wrist and hand. They have their origin on the humerus, and on the radius and ulna, and they are inserted by means of long tendons on bones of the wrist, palms and fingers. When the arm is hanging at the side of the body with the palm facing the thigh, the flexors are at the front and inner parts of the forearm (*see* Fig. 72) and the extensors are at the back and outer parts. In normal grasping, the flexors and extensors of the wrist act as synergists, so that the wrist is kept steady. Then, when the flexors of the fingers contract, the fist is clenched. There are special muscles for flexing and extending the thumb and for rotating it so that it lies opposite the fingers. The mass of muscle at the base of the thumb on the palm side is the one that pulls the thumb inwards and rotates it, so that it comes to lie in the grasping position. The thumb is brought back into line with the fingers by the contraction of a muscle between the palm and the thumb at the back of the hand. There are also small muscles of the hand that enable one to spread the fingers apart and bring them close

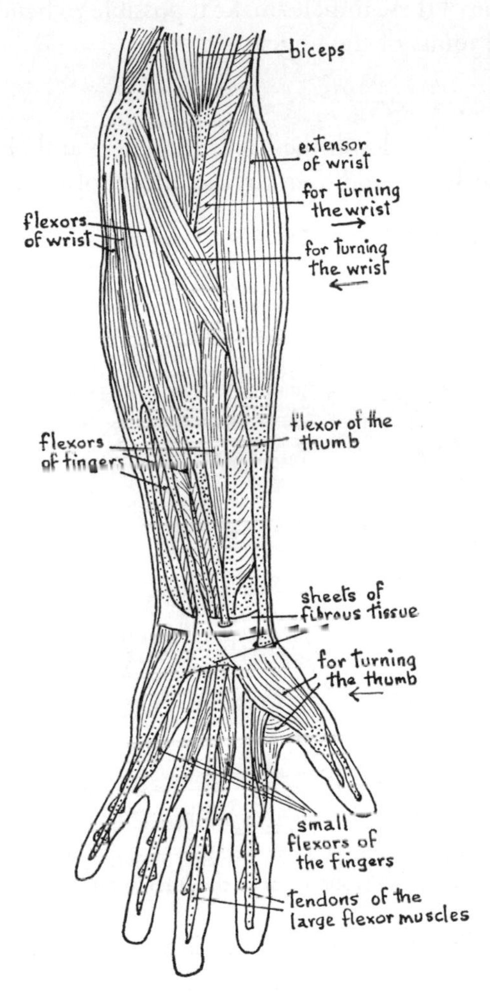

FIG. 72. Front View of the Forearm and Hand Showing the Main Muscles used in Grasping

together. Some of these muscles make it possible to bend and straighten the most distal joints of the fingers.

Breathing Movements

A number of muscles in the back, shoulders and chest assist in the breathing process, in addition to the more obvious action of the

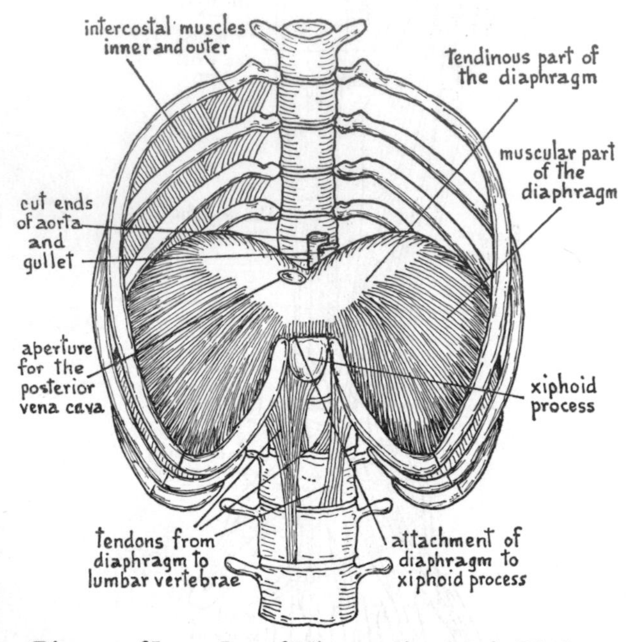

Fig. 73. Diagram of Lower Part of Rib-cage, Showing the Diaphragm and its Attachments and Parts of the Intercostal Muscles

muscle of the diaphragm and of the intercostal muscles of the ribs. Here the action of the prime movers will be described. The diaphragm is a dome-shaped muscular sheet separating the thorax and abdomen. It is attached to the inner surfaces of the xiphoid process and the lower ribs and also to the upper lumbar vertebrae. The muscle-fibres converge towards the apex of the dome, where they are inserted into a strong, white, fibrous area in the centre. When the muscles contract, the dome becomes less arched and the capacity of the thorax is increased (*see* Fig. 73).

At the same time as the upward and downward action of the diaphragm occurs, the intercostal muscles between the ribs act. Between each pair of ribs on each side there are external and internal intercostal muscles (*see* Fig. 73). The former are thick outer sheets and the latter are thin inner sheets. Each external intercostal extends diagonally, and slightly inwards, from the lower border of a rib to the upper border of the rib below. When they contract they pull the ribs towards each other, the upper pair of ribs being held stationary by muscles of the chest and back. Thus the combined action of the external intercostal muscles is to elevate the ribs and thus increase the volume of the thorax. The internal intercostal fibres are attached in a similar way to those of the external, but the fibres extend in a direction almost at right-angles to the other set. When they contract, the lower pair of ribs being held firmly by the **quadratus lumborum** muscles, which extend from the lower pair of ribs to the pelvic girdle, the ribs are pulled together in the downward direction, thus decreasing the capacity of the thorax.

Energy Supply of Muscles and Fatigue

The source of the energy that is used by muscles when they contract is dealt with in the section on "Utilization of food," on p. 154. When a muscle will not contract when it is stimulated by a nerve impulse, **fatigue** is said to occur. Fatigue in a muscle is caused by increase of **lactic acid,** which is produced in the muscle as a result of contraction. Details of fatigue in muscles are given on p. 171.

SUGGESTED EXERCISES

1. Examine, under the high power of a microscope or microprojector, prepared slides of (*a*) teased striated muscle, (*b*) longitudinal section of striated muscle, (*c*) transverse section of striated muscle, (*d*) section of cardiac muscle, (*e*) transverse section of the intestine, to find the two layers of smooth muscle.

2. Make a permanent stained preparation of striated muscle from the leg or arm of a preserved frog, rat or rabbit, by the method described in the Appendix, p. 400.

3. Using an electric torch as the source of light, hold it close to another pupil's eye and watch the alternate contraction and relaxation of the iris when the light is switched on and off.

4. Using a preserved frog, rat or rabbit, dissect the skin from the upper arm and find the origin and insertion of the biceps muscle. If the skin from the forearm and

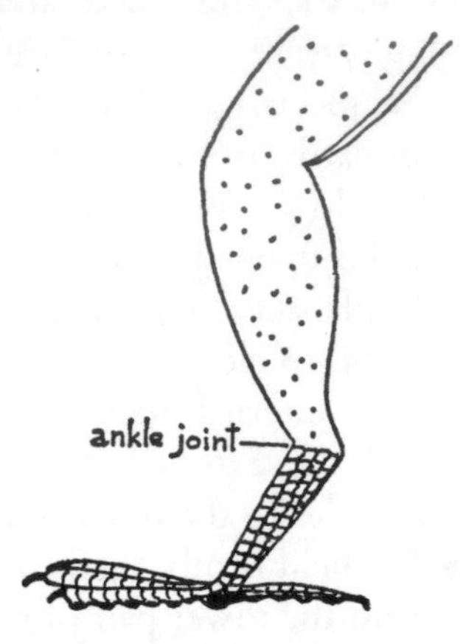

FIG. 74. Leg of a Bird

back of the hand of the animal is removed, some of the muscles and tendons of the fingers can be seen.

5. Try to locate the diaphragm, and the external and internal intercostal muscles, in a preserved rat or rabbit.

6. If the leg of a dead chicken or turkey is cut off at the ankle joint (*see* Fig. 74), the tendons (silvery) can be located at the cut end. Pulling these with a small pair of pliers will cause the toes to move. Find out where the tendons for flexing and extending the toes are located.

7. With a long straight cane, check the standing posture of some of your fellow pupils.

8. With the hand fully extended and the fingers separated as widely as possible, note the tendons at the back of the hand and particularly the two tendons below the base of the thumb.

CHAPTER 7

FOOD AND DIET

THE **diet** of an animal means all the substances it eats as well as those that it drinks. "Food," therefore, includes both solid and liquid substances. To achieve some understanding of the nature of foodstuffs and the purposes for which they are needed, it is useful to compare the requirements of the body with the requirements of a machine such as a motor-car.

COMPARISON OF THE BODY WITH A MACHINE

A motor-car is designed to do work, and if it is to remain in good order, so that it can do its work, it must be supplied with certain materials, the quantities of which will depend on how much work it does. The living body does work of various kinds, and it must be supplied with certain materials, which, here again, depend on the amount of work that is done. The car must have petrol or some other fuel; when burnt, this fuel releases energy. It is the energy stored in the petrol that enables the moving parts of the car to do work, and this energy must be released from the petrol. The living body must have fuel that will release energy when suitably treated; the chief fuel foodstuffs are **carbohydrates** and **fats.**

The working parts of any motor-car eventually become badly worn, damaged or even useless. These parts have to be repaired or be replaced by new parts. Parts of the human body become worn out, damage may be caused by disease or accident, and, in any case, there is a continuous **replacement** of all parts. For this work of **repair** and replacement, the body needs **amino-acids** (*see* p. 106), most of which are obtained from the **proteins** in the diet. Certain kinds of **salts** are also necessary for replacement and repair.

To ease the working parts of a car, various oils and greases are necessary. The living body eases and speeds up its working processes with certain complex proteins called **enzymes.** Certain enzymes are present in every cell of the body; for example, the enzymes concerned with the release of energy from carbohydrates. Others, such as those concerned with digesting food, are produced only in the cells of certain parts of the gut or in glands connected with the gut. The body makes its own enzymes; some of them are present from the beginning of the individual life, in the fertilized egg. Some are represented in the nucleus as a code of instructions that the cell can copy (*see* p. 306). Some enzymes will not function unless certain substances, called **co-enzymes**, are present. Several of the well-known **vitamins** are co-enzymes, and their absence causes certain **deficiency diseases** (*see* p. 110).

The heat produced by the moving parts of a motor-car is absorbed partly by a system of water jackets surrounding the engine. The water is circulated in pipes to and from a radiator, which is specially made to cool the water, usually with the assistance of a fan. Heat produced in the body is absorbed largely by the blood, which is circulated in the blood-vessels, and much of the heat is lost through the skin (*see* p. 54). The main constituent of blood is water.

The waste-products of combustion in the cylinders of a motor-car are mainly carbon, carbon dioxide, carbon monoxide and water vapour. They are eliminated by the exhaust system. The waste-products of the living body are carbon dioxide, water and certain nitrogenous substances. These are eliminated by the excretory organs of the body: the kidneys, skin and lungs.

There are other ways in which a motor-car and a human body may be compared, such as, for example, the fact that both are moved from place to place by the operation of a system of levers, and that in both cases the intake of fuel, oxygen and water can be adjusted fairly precisely. There are, however, several ways in which the motor-car and the living body show marked contrast, the chief one being that the body is alive and the machine is not. The control and regulation of the working of the car lies with the driver, whereas the animal body has within itself its own system of control and regulation. The body grows from a small size to a larger size, while a machine obviously does not. Finally, living creatures can produce others of their own kind, but large cars cannot produce small cars.

From the foregoing paragraphs it should be clear that the body requires six essential classes of food substances. These are **carbohydrates, fats, proteins, salts, vitamins** and **water.** Before considering the purposes and nature of each of these, it is useful to consider the energy relationships of the body.

Energy Relationships in the Body

Every living cell is continually using energy to do its work. Some cells are using energy to divide into two, and after the division, the two daughter cells use energy to make more protoplasm so that they may grow to full size. Nerve-cells use energy in sending out impulses along the nerves. Muscle-cells use energy for contraction. Many types of cells use energy in making some product that is useful to the body. No living cell is ever in a state of complete rest.

The energy used by every cell of the body comes from a complex substance known as **adenosine triphosphate,** usually abbreviated as **ATP.** Part of the ATP consists of three **phosphate** groups. When one of these is removed by enzyme action, considerable energy is released to the cell. A second phosphate group can also be removed, yielding further energy. If the cell is to remain alive and continue to do its work, there must be constant rebuilding of the ATP by adding on the phosphate groups again. This addition requires considerable energy; in fact, it requires the same quantity of energy as that derived from the breakdown of ATP. It is for this purpose that the cell requires constant supplies of substances that can be made to release their energy. These substances are normally carbohydrates or fats. Animals use their carbohydrate first but they cannot store much, so fat supplements the carbohydrate as a supplier of energy. The carbohydrates that enter cells are brought by the blood and must always be simple sugars. In fact, in the blood, the carbohydrate is always **glucose.**

In the cells, the glucose is broken down finally, by enzyme action, to carbon dioxide and water. The energy released from this breakdown process is used to rebuild ATP. Fig. 75 shows, in a simple form, the breakdown and reformation of ATP.

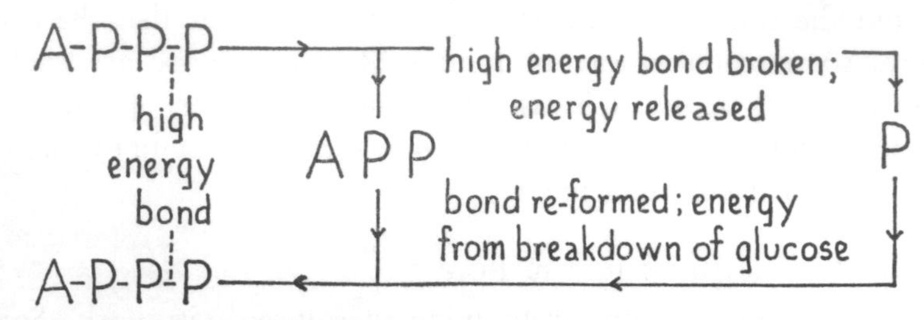

FIG. 75. The Breakdown and Restoration of ATP in the Cells

CARBOHYDRATES

The carbohydrates are all either simple sugars or are formed by the union of two or more simple sugars. The sugars are all soluble crystalline compounds with a sweet taste. There are three main groups of carbohydrates: simple sugars such as glucose; compound sugars such as **sucrose,** formed by the union of two molecules of simple sugars; and long chain compounds such as **starch,** formed by the union of many molecules of simple sugars.

The simple sugars are all called **monosaccharides** and there are two groups important in the body. They are **hexose sugars,** with six carbon atoms, and **pentose sugars,** with five carbon atoms. An important characteristic of carbohydrates is that many of them have the same chemical composition, but a different arrangement of the atoms and also somewhat different properties. All the hexose sugars have the composition indicated by the formula $C_6H_{12}O_6$, which denotes one molecule. The table below shows the monosaccharide sugars that are important in the human being—

pentoses	hexoses
1. ribose $C_5H_{10}O_5$	1. glucose $C_6H_{12}O_6$
2. deoxyribose $C_5H_{10}O_4$	2. fructose $C_6H_{12}O_6$
	3. galactose $C_6H_{12}O_6$

Glucose (grape sugar) is widely distributed in plant and animal tissues and it is present in the blood, normally to the extent of 0·1 per cent. It does not form an important part of the diet, and is produced by digestion of the compound sugars and other higher carbohydrates. **Fructose** (fruit sugar) is present in all sweet fruits and is also produced by digestion of cane sugar (sucrose). **Galactose** is not found free, but combined with glucose in the form of **lactose** (milk sugar). The pentose sugars, **ribose** and **deoxyribose,** form a small but important part of the **nucleic acids** of every living cell (*see* p. 305). We obtain them by digestion of those nucleic acids in the plant and animal cells we eat. Pentose and hexose sugars are the only carbohydrates that can be absorbed from the gut and transported in the blood. Thus, during digestion, all carbohydrates that can be digested are broken down to these simple molecules.

The compound sugars are **disaccharides,** so called because one molecule of each of them is made up of two molecules of hexose sugars. Thus one molecule of sucrose (cane sugar) can be broken down into one molecule of glucose and one molecule of fructose. All the disaccharides have the chemical formula $C_{12}H_{22}O_{11}$. During digestion, or by boiling with a dilute acid, a disaccharide is broken up as in this equation—

$$\underset{\text{disaccharide}}{C_{12}H_{22}O_{11}} + \underset{\text{water}}{H_2O} \rightleftharpoons \underset{\text{hexose}}{C_6H_{12}O_6} + \underset{\text{hexose}}{C_6H_{12}O_6}$$

The arrows pointing in opposite directions show that this is a **reversible reaction.** When a disaccharide molecule is broken down, a molecule of water is added to it and two molecules of hexose sugar are produced. In the other direction, two molecules of hexose combine to form one molecule of disaccharide and one molecule of water. Disaccharide sugar forms an important part of the diet in the form of cane sugar and beet sugar, which are both sucrose, and milk sugar or lactose. **Maltose** is not a normal constituent of the diet but is produced by the digestion of starch. In the gut, these disaccharide sugars are digested as shown below—

$$\begin{array}{lcl} \text{sucrose} & \rightarrow & \text{glucose} + \text{fructose} \\ \text{lactose} & \rightarrow & \text{glucose} + \text{galactose} \\ \text{maltose} & \rightarrow & \text{glucose} + \text{glucose} \end{array}$$

The complex chain carbohydrates, formed by the union of many simple sugars, are **starch, cellulose, glycogen** and others not important in the human diet. Starch forms the largest single component of most human diets. It is obtainable only from plants and is present in considerable amounts in potatoes, cereals (wheat, rye, barley, oats, maize, rice), sago (from the pith of the sago palm), tapioca (from roots of *Cassava*), and arrowroot (from rhizomes of *Maranta*). The chemical formula for starch is $(C_6H_{10}O_5)n$. The letter *n* means that the starch molecule consists of a large number of units of $(C_6H_{10}O_5)$ joined together. The number differs for each type of starch but is at least three hundred.

Cellulose, which forms the cell walls of plant cells, also has the formula $(C_6H_{10}O_5)n$. It forms a part of the diet, being present in all fruits and vegetables. Very few organisms can digest cellulose; examples are earthworms, snails, some protozoa and bacteria. Herbivorous animals such as cows, horses and sheep are able to utilize cellulose because they have large colonies of bacteria in special parts of the gut. If we are to obtain any benefit from eating plants, then it is essential that the material should be thoroughly chewed to break up the cellulose walls, so that the cell contents are available for digestion. Nevertheless, cellulose forms an important part of the diet merely because it constitutes **bulk** or **roughage** and stimulates **peristalsis,** which is the muscular squeezing of material through the gut.

Glycogen, or animal starch as it is sometimes called, is the form in which animals store carbohydrate. It is normally present in the liver and in the muscles. Again it has the formula $(C_6H_{10}O_5)n$. Glycogen and glucose are sources of carbohydrate for carnivora, which exist entirely on a diet of meat. It is noteworthy that many carnivora consume the liver of the prey before the rest of the carcase.

All these complex carbohydrates, starch, cellulose, glycogen and others, are called **polysaccharides,** since they are made up of many monosaccharide units each having lost a molecule of water.

$$n\,C_6H_{12}O_6 - n\,H_2O = (C_6H_{10}O_5)n.$$

FATS

Like carbohydrates, fats consist of carbon, hydrogen and oxygen, but there is relatively less oxygen in the fats. Fats yield more energy per unit mass than any other type of foodstuff. The amount of energy released in the body by complete utilization of any foodstuff is the same as the energy it would yield if it were burnt. In the latter case, all the energy would be obtained as heat, but in the body, although some heat is produced, some of the energy is used for other purposes. Energy as a quantity is expressed in the unit **joule,** which is defined as the work done when a force of one newton acts through a distance of one metre in the direction of the force. The once-used heat unit, calorie, is about 4·2 joule. 1000 joule (J) = 1 kilojoule (kJ).

Thus expressed, we have the following values for the principal food-stuffs—

1 g Monosaccharide yields	15·7 kJ
1 g Disaccharide yields	16·54 kJ
1 g starch yields	17·7 kJ
1 g carbohydrate averages	17·17 kJ
1 g fat averages	38·94 kJ
1 g protein averages	22·61 kJ

Thus one gram of fat yields, on average, more than twice as much energy as one gram of carbohydrate. Many animals and plants have evolved means of storing fat as a reserve of energy-producing material.

One molecule of fat is composed of the union of three molecules of **fatty acids** with one molecule of **glycerine.** There is a whole series of these fatty acids, the simplest being **acetic acid,** CH_3COOH, the acid of vinegar. The next fatty acid in the series adds on two (CH_2) groups, and this process can be continued until a total of fourteen naturally occurring fatty acids is produced. Thus, acetic acid is CH_3COOH and butyric acid is $CH_3(CH_2)_2COOH$, etc. No fats are formed from acetic acid, but some fats, e.g. butter, contain **butyric acid.** The most important fatty acids in this series are **palmitic,** $CH_3(CH_2)_{14}COOH$, and **stearic,** $CH_3(CH_2)_{16}COOH$. The whole series is known as the **stearic series.** There are several other series of fatty acids, the most important being the **oleic series,** which differs from the stearic series in having two (CH) groups replacing of two (CH_2) groups. **Oleic acid,** which is the commonest fatty acid of mammals and the chief fatty acid in human fat, has this formula— $CH_3(CH_2)_{14}(CH)_2COOH$.

In a fat, three molecules of fatty acids are united with one molecule of **glycerol** (glycerine). The chief fat of human beings, **triolein,** is formed of three molecules of oleic acid combined with one molecule of glycerol, with the elimination of three molecules of water.

$$\begin{matrix} CH_2\,\boxed{OH} \\ | \\ CH\,\boxed{OH} \\ | \\ CH_2\,\boxed{OH} \end{matrix} + 3CH_3(CH_2)_{14}(CH)_2COO\boxed{H} \rightleftharpoons \begin{matrix} CH_2\text{—}OCO\cdot CH_3(CH_2)_{14}(CH)_2 \\ CH\text{—}OCO\cdot CH_3(CH_2)_{14}(CH)_2 \\ CH_2\text{—}OCO\cdot CH_3(CH_2)_{14}(CH)_2 \end{matrix} + 3H_2O$$

glycerol + oleic acid triolein (fat) + water

The dotted lines indicate the sources of the three water molecules. The arrows show that this reaction is reversible; when fat is being built up, water is eliminated, and when fat is being broken down, as in digestion, water is absorbed. Often a fat molecule contains two, or even three, different fatty acids, such as—

oleic acid stearic acid palmitic acid	→ glycerol and	oleic acid stearic acid stearic acid	→ glycerol

The fats of animals, unlike the fatty oils of plants, do not have a constant composition but depend to some extent on the fat in the diet.

But if an animal is given a wide selection of fats, it will build up fat of fairly constant composition. Thus we recognize the characteristics of beef fat, mutton fat, bacon fat, etc. The average composition of fatty acids in some fats is given below—

Lard: stearic acid 40 per cent; oleic acid 50 per cent; **linoleic acid** 10 per cent.
Beef and mutton fat: stearic acid 35 per cent; oleic acid 50 per cent; palmitic acid 15 per cent.
Butter: oleic acid 45 per cent; **myristic acid** 20 per cent; palmitic acid 15 per cent; small quantities of five others.

Human fat resembles beef and mutton fat, but has a somewhat greater quantity of oleic acid.

Apart from supplying a store of energy, fats serve several other purposes in the body. Firstly, the fat in the dermis of the skin forms an insulating layer, which helps to reduce loss of heat. Mammals such as the whale and seal of the colder waters, and the polar animals (such as the arctic fox and polar bear), have extra thicknesses of this fat. People of the colder countries, such as Eskimos and Lapps, also store extra fat under the skin. Secondly, **vitamin D** can be made by most animals from their own fat, if they have sufficient sunshine. Also, some of the vitamins will dissolve in fat but not in water, and hence they can be absorbed from the gut only after being dissolved in the fat of the cells lining the gut. Thirdly, fats form the basis of some essential constituents of every living cell.

Common fats in the human diet are butter, margarine, the cream of milk, and the fat of various kinds of meat, such as beef, mutton and pork. Plant fatty oils that are in common use are olive oil and peanut butter, and many foods are cooked in various plant oils.

Proteins

The living material, protoplasm, consists of many substances, but the two present in largest quantity are proteins and water. The proteins of different kinds of animals are never quite the same, and even in different parts of the same body there are differences in the proteins.

The proteins of muscle differ from the proteins of the brain, or liver, or skin or any other part. For growth of the body, more cells of many different kinds are required, and hence different kinds of proteins must be made. Every protein consists of large numbers of **amino-acids;** proteins differ from one another in the numbers and kinds of amino-acids they contain. The body requires about twenty kinds of amino-acids, some of which it can make from simpler substances. Those that cannot be made in the body must be obtained from the proteins of the diet; they are known as the **essential amino-acids.** They are more abundant in animal than in plant proteins. The remaining amino-acids necessary for the body's requirements are known as **non-essential** (*see* table).

ESSENTIAL	ABBREVIATION	NON-ESSENTIAL	ABBREVIATION
Lysine	L	Glutamic acid	Gl
Leucine	Le	Citrulline	C
Phenylalanine	P	Hydroxyproline	Hy
Valine	V	Tyrosine	Ty
Methionine	M	Glycine	G
Threonine	Th	Proline	Pr
Isoleucine	Ile	Cystine and Cysteine	C and Ce
Histidine	H	Aspartic acid	As
Arginine	A	Alanine	Al
Tryptophane	T	Serine	Se
		Iodogorgoic acid	I

When proteins are digested in the gut, they are eventually broken down into their amino-acids. These are absorbed and carried in the blood to the liver, where selection is made of the kinds and numbers necessary for building, repairing and replacing protoplasm. Those that are not required are utilized for release of energy (*see* p. 153).

Most of the amino-acids are derived from fatty acids. For example, if in a molecule of acetic acid ($CH_3.COOH$) one of the H atoms attached to the first C atom is displaced by the amine group (NH_2), the simplest amino-acid is formed, **amino-acetic acid,** or **glycine** (CH_2NH_2COOH). Ten of the amino-acids are derived from **propionic acid,** CH_3CH_2COOH, the simplest being **alanine,** $CH_3CH.NH_2COOH$.

When amino-acids are chemically linked together, they form chains called **peptides.** The type of linkage between the amino-acids is always the same; it is a **peptide linkage** between the groups CO and NH. The example below shows the joining of two glycine molecules to form a **dipeptide.** The glycine molecules are written in a more convenient form to show how the linkage arises.

$$H_2NCH_2CO\underbrace{\boxed{OH + H}}_{\text{water molecule eliminated}}NHCH_2COOH \rightleftharpoons \underbrace{H_2NCH_2CO}_{\text{amino-acid residue}}\underbrace{NH}_{}CH_2COOH + H_2O$$

water molecule eliminated; amino-acid residue; peptide linkage; amino-acid residue

This is again a reversible reaction; in building up proteins from amino-acids, water is eliminated; in breaking down proteins into amino-acids, water is incorporated. In like manner, three amino-acids joined by peptide linkages form a **tripeptide**; many form a **polypeptide.** It is known that many polypeptide chains are coiled like spiral springs. In some proteins, the polypeptide chains are folded and in others they are straight, but in any case, a number of polypeptide chains make up a single protein molecule. The structure of a simple protein, **insulin,** consisting of two polypeptide chains, is shown below—

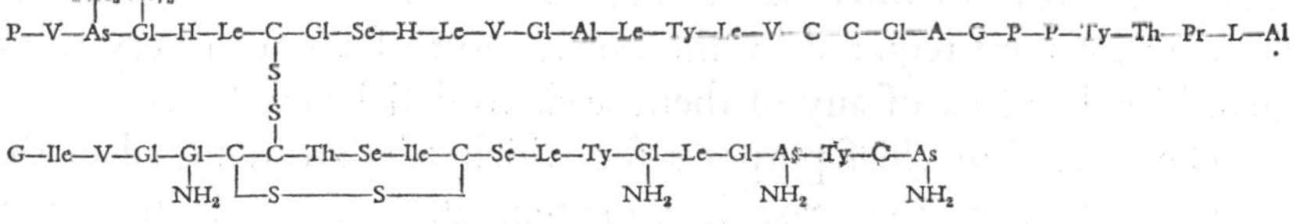

```
     NH2 NH2
      |   |
P—V—As—Gl—H—Le—C—Gl—Se—H—Le—V—Gl—Al—Le—Ty—Le—V—C  C—Gl—A—G—P—P—Ty—Th—Pr—L—Al
                |
                S
                |
                S
                |
G—Ile—V—Gl—Gl—C—C—Th—Se—Ile—C—Se—Le—Ty—Gl—Le—Gl—As—Ty—C—As
              |   |                |            |        |          |
             NH2  └—S————————S—————┘           NH2      NH2        NH2
```

Note. S — S = sulphur linkage.

Proteins are important not only for building up the living protoplasm; they also help to form supporting and protective structures such as connective tissue, bone, cartilage, hair, skin and nails. The enzymes, which bring about almost all the chemical changes in the body, are also proteins. Apart from the value of proteins as food, man uses them for his clothes in the form of wool, silk, fur and leather.

There are no pure proteins in the normal human diet; animal proteins are usually mixed with fat, and plant proteins with carbohydrate. Foods that contain a high percentage of protein are lean meat, fish, cheese and eggs. The plants with the highest percentage of protein are peas, beans and nuts, but even in these, it averages only about half that found in the animal products (*see* table on p. 123).

Salts

These are commonly known as **mineral salts,** since they all come originally from the soil; they are taken into the body either dissolved in water or contained in plant or animal foodstuffs. In dilute solution, salts are completely **ionized**; each salt molecule is split into an **electropositive ion** or **cation** and an **electronegative ion** or **anion.** Some examples of ionization are given below—

1.	$NaCl$ Sodium chloride	→	Na^{+} Sodium ion	+	Cl^{-} Chloride ion
2.	$CaCO_3$ Calcium carbonate	→	Ca^{++} Calcium ion	+	CO_3^{--} Carbonate ion
3.	K_3PO_4 Potassium phosphate	→	$3K^{+}$ 3 Potassium ions	+	PO_4^{---} Phosphate ion

It is in the form of **ions** that salts are absorbed from the gut. Many different ions are required by the body, most of them in very small quantities, but lack of any of them leads to deficiency disease.

There are four chief purposes for which salts are required in the body. Firstly, in the growth, replacement, and repair processes, certain tissues need salts as essential constituents. For instance, bones and teeth consist largely of **calcium phosphate** and **calcium carbonate,** and for correct growth of these organs, the right salts must be present in sufficient quantities. The bone marrow needs constant supplies of **iron salts** for the manufacture of the red corpuscles of the blood. For making more ATP as the body grows, phosphates must be present. Secondly, the body fluids, such as blood, lymph and the coelomic fluid surrounding the viscera, must contain the right kinds and quantities of ions in solution; even slight deviations are very serious in their effects. Therefore, there must be a reasonably constant amount of each of the right kinds of salts. The principal salt in the blood is **sodium chloride,** but small traces of many other salts are present. It is interesting to note that our blood approximates closely in composition to sea-water, which was undoubtedly the medium surrounding our remote ancestors. Now we have the medium, but little altered, as an **internal environment** bathing all the cells. Thirdly, certain salts are required by various types of cells that produce useful materials as their main work. Cells of the **thyroid gland** in the neck produce a hormone, **thyroxine,** which affects the growth of the whole body. For the manufacture of thyroxine the iodide ion, I^-, is required. All cells of the body make enzymes containing iron and therefore the Fe^{++} or Fe^{+++} ion is necessary.

Some of the deficiency diseases due to the lack of salts are well known. Severe lack of sodium chloride will cause **muscular cramp**; lack of iron will cause **anaemia**; lack of calcium or phosphate will cause **rickets** and **dental decay,** both also associated with lack of vitamin D. Lack of iodine will cause overgrowth of the thyroid gland, the large swelling in the neck being called **a goitre.** Serious lack of iodine will cause mental and physical backwardness; children suffering from these are idiots of the type known as **cretins.** The only ions likely to be deficient in the average human diet are those of calcium, Ca^{++}, phosphate, PO_4^{---}, iron, Fe^{++}, and iodine, I^-.

Many other ions are needed for special purposes, e.g. magnesium, potassium, zinc, copper, manganese, cobalt, bromide and fluoride. Sulphur is essential for many purposes but is not absorbed into the body in appreciable amounts as sulphate ions, SO_4^{--}; it is contained mainly in the amino-acids, cystine, cysteine, and methionine.

Foods that are rich in mineral substances are milk, butter, cheese, egg yolk, whole-grain cereals, green vegetables, fruits, meat, fish.

Vitamins

In the early years of this century it was established that young animals fed on a diet of purified carbohydrate, protein, fat, salts and water ceased to grow and soon died. It was known since the time of

Name of Vitamin	Deficiency Diseases	Rich Sources
A	Night-blindness; dry-eye; toad-skin	Fish-liver oils, milk, green leaves, carrots, yellow maize
D	Rickets — malformation of the bones especially during growth	Fish-liver oils, egg-yolk; formed in fat of the dermis with sufficient sunlight, and hence found in many animal fats
E	Sterility in males; early death of the embryo in females	Wheat-germ oil, green leaves, egg-yolk, many seed fats
B_1	Beri-beri—an inflammation of the nerves and paralysis of arms and legs; fatal	Whole cereals, peas and beans, nuts, yeast (the vitamin is now added to all white flour sold in Britain)
B_2	Cheilosis—dry, cracked and scaly skin	Most foodstuffs of plant or animal origin
P–P	Pellagra—red swellings on skin, digestive disturbance, considerable loss of weight; fatal	Yeast, liver, lean meat, milk, fish, whole cereals
B_{12}	Pernicious anaemia—non-production of red corpuscles	Liver, eggs, milk, meat, fish
C	Scurvy—loosening of teeth, bleeding under the skin; fatal	Fresh fruit and green vegetables, especially blackcurrants, blackberries, rose-hip syrup, citrus fruits

the ancient Greeks that certain diseases were due to dietary deficiency and could be remedied by eating certain foodstuffs. During the First World War, 1914–18, the high incidence of nutritional disease stimulated a great deal of investigation into the so-called **accessory food factors,** or vitamins. At present, there are about forty vitamins known; not all have been established as essential for human beings. Because of lack of knowledge of their chemical nature, they were originally named by letters of the alphabet, and even now, when the chemical nature of most of them is known, the letters are still used. The vitamins are conveniently divided into **fat-soluble** (A, D, E) and **water-soluble** (B_1, B_2, P–P, B_{12}, C). The table opposite gives the deficiency diseases and the richest sources of the vitamins known to be important to human beings.

All the vitamins are required in very small but regular quantities and there is not likely to be any deficiency when the diet is suitably varied (*see* p. 117).

Diet

The importance of a well-balanced diet for proper growth and development, and for the maintenance of health, is now generally recognized. Deficiency diseases such as scurvy and rickets, which were common in the past, are now rare among civilized communities. If the available knowledge were more widely applied, there is little doubt that other diseases due to defective diet, such as **beri-beri** and **pellagra,** would soon be eradicated. Perhaps a greater danger in modern times, at least in western civilizations, is that of the accumulation of fat, known as **obesity** or **adiposity.** This condition is a considerable danger to general health and there is definite evidence that in persons of middle age it shortens life.

In considering a balanced diet, a number of factors have to be taken into consideration. They are—

1. Energy requirements.
2. The proportions of carbohydrate, fat and protein.

3. Vitamin requirements.
4. Mineral salt requirements.
5. Water requirements.
6. Other factors such as cost, digestibility, etc.

ENERGY REQUIREMENTS

Human energy requirements, like those of all animals, vary during the twenty-four hours of the day; they vary according to occupation, according to age, size, sex and according to external factors such as temperature. Finally, they vary according to the state of the body, i.e. whether healthy or diseased. Before calculating how much energy-producing foodstuff is necessary, it is important to remember that the body is using considerable energy when in a state of rest or sleep. The resting state, for this purpose, is defined as a period when a person is lying awake and relaxed, twelve hours after the last meal and at a room temperature of 20°C. The energy expended during that state is known as the **basal metabolism,** and the average for men in Great Britain is found to be slightly more than 5000 kilojoule for a sixteen hour day, with somewhat less for sleep; e.g. eight hours sleep requires approximately 2090 kilojoule for metabolism during the sleeping period. This gives an average of 7090 kilojoule for a person lying down all day, sleeping eight hours a night and doing no external work of any kind.

To this figure, there must be added, for the average person, an allowance for extra energy expended on work. Sitting at rest uses 126 kilojoule per hour above the basal metabolic rate; standing uses 167; walking at a moderate pace 580 kilojoule; running averages roughly 2090 per hour. By adding the figures for basal metabolism, sleep and work, we have the kilojoule requirement for twenty-four hours. If the person does not get this requirement in his food, then he will lose weight continuously. Some examples of kilojoule requirements are given opposite. (It is assumed that the person is a healthy male and roughly of average size.) The average figures for women are ten per cent lower.

1. Sedentary work, e.g. office work, typing, managerial.

Basal metabolism (16 hours)	5 024 kJ
Metabolism during sleep (8 hours)	2 094 kJ
Allowance for light work	3 350 kJ
Total for 24 hours	10 468 kJ

2. Moderate work, e.g. garage mechanic, road-sweeper, carpenter.

Basal metabolism	5 024 kJ
Sleep	2 094 kJ
Work	5 862 kJ
Total	12 980 kJ

3. Heavy work, e.g. coal delivery, miner (cutting and loading), stonemason.

Basal metabolism	5 024 kJ
Sleep	2 094 kJ
Work	8 374 (and upward)
Total	15 492 (and upward)

The following are recommended kilojoule requirements for various groups leading an ordinary, everyday life, without manual work, in a temperate climate.

Age (years)	Kilo-joule	Age or Condition	Kilojoule
1–2	3 517	Pregnant woman	10 048
2–3	4 187	Nursing woman	12 561
3–5	5 024	Baby 0–6 months	414 per kg of mass
5–7	6 029	Baby 6–12 months	368 per kg of mass
7–9	7 034		
9–11	8 039	Light work add	314 per hour
11–12	9 044	Moderate work add	314–628 per hour
12–15	10 048	Hard work add	628–1256 per hour
and upwards		Very hard work add	1256 and over per hour
Man	10 048		
Woman	9 044		

It must be emphasized that these average figures have been calculated after measurements involving several thousand people of each group and that they do not necessarily apply to a particular individual. Some people are bigger than average, some are smaller, some are very active, some are not; some spend their leisure in vigorous pursuits such as swimming (2000 kJ per hour), cycling (750 kJ per hour) or active games, while others follow more leisurely pursuits such as reading (130 kJ per hour), painting (180 kJ per hour). Any individual in good health should have sufficient energy for normal activity without losing weight. Any abnormal activity will require a further intake of energy-producing foodstuffs.

To translate these energy-requirement figures into actual foodstuffs, the following table gives the average number of kilojoule per kg of some common foods.

Food	kJ/kg	Food	kJ/kg	Food	kJ/kg
Butter	33 170	Potatoes (chipped)	11 053	Cabbage (boiled)	367
Margarine	33 170	Bread (brown)	10 571	Runner beans (boiled)	295
Chocolate (milk)	23 950	Potatoes (boiled)	3 685		
Bacon (fried)	23 030			Brazil nuts	26 712
Cheese	20 264	Sugar (white)	17 132	Peanuts	25 790
Beef (with fat)	11 975	Jam	11 975		
Eggs	7 045	Treacle	10 547	Bananas	3 500
Milk	3 038			Grapes	2 763
		Butter beans (boiled)	2 763	Apples	2 119
Cornflakes	17 500	Spinach (boiled)	1 104	Plums	1 705
Rice	14 738	Carrots (boiled)	830	Oranges	1 566
Sago	14 276	Lettuce (raw)	506	Tomatoes	627
Bread (white)	11 053	Celery (raw)	367		

As later paragraphs will show, a person cannot simply choose from such a table the foodstuffs that will satisfy his energy requirements, because other factors are also important in the diet. Nevertheless, it is interesting to speculate how much of one particular food a hardworking man (15 500 kJ or more) would need to satisfy his energy requirements.

Proportions of Protein, Carbohydrate and Fat

Apart from energy requirements, growth and replacement have to be provided for, so that even adults who have ceased growing, must have a proportion of protein in their diet. After many observations in various countries, fairly standard figures of the proportions of protein, carbohydrate and fat in the diet are generally agreed. They are: protein ten to fifteen per cent, fat twenty to thirty-five per cent, carbohydrate fifty to sixty-five per cent. The wide spread in each of these quantities indicates that it is impossible to specify to a more precise degree.

The standard recommendation for protein intake for an adult man is 1 g of protein per kilogram of body mass every day; i.e. an average man, 70 kg mass (154 lb.), should have 70 g (about 2½ oz) of protein a day. The table below shows the average quantities of daily protein for various ages and conditions.

AGE (Years)	GRAM PER KG BODY MASS PER DAY	OUNCES PER STONE BODY MASS PER DAY
1–3	3·5	0·78
3–5	3·0	0·66
5–15	2·5	0·55
15–17	2·0	0·44
17–21	1·5	0·33
Adult	1·0	0·22
Pregnant woman	1·5	0·33
Nursing woman	2·0	0·44

Note that greater amounts of protein are needed during the period of growth. It is also recommended that at least half the total protein requirement should be animal protein as found, for example, in eggs, meat, milk and cheese.

Apart from Eskimos, human beings cannot take excessive quantities of fat without nausea and diarrhoea. Also, since large quantities of fat in the diet would imply less carbohydrate, normal utilization of fat would be impossible, because in the absence of a sufficient quantity of carbohydrate the diseases **ketosis** (high acidity of the blood) and **ketonuria** (high acidity of the urine) follow. Most adults can utilize 150 g (5·25 oz) of fat a day; the generally recommended minimum is 50 g a day (1·75 oz). The inclusion of fat in the diet is important for a number of reasons.

1. Fats usually contain the fat-soluble vitamins, A, D and E.
2. Lack of fat causes hunger shortly after a meal, probably because digestion and absorption occur rapidly in its absence (carbohydrate is absorbed in three to four hours, fat in five to six hours).
3. Fat yields large amounts of energy; apart from supplying more than twice as much energy as an equivalent weight of carbohydrate, fat contains little water.
4. The body can store only a very limited amount of carbohydrate but considerably more fat. An average well-fed man can store 6000 g (about 13 lb) of fat, whereas he can store only 500 g (just over 1 lb) of glycogen.
5. Carbohydrate in palatable form is usually bulky, and too much tends to delay digestion and produce acidity in the gut.

For most human beings, carbohydrate constitutes the greatest percentage of the food. It is cheaper than either fat or protein and is readily digested, absorbed and utilized for energy release. Too much carbohydrate is undesirable for the reasons given above and it is not possible to replace it entirely by fat. In any case, normal utilization of fat is not possible without at least 50 g per day of carbohydrate in the diet. It is recommended that carbohydrate should not constitute more than sixty-five per cent of the diet. **Thus for an adult male doing light work and utilizing 10 468 kJ per day, the recommended quantities of the three major foodstuffs are**

Protein. 70 g (about $2\frac{1}{2}$ oz)
Fat. 100 g (about $3\frac{1}{2}$ oz)
Carbohydrate. 300 g (about $10\frac{1}{2}$ oz).

Most of the carbohydrate consists of starches, derived mainly from cereals, potatoes, sago, tapioca and arrowroot; and of sugars, mainly from cane and beet sugar with small quantities from sweet fruits and, in some cases, honey. In any case, the natural product is preferable to highly refined starches and sugars because of the vitamins and salts present. Cereals contain vitamins B and E and salts; potatoes contain vitamins B and C and salts; fruits contain vitamin C and salts. Refined sugar contains no vitamins or salts but is nevertheless of value because it is rapidly digested and absorbed and it relieves fatigue very quickly. Natural starchy foods (e.g. whole cereals and potatoes, fruits and vegetables) contain a great deal of cellulose, which is indigestible, but nevertheless forms an important part of the diet because its bulk stimulates peristalsis and keeps the muscles of the gut in good working order.

VITAMIN REQUIREMENTS

Serious diseases due to vitamin deficiency are now rare in most civilized countries, and mild deficiencies are very difficult to detect. The actual quantities of vitamins needed per day are extremely small; some examples are given below—

Vitamin	A	0·003 g
,,	B_1	0·0015 g
,,	B_2	0·0018 g
,,	P–P	0·015 g
,,	C	0·075 g (American), 0·020 (British) (different standards used)

The actual total of all vitamins needed per day by an adult amounts to less than 0·1 of a gram (i.e. 0·0035 of an ounce). In general, if the diet is sufficiently varied, there is little likelihood of vitamin shortage. To ensure sufficient supplies of all the vitamins, the diet should include meat and fish, milk, eggs, some whole-grain cereal, fresh fruit and fresh vegetables (salad). Children, pregnant and nursing mothers have a special need for vitamins A and D. D is particularly important in these cases, since it is essentially concerned with bone and tooth formation. Finally, it might be remembered that cooking destroys some of the vitamins, especially B_1 and C.

Mineral Salt Requirements

The importance of the presence of small amounts of certain salts in the diet has already been stressed and the effects of certain deficiencies mentioned. Normally, the only ions likely to be deficient in human diets are Ca^{++}, Fe^{++}, Fe^{+++}, PO_4^{---} and I^-. The following table gives the recommended amounts of Calcium, Iron, Phosphorus and Iodine necessary per day.

	Calcium g	Phosphorus g	Iron g	Iodine g
Moderately active men	1·0	1·3	0·012	0·00005
Pregnant women	1·5	1·9	0·015	0·000055
Nursing women	2·0	1·9	0·015	0·000055
Children up to 12 yrs	1·2	1·0	0·01	0·000055
Young people of 12 to 20 yrs	1·4	1·3	0·15	0·000055

The total mineral requirements amount to 3 or 4 g. With a normal good mixed diet there is unlikely to be any shortage except in the case of iodine and then only in certain localities where the soil is deficient and hence local plant and animal products are deficient. In these cases, **potassium iodide** is usually added to the table salt. The table above shows what minute quantities of iodine are necessary, but the consequences of deficiency are drastic—goitre, and in extreme cases, cretinism in children. Again, milk, meat, eggs, cheese, whole cereals, fresh fruits and green vegetables will provide ample mineral salts.

Water

Lack of water is quickly fatal; the average human being could not survive more than a few days with a total lack of water. There are many reasons why water is so vitally necessary; the main reasons are given below.

1. Approximately two-thirds of the human body is water. An average man, i.e. 11 stone or 70 kg, contains 47 litres (about 83 pints)

of water: 20 litres in his muscles, 10 litres in his skin, nearly 5 litres in his blood, and the remainder distributed in the other parts.

2. There is a daily loss of water amounting to 2·7 litres, made up as follows: 0·5 l from the lungs in the exhaled air, 0·7 l from the skin in the perspiration, 1·4 l in the urine, 0·1 l in the faeces. This loss is roughly balanced by the intake of 1·3 l in drinks, 0·9 l in food, and 0·5 l by oxidation of food in the cells.

3. Water plays a part in many of the chemical reactions that take place in the body, and all these reactions take place in aqueous solutions.

4. Digested food is absorbed in aqueous solution and excretory materials are passed out of the body in aqueous solution.

5. Oxygen can be absorbed in the lungs only if in solution, and carbon dioxide is eliminated in solution.

As stated above, the intake of water in drinks is about 1·3 litres per day (about 2¼ pints). This applies to the temperate climates; in hot dry countries, the water intake (and loss) may be over 10·1 litres. The amount of water contained in so-called solid food is somewhat surprising. An average three-course dinner, without drinks, contains about fifty per cent of water. The water content of some foods is given below (average values only).

Cow's milk	87%	Bread	40%
Butter	12%	Nuts	20%
Cheese	over 30%	Potatoes (boiled)	78%
Eggs (raw)	75%	Lettuce (raw)	97%
Meat (roast)	60%	Oranges	90%
Cereals (raw)	10%	Tomatoes	92%

Other Factors

A number of other factors affect diet. Even if all the experts agreed on the perfect diet, the majority of human beings would not keep to it for various reasons. Cost, cooking, digestibility, variety, and a number of psychological factors make the diet, at least of adults, a very individual or family affair.

The **cost** of food is a very important factor and the amount of money that can be spent on food is closely related to the family income. Where incomes are very low, then diet is necessarily poor, because good quality foodstuffs, i.e. those that have protective value as well as energy, are expensive. Milk, meat, fish, eggs, cheese, butter, fresh fruit and salad vegetables are much more costly than bread, potatoes, rice, sugar and cooking fat. In under-developed countries there is often serious lack of protein and of the protective foodstuffs. The problem of raising nutritional standards throughout the world is very complex. There is no doubt that available resources are not fully utilized and even if the present production of foodstuffs were equally shared, the standard of nutrition would be below what is regarded as adequate. With the present rate of increase of population (*see* p. 394), and the improved knowledge of disease and of medical care, drastic steps will be necessary in the next fifty years or the population will rapidly outgrow the food supply.

Most foods undergo considerable change during the process of **cooking.** It may be considered that there are three forms of cooking: boiling, frying or roasting, and baking. Whichever form is used, the primary objects are to destroy harmful organisms, to add flavour and to soften the material so that it is more easily masticated and, in most cases, more easily digested. Cooking almost always adds water to the foodstuff; for instance, bread contains more water than the flour from which it is made. However, in the case of roasted meats, there is a loss of about twenty per cent of the water content. The main disadvantages of cooking are that it usually entails loss of mineral salts and destruction of certain vitamins. During boiling, vitamin B_1 is largely lost in the water, and vitamin C is destroyed. Vitamin A is destroyed in cooking at high temperatures, as in frying and roasting. In any type of cooking involving water, most of the mineral salts content of the food is lost. The practice of using small quantities of water, cooking quickly and using the water for gravy, etc., is strongly recommended. During the cooking of meat, the connective tissue that separates the muscle fibres is largely converted to the protein **gelatin**;

thus the fibres are loosened and it is easier to chew the meat. It is important that starchy foods should be cooked, since raw starch is practically indigestible; cooking bursts the grains and the starch is then easily digested.

Digestibility is also an important factor in diet. Although a particular protein may contain all the essential amino-acids, it does not necessarily follow that it is all digested and absorbed from the gut. In this connection, it is noteworthy that vegetable proteins are not absorbed as well as animal proteins, though fats and starches are largely absorbed. By examining the proportions of a given food that are passed out in the faeces, some idea is obtained of the percentage absorbed. The following table compares some animal and plant products in this respect.

Food	Percentage Not Absorbed		
	Protein	Carbohydrate	Fat
Milk	7	—	6
Beef	3	—	—
Eggs	3	—	5
Bread	22	1	—
Beans	18	4	—
Potatoes	30	8	—

Experiment has shown that there is a greater percentage of absorption when the diet is mixed, i.e. contains a reasonable proportion of both animal and vegetable products. Another complicating factor is **habit.** To most people, the Eskimo diet of meat and fish would at first be indigestible. Likewise a diet of rice and vegetable oils would be indigestible to an Eskimo. The wider the **variety** in the diet, the less risk there is of missing some essential vitamin or salt. Even in the wealthier countries, where cost is not a prohibiting factor, the diet is often governed by custom, prejudice and convenience.

Finally, there are a number of **psychological factors** that affect diet. Everything that contributes towards the enjoyment of a meal is beneficial; the pleasant appearance and smell of food, attractive serving, cleanliness of china and cutlery, comfortable accommodation, pleasant company (or no company) are all beneficial. Monotony, slovenliness, uncleanliness, unattractive preparation and serving, spoil one's enjoyment of a meal. When one is ill, convalescent or in a condition of anxiety, digestion is probably not functioning normally and special care is needed over the diet.

Composition of Common Foods

The table opposite gives the percentage proportions of protein, fat and carbohydrate in various common foods. Column 1 (100 g = 3½ oz) enables the student to compare the energy values of the various foods. The water content can be found by totalling the percentages for protein, fat and carbohydrate and subtracting from 100. The mineral salt content is usually less than one per cent and can be neglected here.

Milk

Of all foodstuffs, milk may be regarded as the most perfect because it contains all the essential substances for a complete diet. The proteins, **caseinogen** and **lactalbumen,** are both first class and easily digested; milk fat is the most digestible of all natural fats; the carbohydrate, milk sugar or lactose, is easily digested but less sweet than cane sugar. Milk contains most essential minerals; it is rich in calcium, potassium, sodium, chloride and phosphate, but rather poor in iron. There are sufficient quantities of vitamins A, B_2, B_1, P–P, but little of C and D.

It is not surprising that such a comprehensive food material also provides a perfect medium for the growth of **bacteria,** and hence it is impossible to market fresh milk in a sterile condition. There are two sources of contamination that may account for the presence of dangerous bacteria. In the first place, bacteria from the cow may be present, and of these the most dangerous is the bovine **tubercle bacillus,** which

Food	kJ Per 100 g	Percentage of Protein	Percentage of Fat	Percentage of Carbo-hydrate
Apples	214	0·3	—	11·5
Bacon (fried)	2 293	24·6	53·4	—
Bananas	348	1	—	19
Beef (roast)	941	26·5	12·4	—
Blackberries	134	1·2	—	6·5
Brazil nuts	2 710	14	61·5	4
Bread (whole)	1 121	8·5	1	54·5
Butter	3 246	0·2	83	—
Butter beans (boiled)	1 190	19	1	50
Cabbage (boiled)	38	0·7	—	1·2
Carrots (boiled)	84	0·5	—	4·3
Celery (raw)	38	1	—	1·3
Cheese (Cheddar)	1 747	25·2	32·1	3·4
Chicken (roast)	794	29·5	7	—
Corn flakes	1 743	11	5	79
Eggs	680	12·3	11·3	1·6
Grapes	277	0·5	—	15·5
Haddock (steamed)	412	22	1	—
Honey	1 235	—	—	71
Jam	1 211	—	—	70
Lettuce (raw)	50	1	—	1·8
Liver (fried)	1 207	29·5	16	4
Milk (cow's, whole)	286	3·3	3·6	4·8
Oranges	160	0·7	—	8·5
Peanuts	2 545	28	49	8·5
Plaice (fried)	991	18	14	7
Potatoes (boiled)	365	1·5	—	20
Potatoes (chipped)	1 058	4	9	37
Rice	1 500	6	0·5	80·5
Runner beans (boiled)	29	0·8	0·4	0·9
Spinach (boiled)	113	5	—	1·5
Sugar (white)	1 722	—	—	100
Syrup (golden)	1 394	—	—	81
Tomatoes	63	1	—	3
Turnips (boiled)	50	0·7	—	2·3

may cause bovine **tuberculosis** in human beings. Secondly, bacteria may be introduced during the milking process or during transport. As a result, the germs that cause **diphtheria, typhoid fever** and **scarlatina** may be present. Hence, various measures are taken to lessen the dangers to human beings.

1. **Tuberculin-tested milk.** In this case, the herd is certified as being free from the tubercle bacillus and the milk is collected and bottled under hygienic conditions.

2. **Pasteurized milk.** The dangerous bacteria are destroyed, though the souring bacteria survive. The milk is heated rapidly to a temperature of 52·5°C (145°F), kept at this temperature for thirty minutes and then cooled quickly. The only loss in nutritive value is that about twenty-five per cent of vitamin B_1 is destroyed.

In the flash method there is a continuous flow of the milk while it is being heated to 85°C for 15 seconds or 71°C for 60 seconds, followed by rapid cooling.

3. **Sterilized milk.** In this case, the milk is heated to a much higher temperature, 110°C, and kept at that temperature for thirty minutes. All the bacteria are destroyed and the milk will keep for a long time, but most people find it unpalatable. Also, the protein is largely coagulated, vitamin C is destroyed, and the fine emulsification of fat is lost.

The average composition of cow's milk is: water 87 to 88 per cent, protein 3 to 3·5 per cent, fat 3·5 to 4·5 per cent, carbohydrate 4·5 to 5 per cent and mineral salts 0·7 per cent. In Great Britain, it is illegal to sell cows' milk with a fat content below 3 per cent. By comparison, human milk has this average composition: water 89 per cent, protein 1 per cent, fat 3 per cent, carbohydrate 6·7 per cent and minerals 0·3 per cent. Milk substitutes for babies are usually made by diluting cows' milk about three times to make the protein content right, and then adding cream and lactose to make the fat and carbohydrate contents right. Iron is usually added, since cows' milk is even more deficient in this than human milk.

Milk is the natural food made in the mother's body for the nourishment of her offspring. If the mother has a correct diet, then the milk will be a perfect food for the baby. In some countries, free milk is given daily to schoolchildren, mainly because of the protective value of its vitamins and mineral salts.

SUGGESTED EXERCISES

Difficult or dangerous exercises are marked with an asterisk; it is inadvisable for young pupils to perform these. (Note: 1 cm^3 = 1 ml.)

1. A test for all proteins. Use dried or fresh egg albumen. Dissolve a small quantity in about 5 cm^3 of water. Add about 1 cm^3 of caustic soda solution, then add one per cent copper sulphate solution, drop by drop, shaking after each drop. A **violet** colour is produced. The copper sulphate solution must not be added too quickly, or the violet colour will be masked by the blue of the copper sulphate. Usually one or two drops are sufficient. The colour is due to the CO.NH groups, which form a violet colour with copper sulphate.

2.* A test for all carbohydrates. Use a solution of any sugar or a suspension of starch. Half-fill a test-tube with the solution, and add a few drops of alcoholic α-naphthol solution. Tilt the test-tube, and down its side pour slowly about 2–3 cm^3 of concentrated sulphuric acid. This will sink to the bottom and at the interface between the acid and the other liquid, a **purple ring** will form. The acid produces a substance called furfural from the carbohydrate. Furfural then forms a coloured product with the naphthol.

3.* A test for all fats. All fats are **blackened by osmic acid. (Care!** This is poisonous and expensive.) Place one drop of the osmic acid on a piece of fat, e.g. suet, lard, dripping, and note the blackening. Any plant fatty oil, e.g. olive oil, may be used.

ADDITIONAL EXERCISES

PROTEINS

Use fresh or dried egg albumen.

1.* Dissolve a small quantity of the albumen in about 2 cm^3 of water. Add concentrated nitric acid and warm gently. **(Care!)** A **yellow** colour is produced. Cool the tube and add a few drops of ammonia; the yellow colour deepens to **orange.** The colours are due to the formation of coloured nitro-compounds.

2.* To the albumen solution, about 2 cm^3, add a few drops of Millon's reagent (mercury dissolved in concentrated nitric acid). **(Care! This is poisonous.)** Heat the test-tube and a **red** colour will gradually appear. The colour is due to the presence of a phenolic compound derived from the amino-acid, tyrosine.

5

CARBOHYDRATES

1. To distinguish monosaccharide sugars from disaccharide sugars, boil about 5 cm^3 of Barfoed's reagent (copper acetate in acetic acid) in a test-tube. Add slowly a small quantity of the monosaccharide solution (glucose or fructose). A **red precipitate** will appear. It is due to red copper oxide produced in the reaction.

Disaccharides need prolonged boiling to break them down to monosaccharides; the red precipitate will then appear. (Use cane sugar solution.)

2. To distinguish starch, add a few drops of a solution of iodine in potassium iodide to a weak suspension of starch in water; a **blue to black** colour will appear. The colour is due to absorption of iodine by the starch, producing a substance of varying composition, depending on how much iodine and how much starch are present.

3.* To distinguish cellulose, use the following test on a small piece of filter paper or cotton wool in a test-tube. Cover it with concentrated sulphuric acid and add a small quantity of weak iodine solution. A **blue** colour indicates cellulose. This is due again to the formation of an absorption substance of varying composition.

4. To distinguish glycogen, use the following test on commercial glycogen. Shake a small quantity of the glycogen in a few cm^3 of water, in a test-tube. Add a few drops of iodine solution. A **reddish-brown** colour will appear, again due to formation of an absorption compound.

FATS

1.* With a piece of fat, e.g. meat fat, suet, make a grease-spot on a piece of white paper. This cannot be washed out with water but can be dissolved out with ether, benzene or chloroform. (Care!)

2. Add a little vegetable oil, e.g. olive oil, linseed oil, or castor oil, to some water in a test-tube and shake. Then add a little of the red dye, Sudan III. The droplets of oil will take up the dye and will retain it, even if shaken vigorously in a large volume of water or alcohol.

3. Shake up 5 cm^3 of olive oil or melted fat with 5 cm^3 of 20 per cent sodium hydroxide. A **white emulsion** forms; the droplets of fat are **dispersed** in the aqueous solution. Immerse the tube in boiling water; the emulsion becomes unstable and separates into a layer of oil on a layer of aqueous solution. Shake the tube occasionally; after 30 min remove and cool under the tap. The contents now show three layers: (1) lower—an alkaline solution of **glycerol;** (2) an intermediate layer of soap; (3) an upper layer of unchanged oil. Pour away the liquids and wash the small cake of soap in cold water.

CHAPTER 8

THE ALIMENTARY CANAL, DIGESTION AND UTILIZATION OF FOOD

THE alimentary canal is the tube from the mouth to the anus in which the food is digested. The digested food is absorbed from the canal and indigestible and excess materials are eliminated at the anus. It is not a simple tube with the same diameter throughout; various parts of it are enlarged, constricted or otherwise modified for the performance of various functions. The canal consists of the following parts—**1. mouth. 2. buccal cavity. 3. pharynx. 4. oesophagus. 5. stomach. 6. small intestine, consisting of the duodenum, jejunum and ileum. 7. large intestine, consisting of the caecum, colon, and rectum. 8. anus.**

Along the canal there are a number of valves, usually called **sphincters.** These valves can be closed so that the food is retained in one particular part until treatment of it is completed. A sphincter consists of a strong, thickened ring of smooth muscle fibres, circularly arranged. The contraction closes the tube, while relaxation allows it to open. The **cardiac sphincter** lies between oesophagus and stomach, the **pyloric sphincter** between stomach and duodenum, the **ileo-caecal valve** at the junction of ileum, caecum and colon, and finally the **anal sphincter** is at the end of the canal. The **mouth** may be considered as a valve at the beginning of the canal. The total length of the canal averages about **9 m** in an adult; the lengths of the various parts are given in the following table. A general view of the main parts of the canal is given in Figs. 29(*a*), p. 43 and 31(*a*), p. 46.

THE MOUTH AND BUCCAL CAVITY

The mouth is the aperture for the entry of food and is controlled by the muscular upper and lower lips. The circular arrangement of the

Sphincters (or Valves)	Name of Part	Average Length
Mouth (lips)	—	—
	Buccal cavity	9·0 cm
	Pharynx	12·5 cm
Cardiac sphincter	Oesophagus	25·0 cm
	Stomach	30·0 cm (fully distended)
Pyloric sphincter	Small intestine	7 m
Ileo-caecal valve	Large intestine	1·5 m
Anus	—	—

muscle fibres is shown in Fig. 12(*c*) on p. 20. The buccal cavity is lined with soft **mucous membrane,** so called because it contains

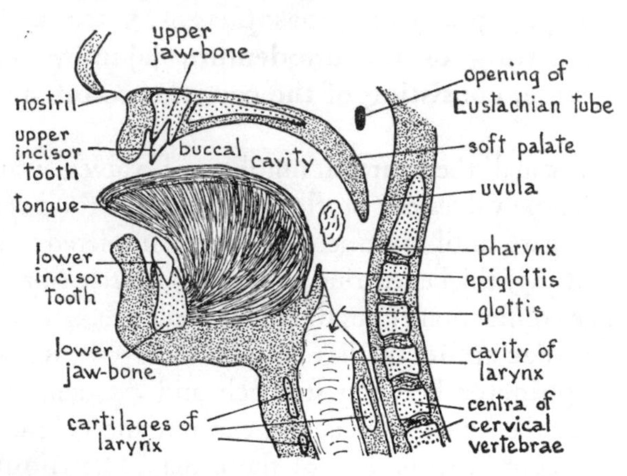

Fig. 76. Median Vertical Section of Part of the Head
One tonsil and Eustachian tube are shown, though they are not in the median plane.

many small **mucous glands,** which secrete the sticky substance, **mucus.** These glands are found throughout the length of the canal; the mucus keeps the surface moist and lubricates the passage of food. The roof of the buccal cavity is the **hard palate** in the front part and the **soft palate** in the back part. The soft palate tapers backwards, forming the **uvula.** At the sides and front of the buccal cavity are the **teeth,** and on its floor is the **tongue.** The ducts of the **salivary glands** open into the buccal cavity. A median section through the buccal cavity is shown in Fig. 76.

The Teeth

The typical structure of a single-rooted tooth is shown in Fig. 77(*a*). A tooth has three distinct regions: the **crown,** which projects above

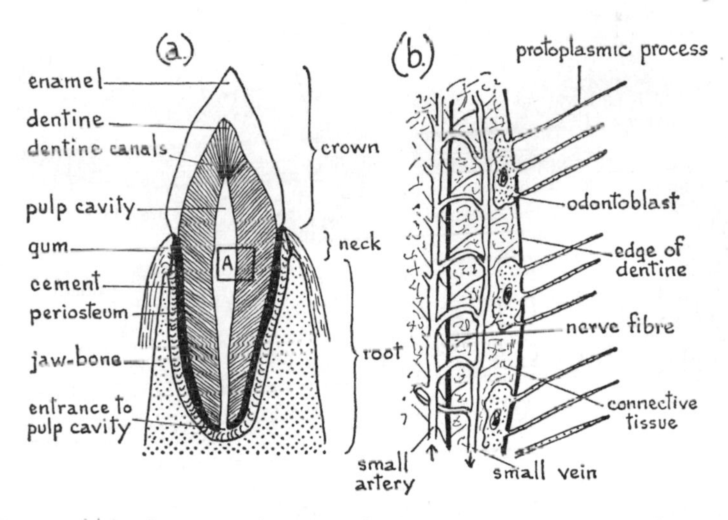

FIG. 77. (*a*) Median Vertical Section of a Canine Tooth *b*) The Portion at A, Considerably Enlarged

the gum, the **root,** which is fixed in a socket in the jaw-bone, and a narrow **neck,** which lies between the root and crown. In the centre of the tooth is a **pulp cavity,** tapering to a narrow entrance at the base of the root. Through this entrance, blood-vessels and sensory nerves extend into the cavity. The bulk of the tooth consists of a yellow, bone-like substance called **dentine,** secreted by cells called **odontoblasts,** which line the pulp cavity (*see* Fig. 77(*b*)). Very fine **protoplasmic processes** from these cells extend into **dentine canals**; these processes are said to perceive pain when a tooth is decayed; there are no nerve-endings in the dentine. Outside the dentine, the crown is capped by **enamel,** which is the hardest substance in the body. The root is covered, except at its aperture, with bone-like **cement** secreted by the **periosteum** of the jaw-bone. The cement and enamel meet just under

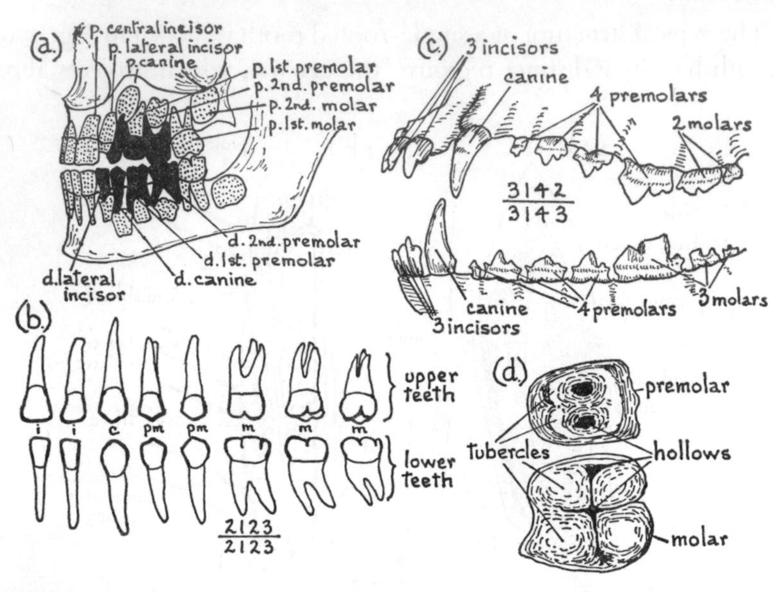

FIG. 78. (*a*) Teeth of a Child of Seven Years, Three Months (*p* = *permanent*, *d* = *deciduous*) (*b*) Permanent Teeth of the Right Side (*c*) Jaws of a Dog (*d*) Upper Surface of One Premolar and One Molar

the edge of the gum. A tooth fits neatly into its socket in the jaw-bone and is secured by the cement and the periosteum.

Human beings, like most mammals, have two sets of teeth. The first set, called **milk** or **deciduous teeth,** are pushed out and replaced by the **permanent teeth** (*see* Fig. 78(*a*)). There are four sorts of teeth in the permanent set and three sorts in the deciduous set (*see* Fig. 78(*b*)). The flattened front teeth, called **incisors,** are adapted for cutting; the pointed **canines,** adapted for piercing, are seen at their best in carnivora (*see* Fig. 78(*c*)); the **premolars** and **molars** are adapted for crushing and grinding, having raised **tubercles** and hollows (*see* Fig. 78(*d*)). There are no molars in the deciduous set.

The new-born baby does not ordinarily have any teeth visible above the gums, but they are forming in the jaw-bones. The table shows the average ages at which the various types of teeth appear above the gums.

Milk Teeth	Lower central incisors	6 to 9 months
	Upper incisors	8 to 10 months
	Lower lateral incisors	15 to 21 months
	First premolars	15 to 21 months
	Canines	16 to 20 months
	Second premolars	20 to 24 months
Permanent Set	First molars	6 years 2 months
	Central incisors	7 years 3 months
	External incisors	8 years 3 months
	First premolars	9 years 3 months
	Second premolars	10 years 3 months
	Canines	11 years 3 months
	Second molars	12 years 3 months
	Third molars (wisdom teeth) (sometimes not formed)	17 to 21 years or later

It is usual to express the total numbers and kinds of teeth of a mammal in a brief way, called a **dental formula.** Thus the pig, which is considered to have the typical complete mammalian set, has, in half the upper jaw, three incisors, one canine, four premolars and three molars. Half the lower jaw has the same kinds and numbers. Thus the pig's dental formula can be expressed as $\frac{3143}{3143}$. To find the total number of teeth, all the figures are added together and the answer multiplied by two. Some examples of dental formulae for adult mammals are given below—

Pig	$\frac{3143}{3143}$	44	Man	$\frac{2123}{2123}$	32	Dog	$\frac{3142}{3143}$	42
Cat	$\frac{3131}{3121}$	30	Sheep	$\frac{0033}{3133}$	32	Rat	$\frac{1003}{1003}$	16

The formula for the milk teeth is obtained by leaving out the last figure, that for molars, except in the case of rats and mice, which have permanent teeth only.

Decay of the teeth (**dental caries**), sometimes early in life, is more widespread among civilized than among primitive peoples. This suggests strongly that diet is an important factor. Undoubtedly, proper diets for the pregnant mother and for the child in its early years are of great importance. Another factor that is generally considered important is the presence of traces of **fluorides** in the diet. **Calcium fluoride** forms a small fraction of the enamel and is said to contribute to its hardness. In some districts in Great Britain, fluoride is added to the water in reservoirs, and the results are somewhat encouraging. One of the causes of the decay of teeth is undoubtedly the slow softening and dissolving of the enamel and the dentine by the products of **acid-forming bacteria.** The junction between neck and root is particularly vulnerable; decay in this region is known as **cervical caries.** Attention to the following factors would undoubtedly lessen the incidence of dental caries.

1. A good diet, rich in calcium and vitamin D, particularly for the pregnant and nursing mother and for the young child.
2. Sufficient fluoride in the diet; only minute traces are necessary.
3. Avoidance of cracking the enamel; for example, by removing screw stoppers or by cracking nuts with the teeth.
4. Frequent firm food, which gives the teeth plenty of work to do, e.g. apples, raw carrots.

5. Regular brushing will remove particles of food and reduce the opportunities for bacteria to flourish. A soft brush should be used and the teeth should be brushed towards the tip of the crown. Tooth pastes and powders clean and whiten the exposed surfaces but cannot be expected to work wonders.

6. Regular attention by a dentist will prolong the life of the teeth. Decay can be arrested by cleaning cavities and filling them; useless teeth can be extracted, and malformations can be treated.

It is rare to find an adult with a perfect set of teeth.

The Tongue

The tongue is a large, muscular organ that occupies most of the buccal cavity. It is covered with soft mucous membrane, which contains numerous glands secreting mucus. Outer cells are constantly shed into the cavity, being replaced by the division of the germinative layer, as in the case of the skin (*see* Fig. 35, p. 49). The rate at which these cells are shed gives rise to different kinds of appearance in certain diseases, and hence an examination of the upper surface of the tongue is useful in a doctor's diagnosis. The back of the tongue is the front wall of the pharynx and at its sides are the **tonsils** (*see* Fig. 79(*a*)).

The upper surface of the tongue is covered by small projections called **papillae.** There are three kinds of these: small pointed ones called **filiform papillae,** which are the most plentiful and give a furry appearance to the front part of the surface; larger **fungiform papillae,** like button mushrooms; and two or three rows, behind the centre, of **circumvallate papillae,** each like a flattened mound surrounded by a moat (*see* Figs. 79(*a*) and (*b*)). The tongue is very sensitive to touch, heat and cold, but most of all it is sensitive to chemicals in solution, and this we call the **sense of taste.** The sense organs that perceive these chemicals are called **taste buds**; they are found in the grooves surrounding the circumvallate papillae, at the sides of the fungiform papillae, and are most numerous in the vertical grooves present at the lateral edges opposite the centre of the tongue (*see* Fig. 79(*a*) and (*b*)). Each taste bud is rounded, and consists of a group of **sensory cells** with projecting, hair-like, protoplasmic processes. At the bases of the cells, fine nerve fibres join with those from other taste buds to form a **sensory nerve.** The nerve carries impulses to the brain, where taste is interpreted. There are four fundamental

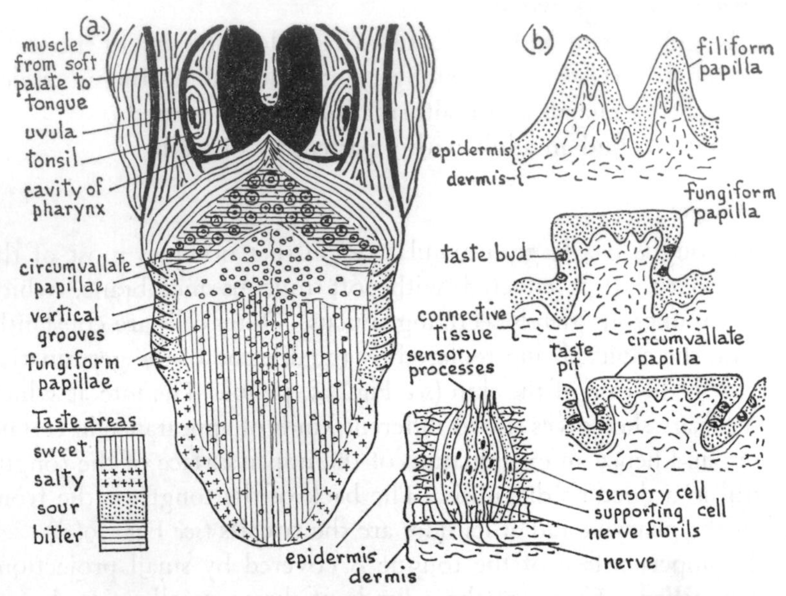

FIG. 79. (*a*) Upper Surface of the Tongue, and the Tonsils (*b*) The Three Kinds of Papillae, × 15 Approximately, and a Taste Bud, × 180

tastes: **sweet, sour, bitter** and **salty,** and the taste buds associated with each have a definite distribution, shown in Fig. 79(*a*).

Apart from the function of taste, the tongue is concerned with moving food about the buccal cavity, with swallowing and with speech.

The Salivary Glands and Ducts

There are three pairs of salivary glands, the **submaxillary, sublingual** and **parotid** (*see* Fig. 29(*b*), p. 43). Each submaxillary gland lies beneath the floor of the buccal cavity, close to the inner aspect of the back of the lower jaw. The ducts from each gland open beneath the tongue (*see* Fig. 80). The sublingual glands also lie beneath the floor of the buccal cavity, nearer the midline; a number of small ducts from these glands open beneath the tongue. The parotid glands, which are the largest of the three pairs, lie one on each side, below and in front of the ear, almost immediately beneath the skin. The duct of

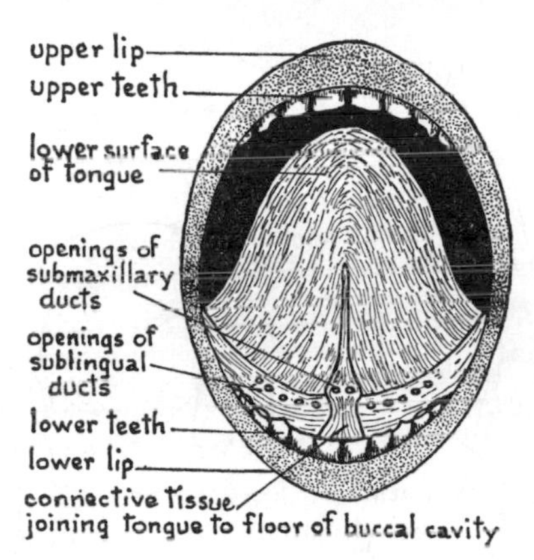

FIG. 80. Wide-open Mouth, with Tongue Raised to Show the Openings of the Salivary Ducts

each opens into the buccal cavity on the inner surface of the cheek, opposite the second upper molar tooth. These glands secrete fluids that together make up the digestive juice, **saliva**, which, apart from its digestive function (*see* p. 148), moistens and lubricates the food material. The saliva is continually swallowed. The disease **mumps**, characterized by painful swellings, which extend from the front of the ears down the sides of the face and under the chin, is due to a **virus infection** of the salivary glands (*see* p. 338).

THE PHARYNX

The extent of the pharynx is shown in Fig. 81(*a*). It is a tubular structure, divisible into three parts: an upper part behind the nasal passage, the **naso-pharynx**, a middle part behind the tongue, the

oropharynx and a lower part behind the larynx, the **laryngopharynx.** The air passages from the nostrils meet in the nasopharynx behind the uvula, and on each side a narrow tube, the **Eustachian tube,** leads outwards and upwards to the middle ear. The functions of these tubes are discussed on p. 159. When swallowing occurs, the uvula is raised to shut off the nasopharynx. On each side of the oropharynx are the

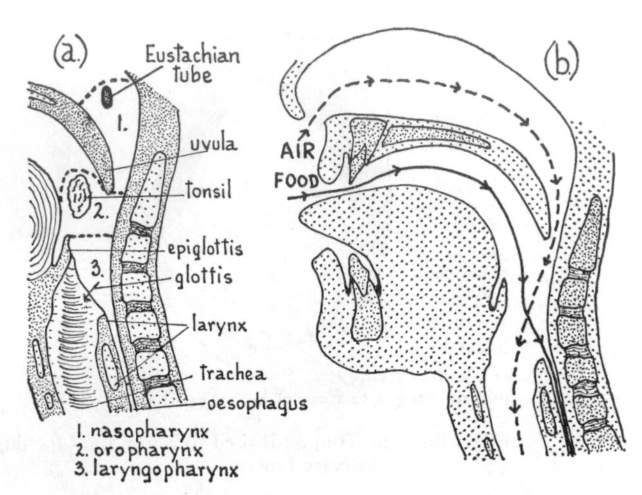

FIG. 81. (*a*) Diagram of the Pharynx in Median Vertical Section (*b*) The Food and Air Passages

tonsils, which are **lymphatic glands** (*see* p. 207), and below it is the **epiglottis,** standing almost erect in front of the air passage into the **larynx.** The **laryngopharynx** lies behind and below the **epiglottis**; in front of it lies the larynx with its entrance, the **glottis,** and below it lies the **oesophagus.**

In the lower chordates, the pharynx is an extensive region with all the gill slits opening out of it. In the land vertebrates, it has become much reduced, the last remnant of a gill slit being the Eustachian tube. The pharynx is essentially a tube where various passages meet; it is a passage-way for food from the buccal cavity and for air from the nasal passages, and from it lead the air-passage to the larynx and the food-passage to the oesophagus. The act of swallowing is essentially the processes of getting food from the buccal cavity through the pharynx and into the oesophagus. The various passages are shown diagrammatically in Fig. 81(*b*).

Because of the access to the pharynx through the nose and through the mouth, it is particularly liable to infection by germs; **sore throat, laryngitis, pharyngitis** and **tonsillitis** are all very common.

The Oesophagus

The oesophagus, extending from the pharynx to the stomach, is one of the most muscular parts of the alimentary canal. It is a tube for conveying the food, after treatment in the buccal cavity, to the next treatment region, the stomach. The food is forced along by a series of wave-like contractions known as **peristalsis.** As soon as a portion of food enters the upper part, having been squeezed in by the voluntary muscles of the pharynx, the circular muscles in the wall of the oesophagus contract just above the food, thus forcing it downwards. This action is repeated by all the circular muscle fibres down the tube, each set contracting just behind the food and relaxing in front. The cardiac sphincter relaxes as the food approaches it, and the food is then squeezed into the stomach. Liquids are swallowed very quickly; one can feel a hot drink passing down; solids are swallowed more slowly; most people who have inadvertently swallowed a boiled sweet will know this. After the process of peristalsis has begun, it cannot be stopped voluntarily.

The oesophagus at rest is closed and flattened by the pressure of the trachea, to which it is attached by connective tissue (*see* Fig. 82). The wall of the oesophagus consists of an outer layer of fibrous connective tissue, then two coats of muscle, an outer longitudinal and an inner circular, then more connective tissue and finally a lining of mucous membrane.

The act of **vomiting** forces the stomach contents upwards through the oesophagus and pharynx and then out through the mouth. It is a **reflex action** (*see* p. 284), controlled from a **vomiting centre** in the brain. Some of the stimuli that cause vomiting are chemical irritants in the stomach, irritation of the back of the buccal cavity or pharynx, disturbance of the semicircular canals of the ears (*see* p. 263), as in motion sickness, and psychological factors such as the witnessing of gruesome scenes.

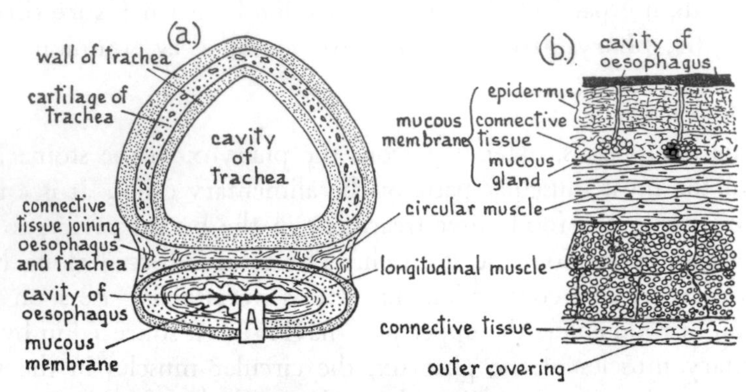

FIG. 82. (*a*) Transverse Section of Oesophagus and Trachea, × 1 (*b*) Vertical Section of Oesophagus Wall at A, × 18

THE STOMACH

The stomach, lying in the abdominal cavity just below the diaphragm on the left side (*see* Fig. 31, p. 46) is the most dilated portion of the alimentary canal. When empty, it is a wide soft tube, and when dilated, it is sac-like and firm. In the stomach, there are not only wave-like peristaltic contractions from the cardiac sphincter towards the pyloric sphincter (*see* Fig. 83(*a*)), but also what are known as churning movements, which are intermittent contractions at various points and in various planes. The muscular wall is able to exercise these different types of contraction because there are three muscle layers, an outer longitudinal, a middle circular and an inner oblique (*see* Fig. 83 (*b*)). Apart from the muscular layers, the wall consists essentially of an outer firm smooth membrane, the **peritoneum,** and an inner soft mucous membrane. When the stomach is empty, the mucous membrane lies in longitudinal folds (*see* Fig. 83 (*a*)); when it is distended, the membrane is smooth and even. This mucous membrane is perforated by very numerous minute openings, which lead into the **gastric pits**

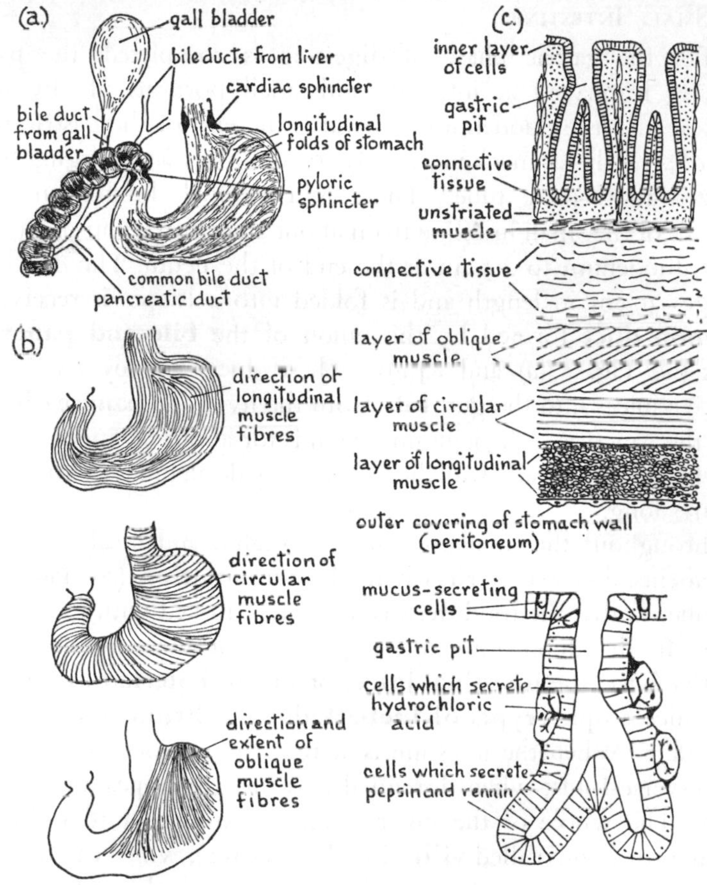

FIG. 83. (*a*) Relationships of the Stomach, and its Internal Folds (*b*) The Three Muscular Layers of the Stomach Shown Separately (*c*) Vertical Section of Stomach Wall, × 35; One Gastric Pit, × 80

(gastric glands) (*see* Fig. 83(*c*)). There are about thirty-five million of these glands in the average adult human stomach; they produce the second digestive juice, **gastric juice.** Food remains in the stomach for a period ranging from one hour to four and a half hours, during which time it is converted into a paste-like mass called **chyme.**

The Small Intestine

After the gastric phase of digestion is completed, the pyloric sphincter is relaxed at intervals and small portions of chyme are squeezed into the duodenum. Towards the end of the gastric phase, chyme leaves the stomach in a steady stream, like that emerging from a squeezed tooth-paste tube. There is a gradual diminution in the diameter of the small intestine from about 5 cm at the commencement of the duodenum to 2·5 cm at the end of the ileum. The duodenum averages 30 cm in length and is folded into a loop; it receives the **common duct** formed by the union of the **bile and pancreatic ducts** (*see* Fig. 83(*a*) and 84(*a*)). These ducts convey two further digestive juices into the gut, **bile** from the liver and **pancreatic juice** from the pancreas. The jejunum extends for about 2·7 metres and has a thicker, redder wall than the relatively pale ileum, which is about 4 metres long.

Throughout the small intestine, peristaltic and local constricting movements, known as **rhythmical segmentation** (*see* Fig. 84(*b*)), continue, so that the food mass is intimately mixed with the digestive juices. In the duodenum and upper part of the jejunum, a final digestive juice, the **intestinal juice,** or **succus entericus,** is produced from microscopic **crypts of Lieberkuhn** and **Brünner's glands** (*see* Fig. 84(*c*)). When the intestine is at rest, the mucous membrane lies in transverse folds; when distended it is furry in appearance (*see* Fig. 84(*d*)). The whole of the inner lining is extended into millions of minute projections called **villi** (singular—**villus**), which can be moved like moving fingers (*see* Fig. 84(*c*)). They increase the surface area for both secretion and absorption. Most of the work of digestion and absorption is completed in the small intestine.

The Large Intestine

The material passed through the ileo-caecal valve is mainly undigested or indigestible, and normally it consists largely of cellulose.

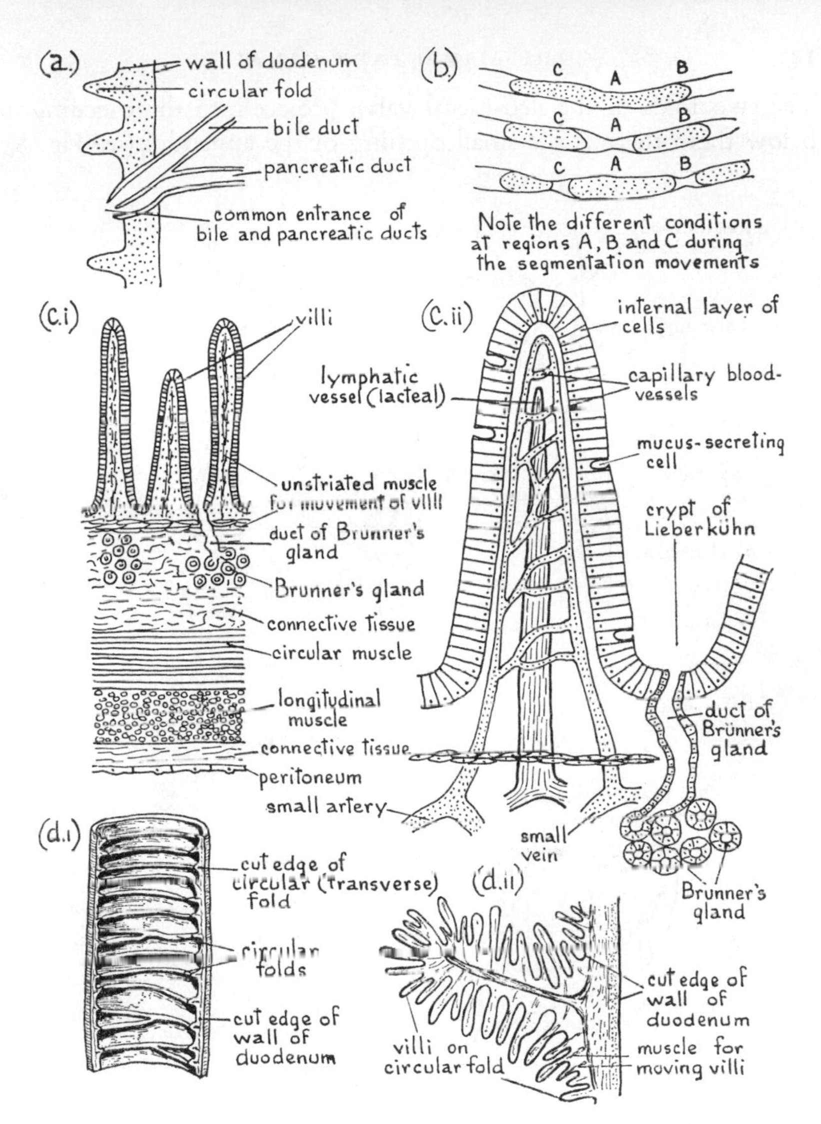

Fig. 84. (*a*) Longitudinal Section of Duodenal Wall at Entrance of Bile and Pancreatic Ducts (*b*) Segmentation Movements (*c.i.*) Vertical Section of Wall of Duodenum, × 75 (*c.ii.*) Vertical Section of One Villus, × 150 (*d.i.*) Longitudinal Half of Duodenum, Showing Circular Folds, × ⅓ (*d.ii.*) Vertical Section through One Circular Fold, × 30

The two folds of the ileo-caecal valve project into the caecum, and below them there is the small opening of the appendix (*see* Fig. 85).

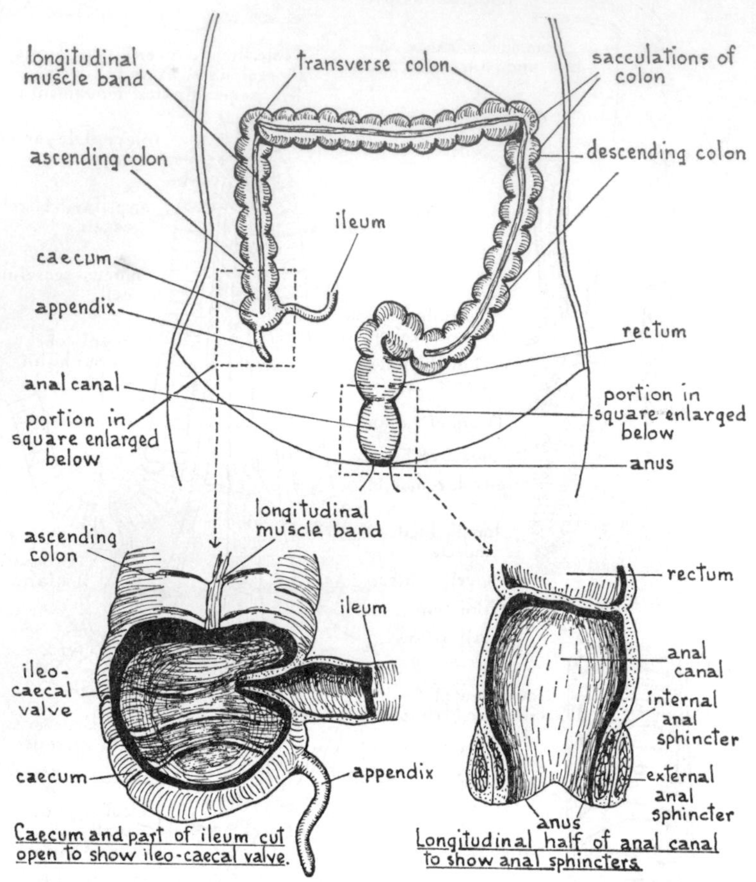

FIG. 85. The Large Intestine: Portion of Caecum and Anal Canal Drawn Below, Opened and Enlarged

The valve prevents the passage of faecal material back into the ileum. In the caecum and colon, there is a great deal of bacterial action on the remaining material. Some of the products of this decomposition are acids, gases and two evil-smelling substances, **indole** and **skatole,** which contribute largely to the offensive odour of faeces. Sometimes, the appendix becomes distended and inflamed by a mass of this putrefying material, one cause of the painful condition of **appendicitis.** It is essential that such an inflamed appendix is surgically removed; if it bursts and its contents are liberated into the abdominal cavity, the serious condition **peritonitis** may result.

The caecum is about six centimetres long and of equal diameter; the appendix averages about ten centimetres in length and six millimetres in diameter. The colon is a long, sacculated tube, about five centimetres in diameter; it has ascending, transverse and descending portions. In its wall, the longitudinal muscle is restricted to three bands (*see* Fig. 85); there are no villi, but numerous, closely-packed, tubular glands, which secrete mainly mucus. The essential function of the colon is the absorption of water, and hence the faecal matter gradually becomes drier and more firmly compacted.

Finally, the faecal material enters the rectum and is eliminated through the **anal canal,** which has internal and external sphincters (*see* Fig. 85). The general shape and disposition of the intestines are shown in Fig. 31, p. 46.

MESENTERIES

The internal surfaces of the abdomen (and thorax) are lined with a thin, smooth membrane known as the peritoneum. All the organs of the abdominal cavity are suspended in the coleomic fluid by a double sheet of peritoneum, known as a **mesentery** (*see* Fig. 86(*a*)). Between the two sheets of mesentery, there is connective tissue, which supports the blood-vessels and nerves supplying the various organs. Two portions of the mesentery need special mention because they are largely responsible for preventing the gut sagging when one stands in an erect position. The **lesser omentum** fastens the stomach firmly to the liver, which is itself attached to the diaphragm and the front wall of the abdomen by another mesentery called the **falciform ligament.** The **great omentum** joins the transverse colon to the stomach; it usually contains a great deal of fat and hangs down over the small intestine like an apron (*see* Fig. 86(*b*)). It gives protection to the front of the abdomen, is an easily movable packing material and acts as a storehouse for fat.

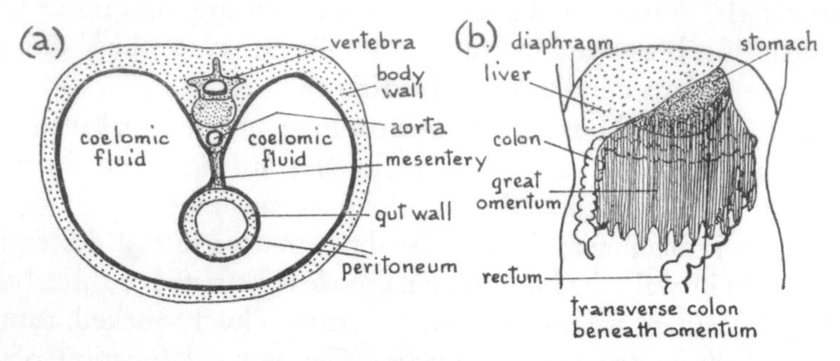

FIG. 86. (*a*) Diagrammatic Transverse Section of the Abdomen, to Show Suspension of the Gut by a Mesentery (*b*) Location of the Great Omentum

Digestion of the Food

While it is in the gut, the food is of no value; it must be absorbed from the gut and circulated round the body in the blood. Before it can be absorbed, the food must be in the form of molecules that are small enough to pass through the cells of the villi into the network of vessels (*see again* Fig. 84(*c*)). This passage from the gut into the villi is known as **absorption.** Thus we can define digestion by saying that it includes all the processes that make the food soluble and absorbable. There are two different aspects of digestion to be considered, the physical and the chemical. In the physical processes, the nature of the food is not changed, but it is reduced to small particles, which enables the later chemical changes to take place more easily.

PHYSICAL PROCESSES OF DIGESTION

Small portions of the food are placed in the mouth direct, or are bitten by the incisors from larger pieces. Then the tongue pushes the food between the upper and lower premolars and molars where the chewing or **mastication** takes place. The tongue is very active during this process, ensuring that all the pieces of food are eventually chewed. Then the smaller particles are collected on the upper surface of the tongue in one mass called a **bolus.** During mastication, which should be as thorough as possible to make the chemical processes more effective, the food is intimately mixed with saliva, which moistens it, lubricates it for its future passage along the pharynx and oesophagus, and performs the first phase of chemical digestion.

Swallowing the bolus is quite a complex process, voluntary until the bolus is well into the oesophagus, and then involuntary. The bolus is lifted by the tongue, pressed against the hard palate and squeezed backwards into the pharynx. In the pharynx, the bolus is gripped by the pharyngeal muscles and squeezed into the oesophagus. While the bolus is being passed through the pharynx, the back of the tongue projects over the larynx thus protecting it; the glottis is closed momentarily to prevent food entering the breathing passage, so that breathing is halted for a second or so; the soft palate and uvula are raised to close off the nasopharynx. The pharynx shortens rapidly, lifting the bony larynx upwards and forwards, so that the bolus can slip past it. The bolus is then in the oesophagus and it proceeds down to the stomach by peristalsis, the journey taking five or six seconds for solid masses.

Then follow the churning movements in the stomach, and when a sufficient degree of liquefaction has been attained, the pyloric sphincter opens at intervals and portions of chyme are squeezed into the duodenum by contraction of the circular muscle of the pyloric region of the stomach.

One of the effects of the bile on the food is to break up fats into minute globules, a process known as **emulsification.** The substances in the bile that have this effect are the two **bile salts, sodium glycocholate** and **sodium taurocholate.** This emulsification of fats renders them easier to digest, because it results in a very great surface area of fat being presented to the chemical agents of digestion.

Throughout the small intestine, peristalsis and rhythmical segmentation movements proceed, so that the food mass is intimately mixed with the digestive juices and ever-changing portions are placed in contact with the villi for absorption.

There are slower peristaltic movements in the colon, which allow sufficient time for water to be absorbed. Defaecation, the expulsion of the faeces, is partly involuntary and partly voluntary. The former is due to strong contraction of the muscular walls of the rectum and relaxation of the internal anal sphincter muscles. The latter is due to controlled relaxation of the external sphincter muscles and the contraction of the abdominal muscles and the diaphragm. These two latter actions press the abdominal viscera against the rectum, thus aiding the evacuation process. Infants are not able to control the voluntary parts of these actions; the act of defaecation is purely reflex, voluntary control being learned at a later stage of development.

The Chemical Processes of Digestion

Many chemical reactions, both in the laboratory and in industrial processes, are carried out more speedily, and at lower temperatures, by the addition of substances called **catalysts.** For example, in the laboratory preparation of oxygen by heating potassium chlorate, the reaction takes place more quickly and at a lower temperature if manganese dioxide is mixed with the potassium chlorate.

The manganese dioxide is present in the same quantity at the end as at the beginning of the experiment. In the living body, practically all the chemical changes that take place are accelerated by catalysts called **enzymes.** These are very complex proteins, produced only in living cells, though they can be extracted and used for experimental work, and they possess certain properties that do not ordinarily apply to catalysts used in inorganic chemistry.

Properties of Enzymes

1. Each enzyme catalyses one reaction or kind of reaction. Thus the enzyme invertase, which changes sucrose to glucose and fructose, will not act on the other disaccharide sugars, maltose and lactose.

2. The reaction catalysed by an enzyme is reversible. For example,

the enzyme lipase will, under one set of conditions, catalyse the breakdown of fat to fatty acids and glycerol, while under another set of conditions it will catalyse the build-up of fat from fatty acids and glycerol. The conditions that determine which reaction shall take place are the **acidity** or **alkalinity** of the medium in which the reaction is taking place, and the relative amounts of the reacting substances present.

3. The amount of enzyme present will affect the rate of the reaction, since one molecule of the enzyme can deal with only one molecule of the other substance at a time.

4. Temperature affects the rate of enzyme-catalysed reactions; the temperature at which the rate is greatest usually lies between 35 and 40°C. Above 40°C and below 35°C, the rate diminishes. All enzymes are destroyed at 100°C, i.e. the boiling point of water. This question of temperature effect is of little importance to mammals, with their constant body temperature.

5. Provided that the temperature is favourable, the maximum rate of any particular enzyme-catalysed reaction occurs at a particular level of acidity or alkalinity.

Acidity and Alkalinity

Here the student should become familiar with the use of the **pH scale** for expressing acidity or alkalinity, without necessarily knowing how it is obtained. **Hydrogen ions, H^+,** determine acidity in a solution, and **hydroxyl ions, OH^-,** determine alkalinity. Pure water, which is neutral, i.e. neither acid nor alkaline, has equal numbers of hydrogen and hydroxyl ions and is given the value of 7 on the pH scale. **Numbers above 7 indicate alkalinity and numbers below 7 indicate acidity.** Thus 8, 9, 10, etc., represent gradually increasing proportions of OH^- ions and relatively less H^+ ions; the numbers 6, 5, 4, etc., represent increasing degrees of acidity with more and more H^+ ions and less and less OH^- ions. From one number to the next on the pH scale represents a very big change; a pH 3 solution is ten times as acid as a pH 4 solution; a pH 9 solution is ten times as alkaline as a pH 8 solution. Note: the pH scale is logarithmic.

Secretion of Enzymes

Most of the enzymes carry out their work in the cells where they are produced; a few, and especially the digestive enzymes, are passed out of the cells that produce them and carry out their work elsewhere. Those enzymes that digest proteins are secreted by the cells in an inactive form, otherwise they would digest the proteins of the cells that produce them. They are activated by substances present in the gut. Thus the pepsin of the stomach is secreted as the inactive **pepsinogen** and is activated by the chloride ions present in the stomach. **Trypsinogen** from the pancreas is activated by an enzyme called **enterokinase,** present in the intestinal juice.

Digestive enzymes are not continually secreted, but only when food is present in the gut. The stimuli that activate the various glands to begin and continue secretion are in some cases nervous and in some cases hormonal. Saliva is secreted by the salivary glands after nervous stimulation arising from the sight, smell and taste of food. Gastric juice is secreted by the gastric pits in response to nervous stimulation at first, but continued secretion is due to stimulation by a hormone produced by the stomach wall. All the glands that secrete into the duodenum are stimulated by hormones.

There are five digestive juices. They are—1. **Saliva** from the salivary glands. 2. **Gastric juice** from the gastric pits. 3. **Bile** from the liver. 4. **Pancreatic juice** from the pancreas. 5. **Intestinal juice** from the wall of the duodenum. Each of these juices consists mainly of water with very small quantities of other important constituents in solution.

Summary of Digestion

The saliva contains the enzyme **ptyalin,** which converts **starch and glycogen** into the disaccharide sugar **maltose.** This reaction begins in the mouth at pH 6·7, continues down the oesophagus and persists for a short time until the increasing acidity of the stomach contents stops it. In general, not more than twenty per cent of the starch is converted. (Note: there is normally very little glycogen in the diet.)

The gastric juice contains two enzymes, **pepsin** and **rennin.** **Hydrochloric acid** is also present; it increases the acidity to a pH of about 2, in which condition the two enzymes work at their best rate. Pepsin breaks **proteins** down into smaller portions, called **peptones.** Rennin **coagulates the protein of milk** as white curds; pepsin is then able to break the protein down to peptones.

The bile contains no enzymes, but has several other important constituents. The **bile salts emulsify fats;** the bile alkali, **sodium bicarbonate,** neutralizes the hydrochloric acid from the stomach so that the pH is raised above 8; all digestion in the small intestine must take place in an alkaline medium. The **bile pigments,** one red and one green, give an orange colour to the bile.

The pancreatic juice contains the enzymes **trypsin** (secreted as **trypsinogen), amylopsin** and **lipase.** Trypsin acts on the **peptones** produced in the stomach and converts them into **single polypeptide chains** (*see* Fig. 87(*a*)). **Amylopsin** acts on the **starch and glycogen** not broken down by ptyalin and converts them to the disaccharide sugar **maltose** (*see* Fig. 87(*b*)). **Lipase** acts on **fats,** breaking them down to **fatty acids and glycerol** (*see* Fig. 87(*c*)).

The intestinal juice, produced in Brünner's glands and the crypts of Lieberkuhn, contains the enzymes **erepsin** and **lipase,** and a number of enzymes that act on the disaccharide sugars, **invertase, maltase** and **lactase. Erepsin** completes the digestion of proteins by breaking down the **polypeptides** into **amino-acids** (*see* Fig. 87(*a*)). **Lipase** again breaks **fats** down to **fatty acids and glycerol,** while the other three enzymes split **disaccharide sugars** into **monosaccharide** (hexose) **sugars,** as shown below—

sucrose	+	invertase	→	glucose and fructose
maltose	+	maltase	→	glucose and glucose
lactose	+	lactase	→	glucose and galactose

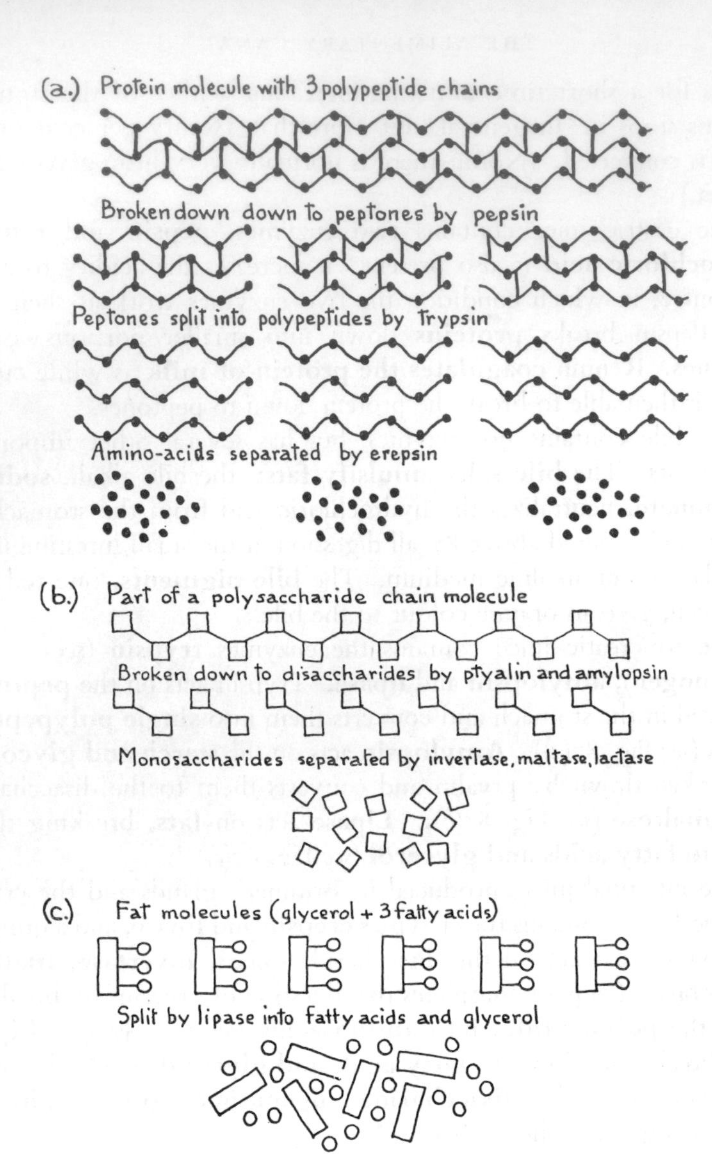

FIG. 87. Diagrams Illustrating the Digestion of Proteins, Carbohydrates and Fats

Thus the final products of digestion are all soluble materials: amino-acids from proteins, hexose sugars from carbohydrates, fatty acids and glycerol from fats. These substances are absorbed by the villi throughout the length of the small intestine.

Absorption of the Digested Food

The great length of the small intestine, and the presence of millions of villi, combine to present an enormous surface for absorption. Each

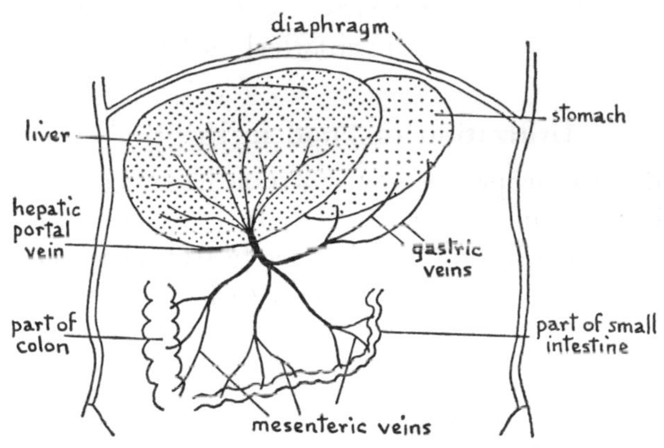

Fig. 88. Diagram Showing the Course of the Blood from the Gut to the Liver

villus has a network of capillary blood-vessels very close to its surface and a single lymphatic vessel, called a **lacteal,** in the centre.

Hexose sugars are absorbed by the cells of the villi, transferred to the blood capillaries and then carried in the **mesenteric veins** to the great **hepatic portal vein,** which leads into the liver (*see* Fig. 88). In the liver, the portal vein leads to numerous smaller veins, which ramify throughout the liver and lead to plentiful capillaries. From these capillaries, the hexose sugars are passed into the cells of the liver, and combined to form the storage carbohydrate, glycogen. In a well-fed adult the liver stores about two hundred grams of glycogen.

The amino-acids are absorbed in a similar manner to the hexoses and they are taken to the liver by the same route.

Fatty acids and glycerol are absorbed into the lacteals and there is also some absorption of minute, unchanged globules of fat. In the lymph of the lacteals the fatty acids and glycerol reform minute globules of fat; these are taken in the lymphatic circulation (*see* p. 206) to enter large veins at the base of the neck. Thus the fat also finds its way into the blood by a roundabout route. In the blood-stream, the fat is conveyed to storage depots: the dermis of the skin, the region of the kidneys, the mesenteries and the great omentum, there to await utilization for the release of energy.

Vitamins, salts and water are absorbed unchanged.

Utilization of Food in the Body

After the various products of digestion have been absorbed from the gut, they are said to be **assimilated;** in other words, they now form part of the body and are not merely passing through it. They are then utilized for various purposes such as growth, repair and energy.

Utilization of Amino-acids

The amino-acids, taken to the liver in the blood-stream, are in some way sorted so that those that are required for building and replacing tissue proteins are passed back into the blood and circulated for use where required. The amino-acid content of blood is small but fairly constant, being 3 to 8 mg per 100 cm^3 of blood. Those amino-acids that are not required for body-building and repair have one of three fates.

1. The amine (NH_2) group is removed and transferred to another molecule and utilized for synthesis of one of the non-essential amino-acids.

2. The amine group may be removed in a complex cycle that results in the formation of **ammonia** (NH_3). The ammonia, which is very toxic, is combined with carbon dioxide to form **urea,** according

to this equation, which expresses the net result of a whole series of reactions in the liver.

$$2\,NH_3 + CO_2 = CO(NH_2)_2 + H_2O$$

Ammonia + carbon dioxide urea + water

The urea is passed into the blood and is almost entirely extracted by the kidneys and passed out of the body in the urine. In a healthy adult, almost all the nitrogen of the protein intake re-appears in the urine, indicating that fate 2 is by far the most common.

3. The amino-acids may be used for synthesis of a number of other nitrogenous products apart from tissue proteins. Examples of these are some of the hormones, **phosphagen** (*see* p. 154) and certain components of the chromosomes (*see* p. 305).

After the amine groups of excess amino-acids have been removed as described above, the remaining parts of the amino-acid molecules can, in sixty per cent of types, be converted into glucose, and in the remaining forty per cent, most are broken down to carbon dioxide and water, which are excreted from the body. The breakdown process releases energy.

Utilization of Glycogen and Glucose

The glycogen of the liver is the only regular source of supply of glucose to the blood. In a healthy body, a balance between glycogen in the liver and glucose in the blood is maintained by a number of factors. The blood glucose is constantly being used by all the cells of the body for energy release, the energy being used to rebuild ATP. More glycogen from the liver is converted into glucose and released into the blood to replenish the supply. This glycogen-glucose balance is maintained by three hormones (*see* Chapter 15). In a normal diet, the glycogen of the liver is kept up to a fairly constant level by the digested products of the carbohydrates in the diet. If these are lacking, the fat stores of the body are brought into use and fat is converted into glycogen in the liver.

In the cells, glucose releases its energy to rebuild ATP in a long series of enzyme-catalysed reactions, known as cell or tissue respiration.

When oxygen is freely available, the end products are carbon dioxide and water, which are excreted. In striated muscle that is performing strenuous exercise, the supply of oxygen soon becomes insufficient for the complete breakdown of glucose (from muscle glycogen) and the substance **lactic acid** is formed after a series of reactions. The net result of these reactions is shown in this equation.

$$\underset{\text{glucose}}{C_6H_{12}O_6} \rightarrow \underset{\text{lactic acid}}{2\ C_3H_6O_3} + \text{energy}$$

The energy for contraction of the muscle is supplied by ATP; the ADP formed is converted back to ATP by the donation of a phosphate group from phosphagen, which is itself regenerated by energy released during the glucose–lactic acid series of reactions. Fatigue is due largely to accumulation of lactic acid in the muscles. Eventually the lactic acid passes into the blood and most of it is reconverted to glycogen in the liver.

Utilization of Fat

An average healthy man will have about six kilograms (about 13 lb) of fat in the fat-storage regions. This fat does not remain as an inert mass, and if normal exercise and a normal diet are customary, it will not accumulate. In any kind of work that needs energy in excess of basal metabolism, glucose will leave the liver at a rate depending on the kind of work. As this rate becomes rapid, the glycogen of the liver is depleted. In the absence of sugars from the gut, the fat-storage depots are used. Fat is converted into fatty acids and glycerol, both of which enter the chain of reactions in the breakdown of glucose to carbon dioxide and water. Until the next meal containing carbohydrates, fatty acids and glycerol can both be built up into glycogen in the liver. Even in conditions of starvation there is a fair amount of glycogen in the liver; it has been obtained by conversion of fat and protein already in the body.

SUGGESTED EXERCISES

1. Examine the teeth in any mammalian skulls available, e.g. pig, dog, rat, rabbit, etc.

2. Examine on the microscope, or microprojector, a prepared slide of a vertical section of a tooth. Draw it.

3. The following observations may be made on your own buccal cavity with the aid of a torch and a mirror, or on someone else, using the torch.

(*a*) Open the mouth wide, press down the tongue with a clean spoon and observe the palate, the uvula, the papillae on the tongue and the tonsils.

(*b*) This time, with the mouth open wide and the tongue lifted and pressed against the palate, locate the central openings of the submaxillary ducts and the two rows of openings of the sublingual ducts.

(*c*) Collect and swallow the saliva in the buccal cavity and then note where you can feel more entering.

(*d*) Examine your own teeth and the teeth of children of various ages and compare them with the table on p. 131.

4. Examine, on a microscope or microprojector, prepared slides of (*a*) a vertical section of the stomach wall, (*b*) a transverse section of the small intestine, (*c*) a transverse section of the large intestine.

5. Cut a preserved rat's or rabbit's head vertically into two, keeping the cut slightly to one side of the midline. For soft structures use a sharp knife or scalpel, and for bone, a fine-toothed hacksaw. Examine.

EXPERIMENTS ON ENZYMES

For satisfactory results, an incubating oven or a water-tank with an electric heater and thermostatic control are almost essential. The experiments should be performed at about 35°C.

6. The action of ptyalin can be shown by using the enzyme obtained from the human mouth—rinse with clean water into a beaker. Use a weak suspension of starch in test-tubes or dishes, and add about 1 cm^3 of the ptyalin solution. Test a different tube at 15 seconds, 30 seconds, 1 minute, 2 minutes, etc., with a few drops of iodine solution. With each successive test, the blue colour will become fainter until no colour at all can be seen. All the starch will have been changed to disaccharide sugar. If desired, commercial ptyalin can be obtained from biological dealers.

7.* The action of lipase on fat can be shown by using commercial lipase with olive oil or castor oil. Shake up a little olive oil with fifty per cent alcohol and test with litmus paper to show that the reaction is neutral. Add a small quantity of lipase and shake vigorously. A litmus test after a few hours will show an acid reaction, due to the presence of fatty acids. To test for glycerine, add a little weak copper sulphate solution and then a few drops of caustic potash solution. A strong blue colour will show the presence of glycerine.

8.* The action of pepsin on protein can be shown by using commercial pepsin with egg albumen. Place a spoonful of fresh egg-white in water about four times its volume. Shake the test-tube until a cloudy solution of fairly even consistency is obtained. Heat gently to coagulate the albumen. Strain through muslin to remove any large lumps and then dilute the solution ten times with water. Use about 5 cm^3 of this liquid and add 1 cm^3 of dilute hydrochloric acid and a small quantity of commercial pepsin powder. Incubate for about an hour. By then the cloudiness will have disappeared because the insoluble proteins will have been converted into soluble peptones.

9. Emulsification can be carried out by shaking a small quantity of olive oil with water in a corked bottle. A milky emulsion is formed, which is not stable; the olive oil will collect on the surface again. But if a pinch of bile salts is added before shaking, a permanent emulsion is formed.

Many other experiments on digestive enzymes can be devised. All the digestive enzymes can be purchased from biological dealers. The effects of different temperatures, different acidity, and different concentrations of the enzymes can be observed. Note. If the incubation is performed in a heated water-tank, ensure that the current is switched off while placing tubes in the tank or taking them out.

Chapter 9

THE RESPIRATORY SYSTEM

It will be recalled that carbohydrates and fats are necessary in the diet because they are sources of energy, which can be released in the cells and then used for rebuilding ATP. Energy is the capacity for doing work; not only physical work in which muscles are used, but many other kinds of work. Building up proteins from amino-acids is chemical work; producing sound is a form of physical work. Thought requires energy and so does heat production. Even when a man is asleep, he uses a great deal of energy in many ways. For example, he must still breathe, the blood must be circulated, the food must be digested, and waste products must be excreted. All these, and all other processes that take place in the body, require energy.

The true biological meaning of respiration is the process of converting the energy of the food into a form that can be used by the cells. This takes place in all living cells, and in most cases, but by no means all, oxygen is needed to complete the process. This release of energy in the living cells is referred to as **tissue,** or **cell, respiration.** The everyday use of the word "respiration" refers to breathing; in biology, this is called **external respiration.** Between external and tissue respiration there are two other essential processes, the absorption of oxygen and the transport of it to the tissues. Therefore, for a clear understanding of the whole sequence, four processes must be known. They are breathing, absorption of oxygen, transport of oxygen and tissue respiration.

Breathing or External Respiration

Oxygen is present in the lower atmosphere to the extent of nearly twenty-one per cent. **Breathing** is the process whereby the air containing the oxygen is brought close enough to the blood capillaries in the lungs for absorption of the oxygen to take place. Furthermore, the air reaching the lungs has to be continually changed so that fresh supplies of oxygen are always available. The pathway for air from the atmosphere to and from the lungs is called the **respiratory passage.**

The Respiratory Passage

The air passes through the nostrils into the nasal cavities, which are lined with soft moist mucous membrane. Certain cells in this membrane secrete mucus, which is essentially a sticky solution of the protein **mucin.** The nature of mucus is familiar to everyone who has had a cold in the nose; there is then excessive production of mucus. In the first part of the nasal passage, hairs moistened by mucus provide a rough filtering mechanism and larger particles in the air are trapped. In a dusty atmosphere, the passage becomes almost blocked by the accumulation of these particles and so we "blow the nose." The upper part of each nasal cavity is partially divided into lower, middle and upper chambers by the three **turbinate processes** (*see* Fig. 89(*a*) and (*b*)). They are covered with the moist mucous membrane and have large, thin-walled veins near the surface. Most of the cells in this region are **ciliated,** i.e. they possess minute, protoplasmic outgrowths called **cilia** (*see* Fig. 89(*c*)). The cilia beat rhythmically towards the nostrils, sweeping mucus with any trapped dust particles outwards. The increased area due to these turbinate processes, together with their moist surfaces and large flattened veins, all help to moisten and warm the inhaled air as it passes through.

Above the upper turbinate process, there is a small, blind pocket containing many patches of **sensory cells,** again interspersed with cells that secrete watery mucus (*see* Fig. 89(*d*)). These cells have short protoplasmic processes, which are sensitive to chemicals in solution. Nerve fibres from the bases of these cells unite to form the **olfactory nerve**; its fibres conduct impulses to the brain, where they are interpreted as various **odours.** Thus we have an **olfactory sense,** or sense of smell, which is very similar to the sense of taste. Odours cannot be subdivided as conveniently as tastes; there are so many distinctly different odours that they cannot easily be classified. **Flavour** is a combination of taste and smell; a good example is the flavour of an

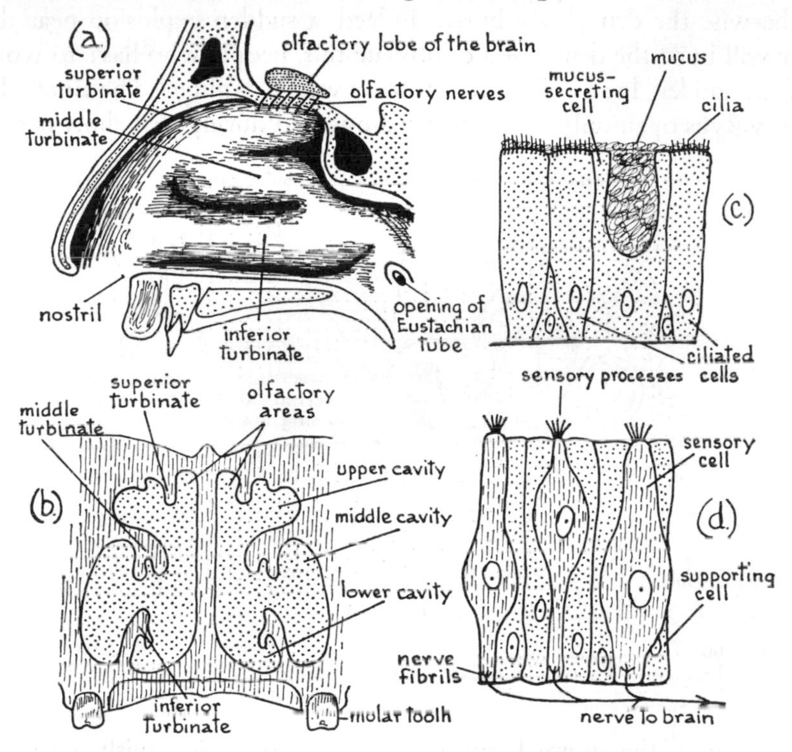

FIG. 89. (*a*) Vertical Section of the Nasal Cavity, Slightly Right of the Mid-line (*b*) Vertical Section Through Both Nasal Cavities, Passing Through the First Molar Teeth (*c*) Vertical Section of Ciliated Epithelium of the Nasal Cavity, × 100 (*d*) Vertical Section of Sensory Cells in the Olfactory Area, × 120

onion. When we have to take unpleasant medicines, they are not quite as objectionable if the nose is pinched while we take them.

After passing the turbinate processes, the air enters the naso-pharynx, at the sides of which are the openings of the **Eustachian tubes,** guarded by valves. These tubes lead outwards and upwards to the chamber of the middle ear (*see* Fig. 90). This chamber is separated from the external atmosphere by the **ear-drum**; and it is essential that the air-pressure on both sides of the drum is equalized, otherwise the drum may burst. Indeed, a sudden explosion near the ear will burst the drum, and to obviate this, people who have to work where sudden, loud noises occur usually wear ear-plugs. The **Eustachian valves** open only when we swallow, and during that short period

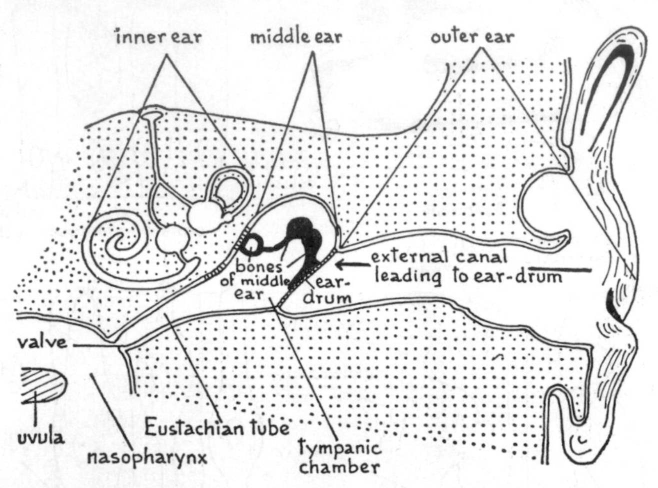

FIG. 90. Diagram Showing the Relationship between the Pharynx, the Eustachian Tube and the External Canal on the Left Side of the Head

of opening, the internal cavity is in communication with the nasopharynx, and thus the internal air-pressure can be equated with the external pressure. When the external pressure is changing rapidly in speedy ascent or descent, as in an aeroplane or a pit-shaft, quick gulping movements take place, so that the necessary adjustments in internal air-pressure can be made.

The air passes on through the oropharynx, through the glottis and into the larynx, which extends from the tip of the epiglottis to the trachea, a distance of approximately 5 cm. The larynx is supported by cartilages, which gradually ossify after maturity (*see* Fig. 91(*a*)). About halfway down the larynx there are the two lateral **vocal folds,** which can be tightened or slackened by muscles attached to the **laryngeal cartilages.** By controlling the exhalation of air from the lungs, these vocal folds can be made to vibrate, thus producing sounds. The **pitch** of the sound produced, i.e. high or low, can be altered by varying the tension of the folds, in the same way as the pitch of a note produced by plucking a guitar string is raised by tightening the string and lowered by slackening it. As the folds are tightened, they are brought closer

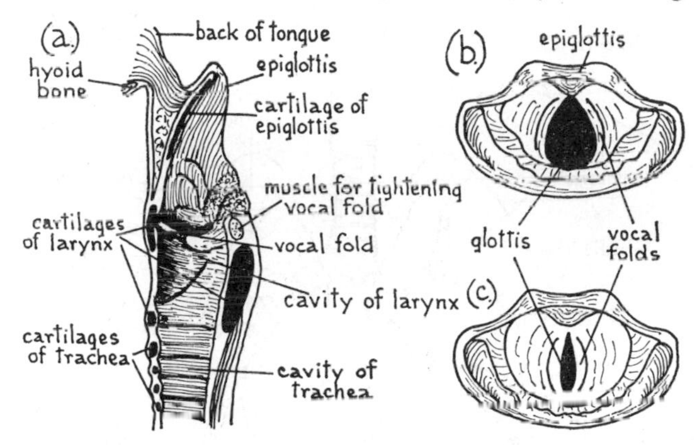

FIG. 91. (*a*) Median Vertical Section of the Larynx (*b*) The Larynx from Above, Vocal Folds Wide Open (*c*) The Larynx from Above, Vocal Folds Nearly Closed

together and the pitch of the sound produced becomes higher and higher (*see* Fig. 91(*b*) and (*c*)). The **intensity** of the sound is varied by altering the force with which the air is exhaled. The vibrations of the folds cause the air in the cavities above to vibrate also. It is the controlled use of these cavities that gives the **quality** of the voice. The larynx, like all parts of the respiratory passage after the nasal hair region, is lined by ciliated mucous membrane; the cilia beat upwards, and trapped particles in the mucus are driven round the sides of the epiglottis into the buccal cavity and are eventually swallowed.

The next part of the respiratory passage is the trachea, which is about 11 centimetres in length in an adult male; it extends down into the upper part of the thorax. The trachea, like the larynx, is kept permanently open by cartilages, which in this case are not complete rings, but C-shaped; the opening of the "C" lies against the oesophagus, which is thus able to dilate when swallowing takes place (*see* Fig. 82 (*a*)). The mucous membrane is ciliated, the cilia beating upwards towards the larynx. In the upper part of the thorax, the trachea

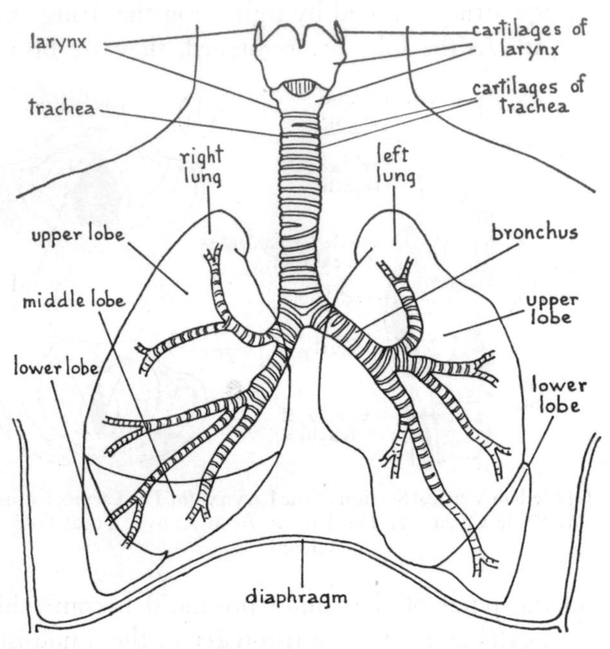

FIG. 92. Larynx, Trachea, Bronchi and Lungs, Front View

forks into two, each branch being a **bronchus** leading into the lung on its own side (*see* Fig. 92).

Each bronchus gives rise to several main branches, which divide repeatedly until, in the final stage of division, microscopic bronchioles are present. Throughout their course, the bronchi and bronchioles have their walls supported by cartilage and their cavities lined with ciliated mucous membrane. In the ultimate branches, the cartilages lose their "C" shape and are present as separated pieces.

Each bronchiole gives rise to three or four branches without cartilage in their walls, and each branch finally leads into an **air-sac** about one millimetre long, with numerous pouches called **alveoli** leading from it (*see* Fig. 93). These alveoli are the ultimate ends of the

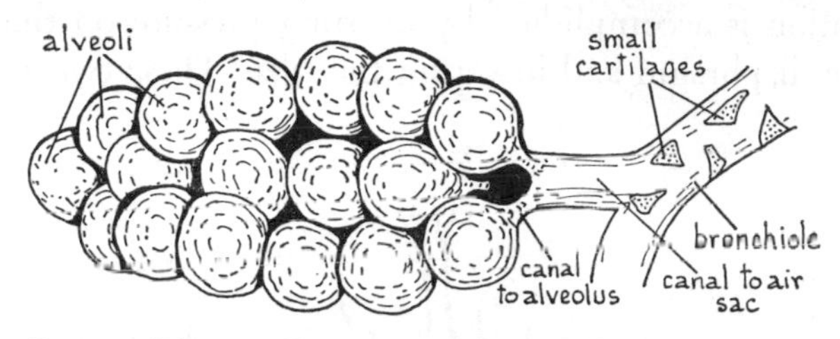

FIG. 93. Diagram Showing an Air-sac with Alveoli, × 45

respiratory passage and it is through their thin walls that gaseous exchange with the blood takes place.

THE BREATHING MOVEMENTS

Breathing in is known as **inhalation** or inspiration, and breathing out is called **exhalation** or expiration. The former names will be used in this account. Since the thorax is a closed cavity packed with organs (the lungs, heart, etc.), and since all the organs except the lungs are filled with fluid and all are surrounded by coelomic fluid, any change in the volume of the cavity will affect only the volume of air in the lungs. During inhalation, the volume of the thoracic cavity is increased by the depression of the diaphragm and the raising of the ribs. The muscular actions that carry out the two processes are described on p. 94. As the volume of the cavity begins to increase, air is forced through the respiratory passages to the lungs by the greater external pressure. This continues as long as the thoracic volume continues to increase. Note that there is never any partial vacuum in the lungs or in the thoracic cavity. The whole process is similar to that of filling a pipette with water, beginning with the teat squeezed flat (*see* Fig. 94). As soon as the pressure on the teat is relaxed, extra volume is available in the pipette and the water is forced in, and will continue to be until the teat is in the fully expanded position. The amount of water that has entered is the same volume as that created by allowing the flattened teat to expand. Similarly, the volume of air that will enter the lungs is equal to the volume created in the thoracic cavity by expanding it.

Exhalation is accomplished by exerting pressure on the lungs by raising the diaphragm and lowering the ribs. These two movements

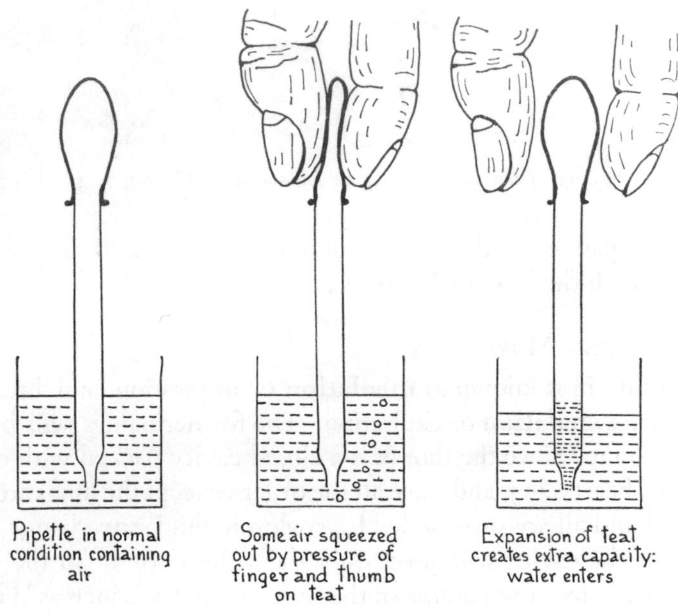

FIG. 94. Water Entering a Pipette, After the Creation of Extra Capacity

are assisted by the elastic tissue of the lungs, which, having been stretched, exerts a force tending to restore it to its smallest size.

In order that the lungs should slide without friction against the thoracic wall or the diaphragm, they are separated from these structures by a very thin layer of coelomic fluid. This fluid is secreted by the peritoneum, which covers the lungs and lines the inside of the thoracic wall and the upper surface of the diaphragm. The two lungs do not touch each other but are separated by a double membrane, the **mediastinum,** which contains the heart (*see* Fig. 95). The sheets of peritoneum covering the lungs and lining the thoracic cavity are known as **pleurae** (singular, **pleura**). Inflammation of the pleurae due to physical damage or to bacterial attack is called **pleurisy**; it may be dry,

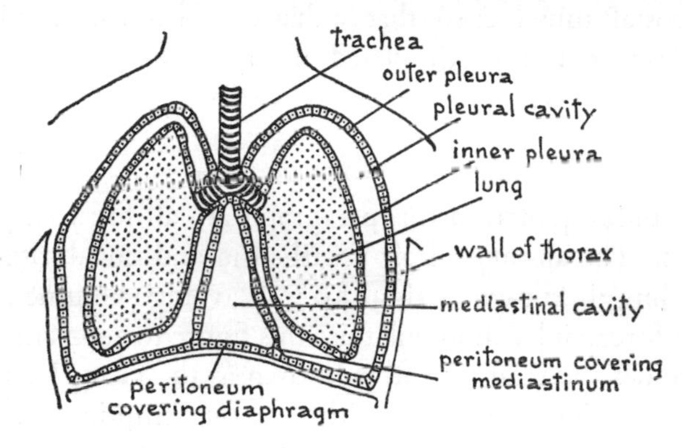

FIG. 95. Relationships between the Lungs, the Pleurae and the Mediastinum
The thicknesses of the pleurae and pleural cavities are exaggerated.

when the cells secrete too little fluid, or wet, when the cells secrete too much.

BREATHING RATE AND FACTORS AFFECTING IT

The average adult breathing rate is about sixteen times a minute; nearly two seconds to breathe in and the same to breathe out. Children breathe much more quickly; the rate for a child of seven years is about thirty per minute, and for a child of one year, about forty per minute. These rates apply to the resting condition. The rate becomes faster when there is a rise in the carbon dioxide content of the air (0·04 per cent normally), or a decrease in the oxygen content (20·96 per cent normally), or during exercise, or when there is increased acidity of the blood. Breathing rate is controlled automatically from a **respiratory centre** in the part of the brain called the **medulla** (*see* Fig. 161), but we can change the rate voluntarily. There are distinct regions in the respiratory centre controlling inhalation and exhalation respectively. For example, the **exhalation centre** is not stimulated by sensory nerves from the lungs until a certain degree of expansion has been achieved. At that point the **inhalation centre** is inhibited and impulses travel out from the exhalation centre to the lungs, diaphragm and intercostal muscles, so that exhalation begins. Similarly, the inhalation centre is not stimulated until the lungs reach a certain degree of contraction.

Breathing Capacity

The volumes quoted in this paragraph are all for young, healthy, adult males. The amount of air breathed in and out during ordinary quiet breathing is known as **tidal air**; its average volume is 500 cm^3. Maximum forced inhalation will raise this figure to 3000 cm^3, which of course includes the tidal air and is known as the **complemental air**; it represents an increase of 2500 cm^3. If one empties one's lungs as completely as possible, i.e. forced exhalation after tidal exhalation (500 cm^3), an extra 1000 cm^3 of air can be expelled from the lungs. This volume, expelled by extra effort, is known as **reserve air.** Even the most forcible exhalation cannot empty the lungs completely; a considerable volume of air, amounting to about 1200 cm^3, remains in the remoter parts. This quantity is known as the **residual air** and its composition is little changed by the flow of air during breathing, but only by **diffusion,** which is random movement of the various molecules present. From these figures, an important quantity known as **vital capacity** can be calculated. The calculation, for the average man, is shown in the table below.

1. Complemental air (which includes tidal air)	3000 cm^3
2. Reserve air	1000 cm^3
3. Vital capacity (total of 1 and 2)	4000 cm^3

Vital capacity is important as a standard of physical fitness. Persons who are active and physically fit have, in general, greater vital capacities than those who are unfit or lead sedentary lives. It is also important in medical diagnosis of certain chest, lung and heart diseases that reduce vital capacity considerably. The volume is determined by using a **spirometer,** the simple basis of which is shown in Fig. 96. There are expensive and very accurate forms of spirometer but the

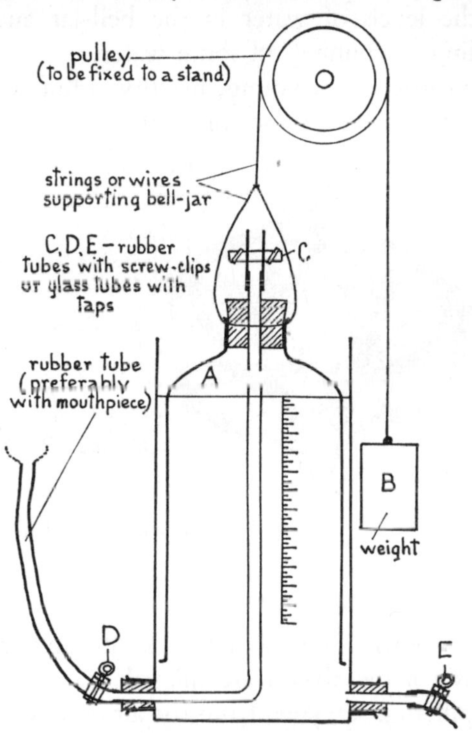

FIG. 96. A Simple Spirometer

one shown in the diagram can be made quite simply and will be accurate enough for obtaining comparative values. The weight or weights at B should balance the weight of the bell-jar, A. The graduation marks should be made at every 50 cm^3, by inverting the bell-jar and adding successive 50 cm^3 portions of water. In use, the bell-jar, with the clip, C, at the top open, should be immersed to the zero graduation mark and then the clip closed. The person to be tested takes a maximum inhalation, then empties his lungs as completely as possible through the rubber tube, with the clips at D and E open; E sufficiently open to prevent overflowing. As soon as he finishes his exhalation, these clips are closed and the volume read on the graduated bell-jar, after the levels of water in the bell-jar and container are equated by adding or running off some water.

The average capacities of young, healthy, adult women are about twenty per cent lower than those for men.

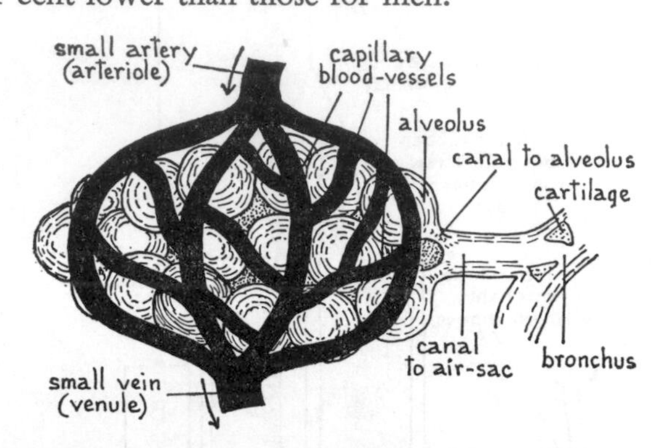

FIG. 97. Diagram Showing a Capillary Network around an Air-sac, × 45

THE ABSORPTION OF OXYGEN

Each air-sac is enclosed in a network of capillary blood-vessels (*see* Fig. 97). When the alveoli are dilated, they are pressed tightly against these capillaries. Oxygen from the air dissolves in the moisture lining the alveoli and passes in solution through their thin walls and then through the thin walls of the capillaries into the blood (*see* Fig. 98). In the blood, the oxygen is absorbed by the **red corpuscles** as they pass through the capillaries and is carried by them to the tissues of the body. At the same time as the oxygen is passing from the alveoli into the blood, carbon dioxide is passing from the blood into the alveoli. This carbon dioxide is then breathed out. The processes of absorbing oxygen and giving out carbon dioxide constitute the **exchange of gases.** The absorption of oxygen is a part of external

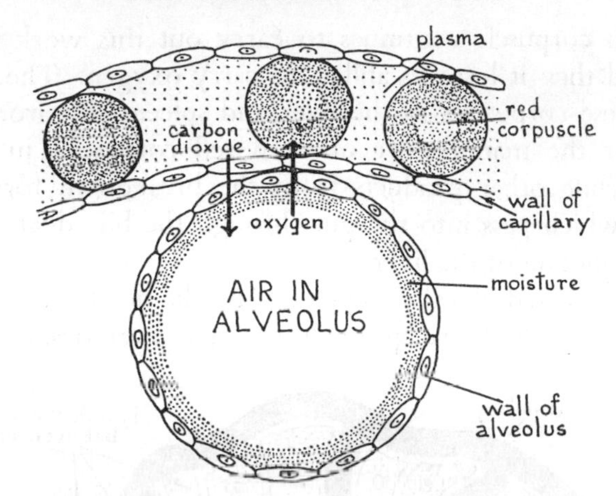

FIG. 98. Diagram Showing Contact between an Alveolus and a Blood Capillary

respiration, while the giving out of carbon dioxide is excretion. The lungs are therefore both respiratory and excretory organs.

TRANSPORT OF OXYGEN TO THE TISSUES

The red corpuscle consists mainly of a jelly-like substance called **haemoglobin,** which contains in its molecule one atom of iron. When there is plenty of oxygen available, as in the lung capillaries, a molecule of oxygen combines loosely with the **iron** atom of the haemoglobin, forming **oxyhaemoglobin.** This is not an oxide so we do not say that the corpuscles are oxidised but **oxygenated.** The blood from the lungs then flows back to the heart to be pumped to the tissues in the arteries. When there is little oxygen available, as in the tissues of all parts of the body except the lungs, the oxyhaemoglobin gives up its oxygen and haemoglobin is formed again. The blood is then said to be **deoxygenated.** The red corpuscles are to be regarded as vehicles for carrying oxygen. They return to the lungs empty, possessing haemoglobin, and they leave the lungs full, containing oxyhaemoglobin. The oxygen is delivered to the tissues by the oxyhaemoglobin and haemoglobin is produced ready for the next refill. Each corpuscle continues to carry out this work for several months and then it loses its ability to carry oxygen. The body then destroys these corpuscles in the liver and spleen; the iron is largely retained for the manufacture of fresh red corpuscles in the bone-marrow, while other products from the breakdown form the bile pigments, which pass into the gut through the bile duct. This is an excretory function of the liver.

The oxygenated blood pumped out by the heart travels in arteries to all parts of the body except the lungs. These arteries, after repeated

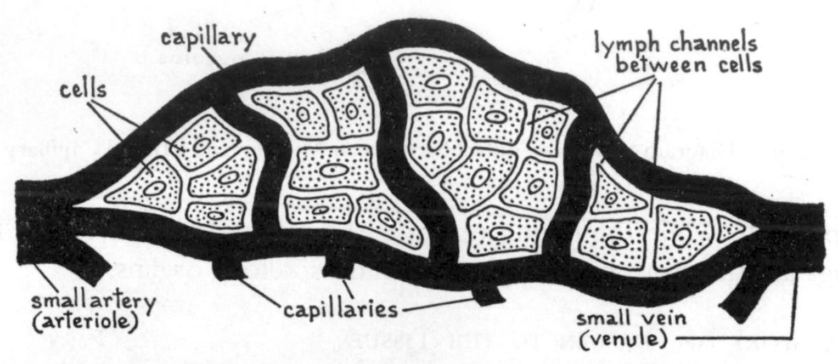

FIG. 99. Diagram Showing Capillary Circulation to the Tissues

branching, finally lead to networks of extremely fine capillaries (*see* Fig. 99). Through the thin walls of the capillaries, a colourless fluid, **lymph,** leaves the blood and circulates freely round all the living cells. The lymph contains the dissolved oxygen and other substances from the blood (*see* p. 203); the oxygen is absorbed by the cells and utilized for respiration. Carbon dioxide, produced by the cells as a result of respiration, passes into the lymph, finally enters the blood stream via the **lymphatic circulation,** and is excreted from the body mainly by the lungs.

Internal or Tissue Respiration

It must now be recalled that the energy required by the cell to perform its work is supplied by breaking down ATP (*see* p. 99).

The process of respiration in each cell is the breakdown of glucose (or fat or protein) so that the energy stored in the glucose is released for the purpose of rebuilding ATP. The breakdown of glucose takes place in a series of reactions, so that the energy is released in small quantities at a time. The process is known as **glycolysis**; it results in the rebuilding of two ATP molecules and the resulting chemical substance is **pyruvic acid,** $CH_3.CO.COOH$. The further breakdown of pyruvic acid depends upon whether aerobic or anaerobic respiration is carried out (*see below*). If the respiration is aerobic, then pyruvic acid is completely broken down to carbon dioxide and water, after a long cyclic series of reactions known after its discoverer as **Krebs cycle.** The net result of the whole process of aerobic respiration, i.e. glycolysis + Krebs cycle, is that 38 ATP molecules are built up.

The whole process is summed up in this equation—

$$\underset{\text{glucose}}{C_6H_{12}O_6} + \underset{\text{oxygen}}{6\,O_2} = \underset{\text{carbon dioxide}}{6\,CO_2} + \underset{\text{water}}{6\,H_2O} + \underset{\text{energy}}{2872\text{ kilojoule}}$$

In actual quantities, the equation indicates that 180 g of glucose will yield 2872 kJ of energy. The energy utilized in building 38 ATP molecules is 1250 kJ. The loss of energy, 1622 kJ, represents small quantities of heat liberated during various stages of the process. The efficiency of the process of transferring the energy of glucose into ATP is thus $\frac{1250}{2872} \times 100$, i.e. forty-five per cent. The heat energy lost during the whole process, thirty-four per cent, is not to be regarded merely as waste since it is used to some extent in the maintenance of body temperature.

In muscle tissue during exercise, glucose is broken down by glycolysis to pyruvic acid, which is then converted into **lactic acid.** Two molecules of ATP are built and the energy yield is 100·5 kJ per 180 g of glucose, an efficiency between three and four per cent. However, since most of the lactic acid is built up into glycogen in the liver, some of the remaining energy is still conserved.

More details on energy yields and energy requirements are given in Chapter 7.

Comparison Between Inhaled and Exhaled Air

The table below shows the composition of inhaled and exhaled air. It can be seen that about four per cent of the atmospheric air is

	Nitrogen	Oxygen	Carbon Dioxide	Water Vapour
Inhaled	79%	nearly 21%	0·04%	Variable
Exhaled	79%	17%	4%	Saturated

absorbed by the body; this is the oxygen taken in by the red corpuscles of the blood. In the exhaled air, there is about four per cent of carbon dioxide, which takes the place of some of the oxygen; this carbon dioxide is produced in the body as a result of respiration. The amount of water vapour in inhaled air varies considerably; sometimes the air is very dry, and sometimes it is very damp. Exhaled air is always saturated with water vapour.

Respiratory Quotient

If the figure for oxygen absorbed is divided by the figure for carbon dioxide given out, the result obtained is called the **respiratory quotient.** Taking the figures from the table: Oxygen absorbed = 4 per cent, carbon dioxide given out = 4 per cent. 4/4 = 1, which is the respiratory quotient or R.Q. The importance of this respiratory quotient is that it shows what type of food material is being used for respiration. If it is carbohydrate alone, the R.Q. is 1; if it is fat alone, the R.Q. is about 0·7. Values between 0·7 and 1 indicate that some carbohydrate and some fat are being used. Protein is not normally used for respiration, except in extreme starvation, since the body does not store protein. The R.Q. for different proteins varies between 0·5 and 0·8.

Artificial Respiration

The cells of the human body cannot survive very long without oxygen. The central nervous system is particularly susceptible to lack of oxygen; if the blood supply to the brain is cut off for five minutes, irreparable damage is done. When breathing is suspended, as in apparent drowning, gas poisoning or severe electric shock, there is still enough oxygen in the blood to maintain life in the tissues for a considerable time. In such cases, it is imperative to begin **artificial respiration** before the heart muscle fails through lack of oxygen. If the beating of the heart is stopped, artificial respiration is useless, unless the heartbeat can be restarted by massage or by other methods.

The object of artificial respiration is to imitate the natural breathing movements by causing rhythmical alterations in the internal volume of the thoracic cavity. Many mechanical methods have been devised, particularly that of Schäfer, and they have resulted in the saving of many lives. Nowadays they are all superseded by a much easier and more effective method, **mouth-to-mouth respiration,** first described in the popular press as "the kiss of life." This method is very easy and everyone should be familiar with it.

Mouth to Mouth Respiration

The essentials of the method are described here in a brief sequence, which should be memorized. The subject (patient) is lying flat on his back (*see* Fig. 100).

1. Push the head back so that the chin points upwards.
2. Open the subject's mouth, place your mouth over his and pinch the nostrils to close them and thus prevent air leakage.
3. Blow into the subject's mouth until you see his chest rise.
4. Then continue to breathe into his mouth regularly about twelve times per minute for an adult; about twenty times per minute for a child. Detach your mouth from his between each breath.

Pointing the chin upwards tends to prevent the back of the tongue blocking the respiratory passage. In the case of drowning, water should be allowed to run out by supporting the subject with the head downwards and the trunk above it. Foreign matter in the buccal cavity can be cleared with the fingers, or if in the throat, slapping on the back may be effective. If the teeth are clenched and cannot be forced apart, it may still be possible to blow through the teeth or through the nose.

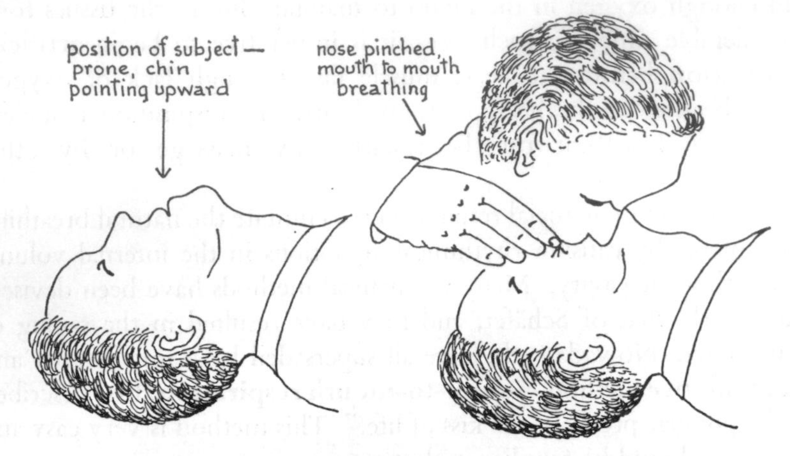

FIG. 100. Artificial Respiration; Mouth to Mouth Method

Various Modifications of Breathing

The respiratory passages, especially in human beings, are used for a variety of other purposes.

LAUGHING

This is essentially a deep inhalation followed by short and rapid exhalations. The characteristic sound is caused by vibration of the vocal folds, which are held tense in the path of the outgoing air. If the laughter is loud and hearty there is vigorous action of the diaphragm; such hearty laughter is a good tonic to the diaphragm muscle.

CRYING

In this case, the respiratory movements are similar to those of laughing, i.e. a long inhalation followed by short and sudden exhalations. These are accompanied by tears and facial contortions.

Coughing

This is caused by stimulation of sensory nerve-endings in the mucous membrane of the larynx, trachea or bronchi. The stimulation may be due to excessive mucus, to inflammation after infection or to the presence of foreign particles, e.g. bread-crumbs. There is a short inhalation followed by a forcible exhalation with the glottis closed. As a high pressure is built up, the glottis is suddenly opened, allowing the air to escape in a blast which may dislodge the source of irritation.

Yawning

This is an indication of fatigue, of sleepiness, of lack of fresh air, or sometimes of sheer boredom. It may also be induced by seeing other people yawn. It consists of a deep inhalation with the mouth wide open, followed by a prolonged exhalation. Yawning is a reflex action that cannot be suppressed, though often it is concealed by the hand.

Sighing

Stimuli due to thoughts of sad happenings, old memories or self-pity arise in the brain and evoke the response of sighing. External occurrences that evoke sympathy have the same result. The physical response is a prolonged and often audible exhalation.

Sneezing

This is another reflex action, initiated by stimulation of sensory nerve-endings in the nasal passages. The response is a short inhalation followed by a forcible exhalation with the glottis open. In this case, the tongue is raised against the soft palate to divert the air stream out through the nose, thus tending to clear the source of irritation.

Hiccoughing

This is usually due to irritation of the mucous membrane of the stomach. There is a sudden inhalation produced by the spasmodic action of the diaphragm. Because it is so sudden, the aperture of the glottis is not prepared for the rush of air and the typical sound is produced by vibration of the vocal folds.

Sniffing

Sniffing consists of a series of short, sharp inhalations, but the mouth being closed, all the air inhaled is made to pass through the nostrils. The sharp inhalations cause air to circulate more readily in the olfactory region and hence the sense of smell is made more acute.

Sobbing

This consists of a series of convulsive inhalations, each taking place with the glottis closed or nearly closed.

Snoring

This occurs while sleeping with the mouth open. The current of inhaled air causes the uvula to vibrate. The position of the lips determines whether there will be a whistle produced by the exhalation.

SUGGESTED EXERCISES

1.* Dissect the neck and thorax of a freshly killed or preserved rat to display the respiratory organs. Make a large, fully labelled drawing of the dissection.

To open the thorax, cut up the mid-line of the sternum with strong scissors; then cut away the diaphragm from the lower ribs. Using the hands, bend the rib-cage outwards and pin down flat with strong spikes. (The heart and thymus gland will obscure part of this dissection and therefore they must be removed. If, however, the same rat specimen is to be used for the blood system, defer this dissection until later.) Note and label the larynx, the thyroid gland, the trachea, bronchi, lungs, intercostal muscles, diaphragm.

2.* Examine, if available, a prepared slide of a lung section. Try to find an air-sac, an alveolus, a small bronchus, a blood-vessel, and elastic fibres. Injected sections, obtainable from dealers, will show the rich blood supply of the lung.

3. Breathing out through the mouth on a mirror or a window will show that water vapour is exhaled. Breathing through a glass tube into lime water will show that carbon dioxide is exhaled. (The lime water will turn milky after you have been blowing into it for a short time.)

4. Find out the rate of breathing per minute for a young child, a mature adult, a dog, a cat, a rat, and any other animals you can observe. Watch the rise and fall of the

thorax and count one for each inhalation, that is, when the thorax expands. If these rates are to be compared, they should all be measured when the animal is at rest. When a number of results have been obtained from a variety of animals, try to find out what general conclusions can be drawn from them.

5. Using a spirometer, determine the vital capacity for various people, e.g. young children, adults, athletes. Note that a simple spirometer is easily constructed if the two large water containers are available.

Chapter 10

THE BLOOD VASCULAR SYSTEM

Living cells in every part of the body must have constant supplies of food, water and oxygen, and they must have their waste products removed. These, and other materials, are transported in the blood vascular system. The word "**vascular**" refers to vessels or tubes and the system is so called because the blood flows in a series of tubes. The system consists of the blood, the blood-vessels and the heart.

The Blood

The blood is a somewhat sticky red fluid with a pH of 7·4, i.e. slightly alkaline. The total quantity of blood in an average man is about 5·7 litres. It is made up of four main constituents, the **plasma,** the **red** and **white corpuscles** and the **blood platelets.**

The Plasma

This is the fluid part of the blood and it occupies fifty-five per cent of the total volume, the remaining forty-five per cent consisting of the various kinds of corpuscles. Plasma is faintly yellow in colour and is composed mainly of water (ninety-two per cent); the remaining eight per cent consists of various dissolved substances. The chief of these are the blood proteins, the salts, the food and the waste materials.

The blood proteins are not to be confused with the proteins from food, which are carried in the blood in the form of amino-acids. Three kinds of proteins are present, **albumen, globulin** and **fibrinogen.** They make the blood slightly sticky and help to prevent rapid changes of pH, which cause the death of cells very quickly. In addition, fibrinogen plays an important part in the **clotting** of the blood (*see* p. 186) while one kind of globulin, **gamma globulin,** contains **anti-bodies** that protect the body against certain diseases (*see* p. 184).

Together, the blood proteins make up about seven per cent of the plasma.

The salts in blood are very similar to the salts dissolved in sea-water, both in kinds and in quantities. The cells of the body cannot continue to live if the fluid surrounding them is not kept fairly constant in composition. The fluid, which constitutes an internal environment, is **lymph,** derived from the blood and returned to the blood (*see* p. 203). The lymph brings back into the blood the waste products from the cells; as the blood passes through the kidneys, waste products are removed and the proportions of salts and water in the blood are regulated (*see* p. 216). The ions present in greatest quantity are sodium and chloride; also present are the following ions: calcium, potassium, magnesium, bicarbonate, phosphate and sulphate. Together, the salts make up less than one per cent of the plasma.

Of the food substances in the plasma, glucose is normally always present, being carried to the cells for use in respiration. The store of glycogen in the liver is gradually changed to glucose and released into the blood; the concentration of glucose is normally about 0·1 per cent, i.e. 100 mg per 100 cm^3 (0·0035 oz). The total amount of glucose present in the blood is about six grams, or one fifth of an ounce. If a meal has contained fat, then for some time afterwards, fat droplets are to be found in the plasma, but they soon disappear as they are deposited in the fat-storage tissues. Small quantities of amino-acids, required by the cells for protein manufacture, are always present.

The two chief waste substances present are urea and carbon dioxide. The urea is made principally in the liver and is derived from the excess amino-acids absorbed from the gut (*see* p. 152). The blood carries the urea to the kidneys, where it is extracted and passed out of the body in the urine. Carbon dioxide is a waste product derived from respiration in the cells; some is dissolved in the plasma, about five per cent, forming carbonic acid; twenty per cent is carried back to the lungs in the red corpuscles, and the remainder, seventy-five per cent, is carried as bicarbonate ions (HCO_3^-). In the blood capillaries of the lungs, an enzyme called **carbonic anhydrase** rapidly splits up both the carbonic acid and the bicarbonate, releasing carbon dioxide. Also, in the presence of the relatively large amount of oxygen in the lung capillaries, the haemoglobin releases its carbon dioxide. In all three cases, the carbon dioxide passes from the plasma into the alveoli and forms four per cent of the exhaled air.

The Red Corpuscles

These are by far the most numerous of the blood cells; there are about five million per cubic millimetre of blood in the average adult

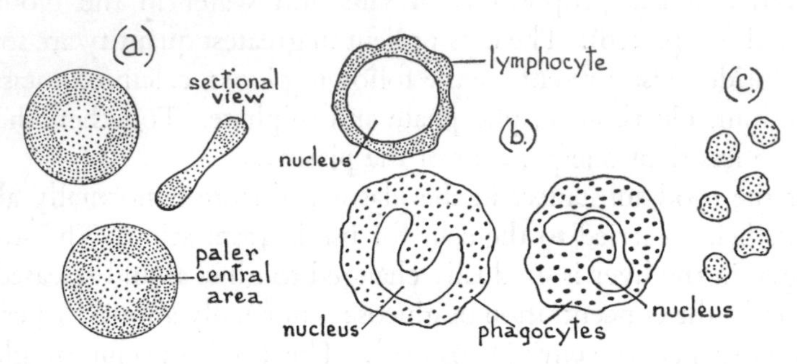

Fig. 101. Blood Corpuscles, All Drawn to the Same Scale, × 1600 (*a*) Red Corpuscles (*b*) Three Kinds of White Corpuscles (*c*) Platelets

man, and about ten per cent less in the average adult woman. These cells are formed in the bone-marrow, and though each cell has a nucleus when formed, it disintegrates before the cell is released into the blood. After the loss of the nucleus, the broader surfaces of each cell sink inwards, thus forming a biconcave disc. The diameter of the disc is about 8μm and its thickness 2μm (*see* Fig. 101(*a*)).

The protoplasm of the red corpuscles consists mainly of a yellow jelly called haemoglobin, composed of a coloured substance, **haem,** containing iron, and the protein **globin.** Haemoglobin combines with oxygen in the lung capillaries to form oxyhaemoglobin. This releases oxygen to the lymph, which carries it to the tissues. The corpuscles return to the lungs containing haemoglobin and some carbon dioxide combined with it. After two to three months of this work, the corpuscles gradually lose their ability to carry oxygen; they are then destroyed in the liver and spleen. The iron is retained for further use, while the yellow colouring material passes out of the liver in the bile. It is estimated that in a healthy young adult, about one million are destroyed per second; it follows that the same number must be made

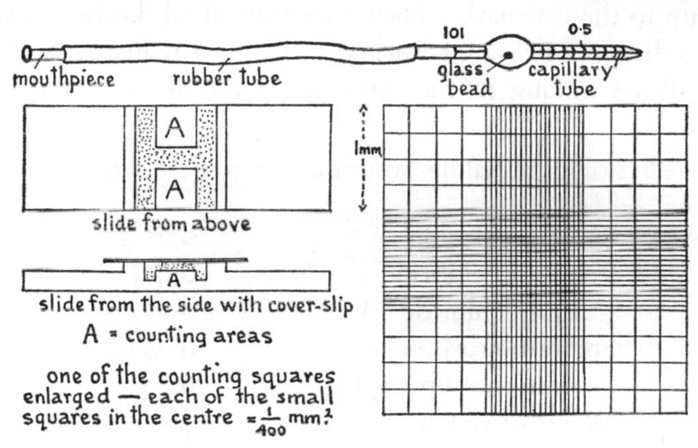

FIG. 102. A Haemocytometer, Used for Blood Cell Counts

per second. In the whole volume of the blood there are about thirty-five billion (35 000 000 000 000) red corpuscles.

Two of the diseases associated with the red corpuscles are **common anaemia** and **pernicious anaemia.** The former is usually due to the inability of the body to manufacture enough haemoglobin because of a lack of iron in the diet or inability to absorb enough iron. This condition is usually remedied by giving the patient a suitable iron salt or by improvement of the diet. Pernicious anaemia is more serious; it is the failure of the marrow to manufacture enough corpuscles. The discovery of an **anti-anaemic** substance in liver has been of great benefit to victims of this disease. In diagnosis of anaemia, a red-cell count is essential. With the correct apparatus, it can be carried out quite easily. The apparatus required is a special microscope slide and a special small pipette (*see* Fig. 102). The slide has a groove shaped like a letter H and two small counting areas, with the centre of each divided into small squares, each having sides of 1/20 mm, i.e. an area of 1/400 mm^2. The pipette has a mouthpiece and a rubber tube and the glass part consists of a capillary tube dilated into a bulb, which contains a glass bead. The thumb is pricked with a sterile needle, between the base of the nail and the knuckle, and blood is sucked up to the 0·5 mark. Then a diluting fluid, known as **physiological saline** (*see* below), is sucked in to the 101 mark. Shaking the pipette allows mixing in the bulb, with the aid of the glass bead.

Physiological Saline Solution (Ringer's Solution)

Sodium chloride	0·65 g
Potassium chloride	0·014 g
Calcium chloride	0·012 g
Sodium bicarbonate	0·02 g
Sodium monophosphate	0·2 g
Water to 100 cm^3	

This is a useful solution for other biological work and it may be worth while to make one litre, i.e. ten times the above quantities.

The liquid, which is blood diluted two hundred times, is squeezed on each counting platform and a cover-slip placed over the slide. The excess fluid runs into the grooves, and the depth above the counting squares is 1/10 mm. The slide is then placed under a microscope, the number of red cells counted in a number of squares and the average taken. Suppose the average is seven, then the calculation proceeds as follows. The volume of fluid above each square is 1/4000 of a cubic millimetre, and the dilution is two hundred. Therefore the number of red cells per cubic millimetre is—

$$4000 \times 200 \times 7 = 5\,600\,000.$$

It is perhaps easier to remember that the average count for ten squares or so is to be multiplied by 800 000.

The White Corpuscles

There are about seven thousand white corpuscles, on the average, in a cubic millimetre of blood; they are therefore about one-eight-hundredth the numbers of the red cells. There are several kinds of these corpuscles, some of which are manufactured in the bone marrow and some in the lymphatic glands (*see* p. 207), but they all fall into two main groups, **phagocytes** and **lymphocytes** (*see* Fig. 101(*b*)). Phagocytes are so called because they ingest bacteria or any other foreign particles; the word means "eating-cells." They can creep about like amoebae and can migrate through the capillary walls into the body tissues (*see* Fig. 103), being especially plentiful at the site of any bacterial

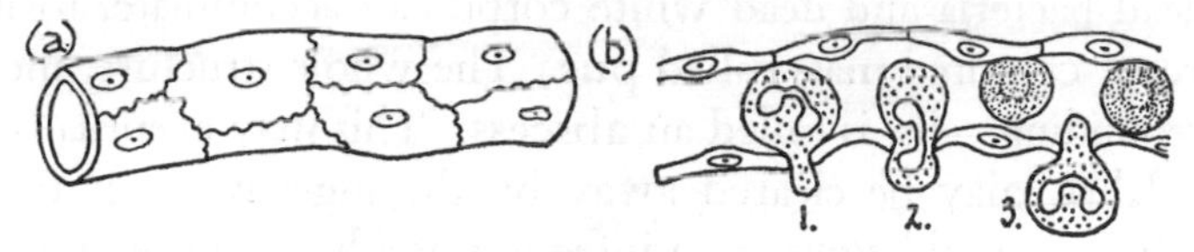

Fig. 103. (*a*) Small Portion of a Capillary (*b*) Migration of a White Corpuscle through the Wall: Both × 700

infection. It is probable that white corpuscles do not last more than a few days, being destroyed mainly in the spleen and lost by migration into the gut.

The lymphocytes are so called because they are made in the lymphatic glands and are therefore plentiful in lymph. The word means "lymph cells." They produce protective substances of four kinds: one causes bacteria to clump together and thus prevents their multiplication; a second causes bacteria to break up, and a third immobilizes bacteria so that they are more easily ingested by the phagocytes. The fourth group consists of a number of substances called **antitoxins,** which neutralize the **toxins** or poisons produced by the bacteria. It is thought that some of the leucocytes produce gamma globulin and then disintegrate, supplying the globulin to the plasma.

Together, the phagocytes and leucocytes combine to protect the body from bacteria. Any infection is an invasion of the body by small organisms usually known as **germs.** In this sense, the word germ denotes any small creature (animal, plant or virus) that can cause a diseased condition. Here we shall consider the case of bacteria introduced by piercing the skin with a thorn or a nail. The damaged cells release substances (**acetylcholine** and **histamine**) that attract the white corpuscles and also cause capillaries in the vicinity to dilate. A condition known as **inflammation** soon arises; it is recognized by its red colour, its swelling and its temperature. These features are due to increased blood supply in the dilated blood-vessels; the swelling is due to more rapid exudation of lymph from the capillaries. The white corpuscles attack the bacteria, ingesting and possibly destroying them or being destroyed themselves by the toxins from the bacteria. Dead cells, dead bacteria and dead white corpuscles accumulate, forming a soft, cream-coloured mass called **pus.** The whole structure, including the surrounding cells, is called an **abscess.** This may eventually burst, or the debris may be cleared away by the ingestive activity of the phagocytes. A boil begins as an infection of a hair follicle. Gradually, the phagocytes remove the overlying cells until the pus has only a paper-thin covering of skin. Soon afterwards, the boil bursts.

If the body is unsuccessful in destroying the invaders, they multiply rapidly and very serious illness, sometimes even death, may follow. Human beings have the great advantage of assistance from **antibiotics** such as **penicillin,** from various drugs and from **immunization.**

Immunity

We say that a person is immune to a disease when it is shown that he cannot contract it although he has been infected by the organisms that cause the disease. Any foreign material that stimulates the production of antibodies in the blood is called an **antigen.** Most of the antibodies are produced by the lymphocytes and a number are present in the gamma globulins, themselves probably derived from certain kinds of lymphocytes. A person who is immune to a particular disease already has the correct antibody present in his blood. Some antibodies are anti-toxins, which render bacterial toxins harmless; some cause the destruction and disintegration of bacteria and some cause bacteria to clump together, preventing their multiplication. These clumped bacteria die and are eventually ingested by phagocytes.

There are three kinds of immunity: natural, active and passive. **Natural immunity** to a disease is inherited; the possessor has the necessary antibody although he has not had the disease. In this way, some people are immune to poliomyelitis, typhoid fever, etc. Although natural immunity is very desirable for the individual, it may be dangerous for other people because the immune person may carry the germs and convey them to others; in other words, he may be an unsuspecting carrier.

When a person has **active immunity,** it means that he has produced the correct antibody either by having had the disease or by having taken a **vaccine.** There are several kinds of vaccines. Some consist of a mild strain of the disease-causing organism, or a related strain, that have relatively little effect on the body. For example, most children are protected against small-pox by vaccination with the milder cow-pox virus. Another form of vaccination consists of injection of dead bacteria; the toxin that evokes production of the required antibody is still present in the dead bacteria. This method is employed in the case of typhoid fever. Yet another type of vaccination consists of injection of bacteria that have been weakened by heat or chemical treatment so that they are non-toxic. This is used for prevention of diphtheria. In all cases of active immunity, actual bacteria or viruses are present in the body, either by accident or design; they evoke the production of the required antibodies and so the person will not contract the diseases.

Utilization of antibodies produced by other people or by animals constitutes **passive immunity.** This method is usually used to reduce the severity of a disease that a person has already contracted. For example, diphtheria anti-toxin is produced by horses into which the toxin of diphtheria has been injected. The horses do not suffer, and small quantities of their **blood serum** (*see* p. 187) are used in the treatment of diphtheria patients. Similarly, serum from patients who have recently recovered from a disease such as scarlet fever, measles or poliomyelitis, is used for passive immunization of other persons suffering from the same disease.

Our increasing knowledge of these chemical processes involving antibody-antigen reactions has proved to be of immense benefit. Diphtheria was once a great killer; now deaths from this disease are extremely rare. Small-pox was once very common in Great Britain; since the discovery of vaccination with cow-pox by Jenner in 1796, cases are very rare. The distressing disease whooping-cough has been almost eradicated by immunization at a very early age, and there are encouraging signs that poliomyelitis vaccines are providing immunity for an ever-increasing proportion of the population.

In a thin film of blood on a **haemocytometer** slide, the white corpuscles are few and far between. Hence a count is made using several of the larger squares. Reduction of the numbers to well below normal brings about reduced resistance to infection and is known as **leukopenia.** During infection, the body responds by increased production of these cells, a condition known as **leukocytosis.** But if they are produced in enormous numbers when there is no known infection, a serious disease called **leukaemia** results; it is associated with a deficiency of red corpuscles.

Blood Platelets (Thrombocytes)

These are very small, disc-like objects about 2μm in diameter, produced in the bone-marrow (*see* Fig. 101(*c*), p. 180). It is difficult to make a reasonably accurate count of them because in drawn blood they tend to clump together in groups, but it is estimated that they average about 250 000 per cubic millimetre. Their sole function, as far as is known, is concerned with the clotting of the blood (*see* below).

The Clotting Mechanism

The ability of the blood to form a clot protects the individual from excessive bleeding from minor wounds. Clotting is an extremely complicated process and is not yet thoroughly understood. The essential facts are these.

1. A substance that promotes clotting is always present in the blood, but normally it is inactive and is then called **pro-thrombin.**

2. When a blood-vessel is damaged, blood platelets break up and release an enzyme (**thrombokinase**), which converts pro-thrombin into the active **thrombin.**

3. In the presence of calcium ions, thrombin causes the fibrinogen of blood to be precipitated as fine threads of **fibrin.**

4. The tangled threads of fibrin, with the meshes blocked by corpuscles, form a sort of soft plug over the wound. This hardens

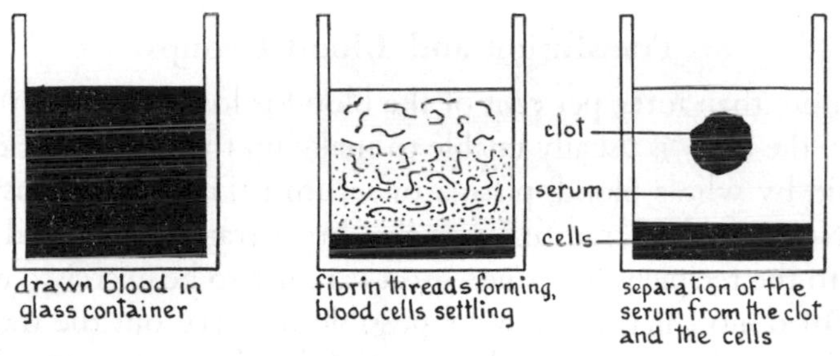

FIG. 104. Separation of Clot and Serum from Drawn Blood

into a **scab** beneath which the repair processes are carried out. Eventually the scab is lost, usually leaving a **scar.**

If drawn blood is placed in a glass vessel, a clot forms and the corpuscles gradually sink (*see* Fig. 104). If the clot and the corpuscles are removed, the remaining straw-coloured fluid is blood serum. It differs from plasma because the fibrinogen has been removed.

Sometimes a blood clot (**thrombus**) forms within a blood-vessel. This condition becomes very serious if it occurs in a blood-vessel in the brain or the heart, causing blockage of the blood supply. In the former case the disease is **cerebral thrombosis** and in the latter case it is **coronary thrombosis.**

HAEMOPHILIA

Individuals whose blood will either not clot at all, or will clot only very slowly are known as **haemophiliacs.** The condition is known only in males; though it is theoretically possible in females, no authentic case has been recorded. It is transmitted by a carrier mother to her sons (*see* p. 319), and these unfortunate persons rarely survive to maturity. The disease is apparently due to deficiency of an unidentified factor in the globulin of the plasma, since normal clotting can be brought about in these haemophiliacs by transfusion of normal blood, or by injection of globulin extracted from normal blood. The trouble is that there is rarely time to carry out the operation before the individual has lost too much blood.

Transfusion and Blood Groups

If more than forty per cent of the blood is lost over a short period of time, the body is usually unable to make up the loss. Replacing the lost fluid by whole blood, plasma or serum is known as **transfusion.** The best method of transfusion is the direct transfer of blood from a donor to the recipient, but since great care has to be exercised over the choice of donor, it is not always possible to carry out the transfer in good time. Blood may not be transfused indiscriminately from any one person to any other; their blood groups must be compatible. If incompatible blood is transfused, the recipient's plasma contains a substance that will cause the donor's red corpuscles to clump together and block the smaller blood-vessels. These clumps disintegrate and the haemoglobin is liberated into the blood; the kidneys attempt to excrete it, but their delicate tubules (*see* p. 213) are severely damaged and the patient dies from suppression of urine.

Whole blood can now be preserved in a refrigerated condition after the addition of sodium citrate, which prevents clotting. Methods have also been developed for preserving plasma and serum almost indefinitely. Every hospital now maintains a "blood bank" ready for any emergency, and healthy individuals are constantly urged to become donors and contribute towards this life-saving store.

It was discovered in 1901 that the human population can be divided into four groups according to the reactions of their bloods when mixed together. The groups are known respectively by the capital letters A, B, AB and the figure O. Interaction between two sorts of blood when mixed depends on two factors; these are the antibodies in the plasma and the antigens in the red corpuscles. An individual may possess in his plasma one, both, or neither of two antibodies known as α and β (Greek *a* and *b*), and he may possess in his red corpuscles one, both, or neither of two antigens A and B. Of these four substances, A, B, α, β, an individual can possess only two. Since the antibody α will clump corpuscles containing A, and the antibody β will clump corpuscles containing B, then it follows that no person can be Aα or Bβ. Therefore the only possibilities left are—Aβ, Bα, AB, $\alpha\beta$. These are respectively known as Groups—A, B, AB, O.

In considering donors and recipients, it is important to remember that it is the effect of the plasma of the recipient on the red cells of the donor that might cause trouble. The plasma (serum is usually used) of the donor is ignored since it is soon diluted by the much greater bulk of fluid of the recipient.

We have, then, the following combinations—

	Antigen in Corpuscles	Antibody in Plasma
Group A	A	β
Group B	B	α
Group AB	AB	O
Group O	O	$\alpha\beta$

The table on p. 190 shows the effect of each kind of blood on the others, a tick indicating that transfusion will be successful and a cross indicating that it will not.

From the table, it can be seen that Group AB persons are **universal recipients,** i.e. they can receive blood from any other group, and Group O are **universal donors,** i.e. they can give blood to any other group.

Testing for blood group is a very simple process. A drop of Group A serum and of Group B serum are placed on a white tile. Then a drop of the blood to be tested is added to both sera. Group A blood will clump only in 2; group B blood only in 1; group AB blood in

7

Donor	Recipient: Group A (Aβ)	Group B (Bα)	Group AB (AB)	Group O ($\alpha\beta$)
Group A (Aβ)	✓	×	✓	×
Group B (Bα)	×	✓	✓	×
Group AB (AB)	×	×	✓	×
Group O ($\alpha\beta$)	✓	✓	✓	✓

both, and group O blood in neither. These results are shown in Fig. 105. Human beings fall into the groups in these proportions—A, 43 per cent, B, 9 per cent, AB, 3 per cent, O, 45 per cent.

Other types of blood-group systems are known, which interact because of the presence of other antigens and antibodies. Of these, the

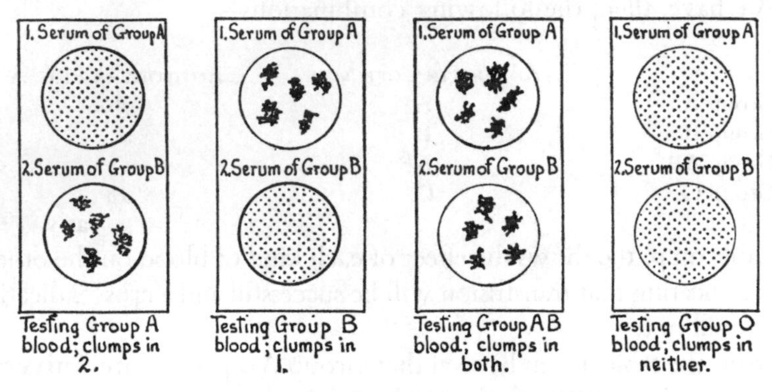

Fig. 105. Testing for Blood Groups in the AB System

best known is the **Rhesus (Rh) system.** The Rh antibody was originally developed by injecting blood of the Rhesus monkey into a rabbit. The rabbit's blood produced an antibody in response to an antigen in the Rhesus red blood cells. Rabbit serum, when extracted, then contained the antibody and when used with a drop of human blood either causes clumping of the red cells (Rh positive) or no clumping (Rh negative). About eighty-five per cent of human beings are Rh positive, i.e. they possess the Rh antigen in their red cells, while the other fifteen per cent, Rh negative, do not possess the Rh antigen.

The Rh factor does not normally cause any trouble in a first transfusion but if repeated transfusions of Rh+ into Rh− are made, then the injected antibody will become strong enough to give a serious reaction. More important is the interaction of the embryo's and mother's blood during pregnancy. If the father is Rh positive and the mother Rh negative, then it is likely that the embryo will be Rh positive like the father (*see* p. 315). This means that the embryo's red cells contain the Rh antigen and if, as occasionally happens, some of the embryo's blood gets into the mother's circulation, then the mother's blood will develop the Rh antibody in the plasma. If in the later stages of pregnancy some of the mother's plasma migrates into the embryo, clumping occurs, resulting often in stillbirth or at least serious blood disturbance. If such an embryo survives birth, then the whole of its Rh positive blood is replaced by Rh negative in an exchange transfusion. This process stops the destruction of red cells and the symptoms disappear. Gradually the infant replaces the transfused blood with its own but since the antigen is not present in Rh negative blood and the antibody is removed by the transfusion, then no further trouble arises.

Apart from the Rh and AB systems, various other blood group systems are known, which are important in special cases.

Functions of the Blood

Most of the functions of the blood have already been mentioned in this chapter. For the student's convenience, they are summarized here.

1. Oxygen is transported from the lungs to the tissues by the red corpuscles.
2. Carbon dioxide is transported from the tissues to the lungs in the plasma and in the red corpuscles.

3. Food materials absorbed from the intestines are carried to the tissues in the plasma.

4. The excretory product urea is carried from the liver to the kidneys to be eliminated.

5. Hormones are distributed from their origins in various glands (*see* Chapter 15).

6. Heat produced in the more active and deeper-seated parts of the body is distributed more or less evenly. Control of loss of heat through the skin does much to keep a constant temperature.

7. The pH of the tissue fluid (lymph) and of the tissues themselves is maintained by interchange with the blood.

8. The ability of the blood to form clots and thus reduce bleeding, has been, and is, of great survival value.

9. The blood plays an important part in protecting the body from bacteria and other organisms that can cause disease. The blood can carry substances that are detrimental to the body, e.g. alcohol, some poisons such as snake venoms, cancerous cells, certain blood parasites, etc.

The Blood-vessels

The blood-vessels are of three types: arteries, veins and capillaries. Each type has its distinctive features and together they form a closed system of tubes around which the blood is pumped by the heart.

Arteries

These contain blood pumped from the heart to the tissues. Only two arteries actually leave the heart; they are the **pulmonary artery** and the **aorta** (*see* Fig. 109, p. 197). The pulmonary artery divides into right and left branches, which supply the lungs; the aorta divides into branches that supply blood to all other parts of the body. The various arteries subdivide, the branches becoming smaller and smaller, and finally the most minute are the **arterioles.** Each arteriole leads into a network of capillaries from which the blood emerges into a small vein called a **venule** (*see* Fig. 99, p. 170).

The walls of arteries contain more muscle and more elastic tissue than the veins and are thus better able to expand and contract and to exert pressure on the blood (*see* Fig. 106). Each time the ventricles of the heart contract, blood is driven along the arteries, and in every artery a wave of contraction followed by expansion passes along in time with the beat of the heart. This wave can be felt with the finger-tip wherever an artery passes near the surface of the body. The regular

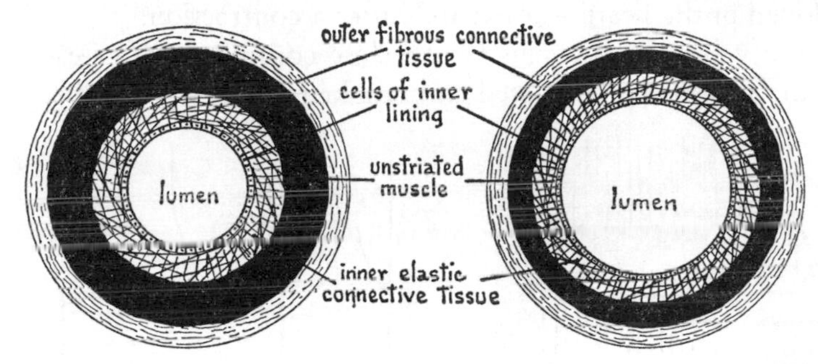

FIG. 106. Transverse Sections of an Artery and a Vein of the Same External Diameter

waves of expansion that are felt are called **pulses,** and pulse-rate is the same as heart-rate, which is about seventy times per minute in the average adult.

VEINS

The blood in veins always flows towards the heart. They begin as minute venules, which collect the blood from the capillaries; the venules unite to form small veins, which lead to larger veins until finally only four veins lead to the heart. They are the **two pulmonary veins,** which contain blood returning from the lungs, and the **two venae cavae, superior and inferior,** which contain blood returning from all other parts of the body. Veins have less muscle and less elastic tissue than arteries have and they exert little pressure on the blood (*see* Fig. 106). In addition, the blood pressure due to the heart and arteries has been considerably reduced in forcing the blood through the vast network of capillaries. Whereas the average arterial pressure is about 96 mm of mercury, i.e. the blood pressure will support a column of mercury 96 mm high, the average venous blood pressure is about 10 mm of mercury. Thus the blood flow in the veins is very sluggish and, to ensure that flow in the right direction is maintained, there are valves like small pockets, which allow the blood to flow one way only (*see* Fig. 107). To aid this sluggish flow, reduced pressure is developed in the heart as it expands after a contraction.

If the valves of the veins fail to close completely after each heart beat, the veins become dilated and crooked in their course because of

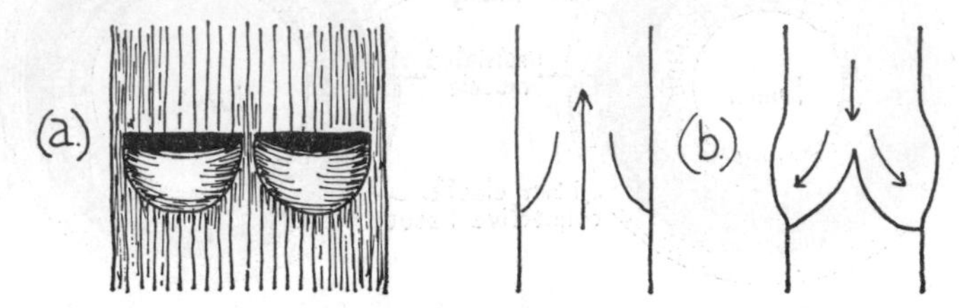

FIG. 107. (*a*) A Vein Opened to Show the Valves (*b*) Longitudinal Sections of a Vein, to Show How the Valves Maintain a One-way Flow

the increased volume of blood in them. This condition is known as **varicose veins**; it may become very painful. Veins in the back of the legs become varicosed most frequently; this may be due to frequent long periods of standing, to lifting of heavy weights, to pregnancy, or in some cases, the condition is inherited. Avoidance of long hours of standing in more or less one position, resting with the legs up and exercise of the legs, all tend to reduce the possibility of contracting varicose veins. Remedies include the wearing of elastic stockings, venous injections, or surgery, when portions of veins are completely removed, the blood flowing via other veins.

CAPILLARIES

These are the smallest of the blood-vessels, sometimes so narrow that the red corpuscles are distorted in shape as they pass through. Their diameter varies from 6μm to 12μm and they average about 1 mm in length. The walls have no muscle or elastic tissue but consist of a single layer of flattened cells (*see* Fig. 103, p. 183). Any constriction occurs in the arterioles, which precede the capillaries in the course of the circulation (*see* p. 196). With each pulse, the capillaries are dilated and their walls stretched. Under the increased pressure some of the fluid part of the blood is forced through the walls to circulate around the cells. The fluid, called **lymph,** differs from blood plasma in that it does not contain the blood proteins. The white corpuscles are also able to pass between the capillary wall cells and migrate into the tissues. The lymph is collected and restored to the blood in the lymphatic system (*see* p. 206).

The rate of blood-flow in the arteries averages 20 cm per second and in the veins 12 cm per second, but in the capillaries, the mean rate of flow is 0·5 mm per second. This astonishing decrease in velocity of the blood is due to the fact that the total cross-section area of all the capillaries is several hundred times that of all the arteries; it is comparable with the effect produced when a rushing river flows into a large lake; the velocity of the river water is soon considerably reduced. This decrease in velocity of the blood in the capillaries also allows more time for exudation of lymph into the tissue spaces.

Vasoconstriction and Vasodilation

The quantity of blood entering the capillaries is regulated by the arterioles. At the point where a capillary branches from an arteriole there is a small band of unstriated muscle called a **precapillary sphincter** (*see* Fig. 108). This is contracted or relaxed after stimulation by impulses carried by **vasoconstrictor** and **vasodilator nerves** from centres in the medulla called the **vasomotor centres.** Various stimuli, particularly rising external temperature, activate the vasodilation centre, and impulses along vasodilator nerves cause the precapillary sphincters to relax and hence more blood enters the capillaries. On the other hand, falling external temperature and other stimuli bring about the opposite reaction and blood-flow in the capillaries is reduced. This reciprocal innervation to the small sphincters controls loss of heat by sweating and is thus one of the main factors in regulating body temperature (*see* p. 54).

Most of the capillaries, especially those in the skin, can be completely closed by contraction of the precapillary sphincters, but in any network of capillaries there are always some in use, the outlets from the arteriole

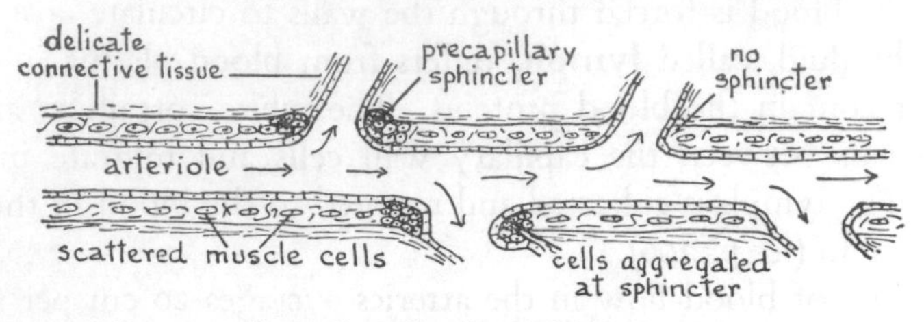

FIG. 108. Diagrammatic Section of an Arteriole, to Show Capillaries with Precapillary Sphincters and without Sphincters

lacking the sphincters. Otherwise the circulation would be completely blocked and the cells in the area would suffer severely (*see* Fig. 108).

The Heart

The heart is a muscular organ located in the mediastinum between and in front of the lungs (*see* Fig. 30, p. 45 and Fig. 95, p. 165). The apex of the heart is inclined downwards and to the left and the beat can best be felt between the fifth and sixth ribs, about 8 centimetres to the left of the mid-line. In size, the heart is about 13 centimetres long and 9 centimetres wide, about the size of a closed fist.

At the upper end of the heart are two thin-walled chambers called **auricles** or **atria** (singular, atrium), and beneath these are two thick-walled chambers called **ventricles** (*see* Figs. 109 and 111). The auricles receive blood from the veins; the ventricles pump blood into the arteries. Each chamber, when fully dilated, has between sixty and seventy cm^3 capacity.

The rate of beat averages seventy times per minute, for resting healthy young adults. This rate increases in the standing position to about eighty and with fairly strenuous exercise, such as riding a bicycle, the rate increases to about a hundred and twenty-five per minute. It is faster in children and slightly faster in old people (*see* table below).

Age	Rate of Heartbeat per Minute
At birth	140
3 yrs.	100
11 yrs.	90
21 yrs.	70
60 yrs.	75
70 yrs.	80

Other conditions, such as emotional excitement, disease, and extremes of external temperature, also affect the rate of beat.

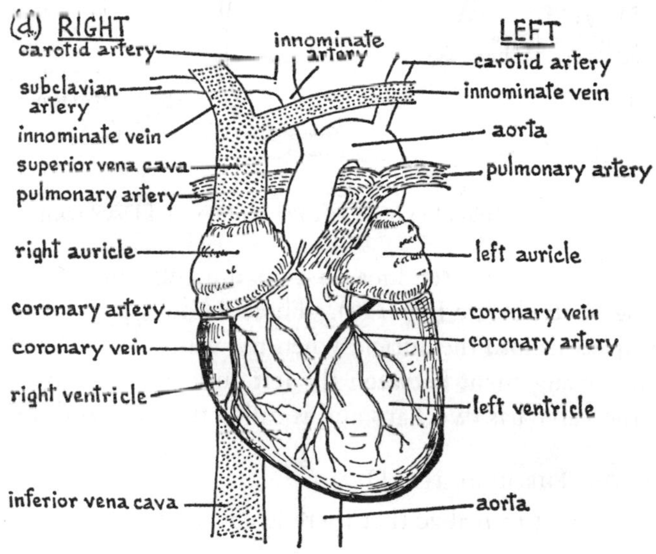

Fig. 109. Anterior View of the Heart and Great Vessels

As in the case of the lungs, the heart is surrounded by two sheets of peritoneum separated by a very thin layer of fluid, which reduces friction as the heart expands and contracts. This double sheet of membrane is called the **pericardium** and the fluid-filled space is the **pericardial cavity** (*see* Fig. 110). Outside the outer sheet of peritoneum, there is a tough, fibrous membrane, which firmly attaches the heart to the sternum and to the diaphragm.

Between the right auricle and right ventricle there is a valve consisting of three triangular flaps called the **tricuspid valve.** Its flaps

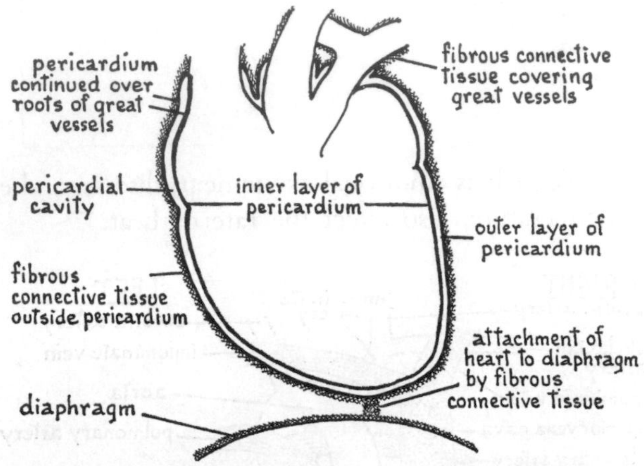

FIG. 110. Relationship of Heart, Pericardium and Diaphragm

are attached by slender tendons to two muscular projections at the base of the ventricle (*see* Fig. 111). These tendons prevent the valves opening upwards into the auricle when the ventricle contracts. There is a similar arrangement between the left auricle and ventricle, but in this case the valve has two flaps and is called the **bicuspid valve.**

Circulation Through the Heart

It is important to realize that there is never any empty space in the heart or indeed in the whole blood system. The heart does not fill and empty, then fill again. The action is best compared with squeezing a rubber bulb connected to a rubber tube at both ends (*see* Fig. 112), the whole system being full of water. Squeezing the tube at any point will not create any empty space but merely force the fluid into other parts. For instance, if the bulb is pinched at A, some water will be squeezed into B and D, and some into C. But if the region B is

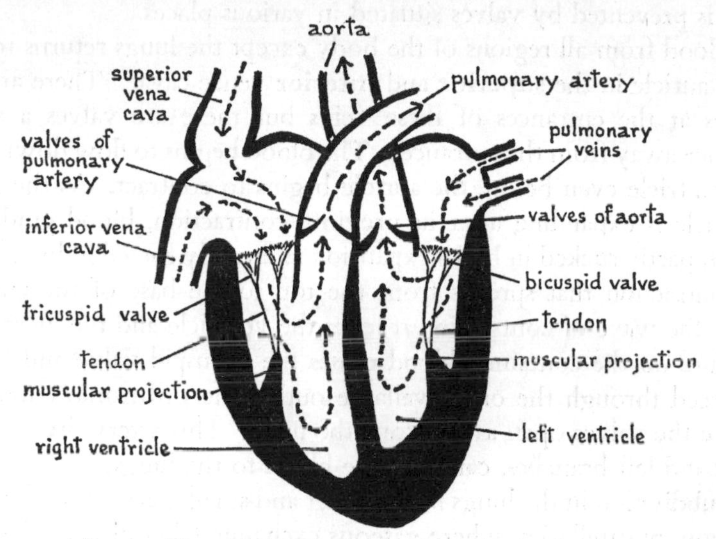

FIG. 111. Interior View of the Heart, Showing Chambers, Valves and the Course of the Circulation

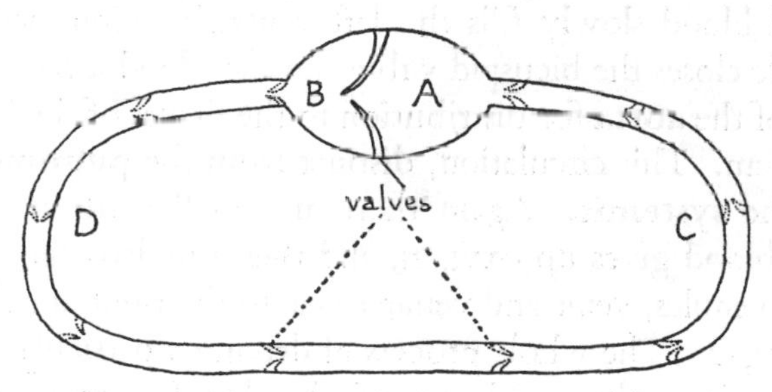

FIG. 112. Model Showing the Principles of the Blood Circulation

squeezed, closure of the valves will stop water flowing back into A, and it must therefore be forced along D. If now some valves are placed at the BD junction, opening from B to D, pressure on D can only force water round the system and back into A.

Like our ABCD system, the blood system is entirely full of blood all the time; it is merely squeezed from place to place and backward flow is prevented by valves situated in various places.

Blood from all regions of the body except the lungs returns to the right auricle in the **superior** and **inferior** venae cavae. There are no valves at the entrances of these veins but there are valves a short distance away from the entrances. The blood begins to flow down into the ventricle even before the auricle begins to contract. As the right ventricle is expanding after its previous contraction, blood gradually fills it, partly sucked in by its expansion and partly forced in by a wave of contraction that spreads from the top to the base of the auricle. Then the wave of contraction reaches the ventricle and the increasing pressure on the contained blood closes the tricuspid valve and blood is forced through the only available outlet, the **pulmonary artery,** where the valves open away from the heart. This artery divides into right and left branches, carrying the blood to the lungs.

Subdivision in the lungs into smaller and smaller arteries ends in the pulmonary capillaries, where gaseous exchange takes place. Then the blood is moved on into venules, veins and finally the two great **pulmonary veins** from the right and left lungs join the left auricle and the oxygenated blood slowly fills the left ventricle. Contraction of the left ventricle closes the bicuspid valve and the blood is forced through the valves of the **aorta** for distribution to the tissues of the body in the arterial system. This circulation, distinct from the **pulmonary** circulation, is the **systemic.** Again there are smaller arteries, capillaries where the blood gives up oxygen, and then the deoxygenated blood returns via venules, veins and venae cavae to the right auricle to begin the circuit again. The whole process of driving a particular portion of blood from the right auricle round the double circuit takes about sixty seconds.

This **double circulation,** i.e. **pulmonary** and **systemic,** was first evolved in some of the reptiles (it is present in crocodiles), and perfected in the birds and mammals. The circulation is here laid out again in brief form. (*See also* Fig. 113.)

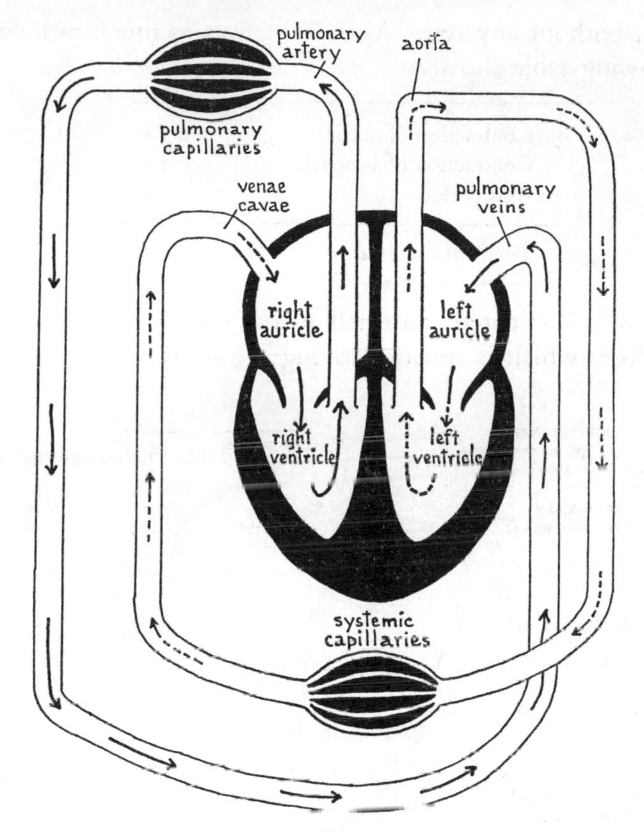

FIG. 113. Diagram Showing the General Course of the Blood Circulation

Venae cavae → right auricle → tricuspid valve → right ventricle → pulmonary artery → lung capillaries → pulmonary veins → left auricle → bicuspid valve → left ventricle → aorta → arteries → tissue capillaries → veins → venae cavae.

CONTROL OF THE HEART

The beat of the heart, which may be defined as the interval of time between one contraction of the auricles and the next, takes about 0·8 seconds. It must not be thought that the heart muscle is always working, without any rest. Actually it rests as much as it works, as the following table shows.

Contraction of auricles	0·1 s
Contraction of ventricles	0·3 s
Rest period	0·4 s
Total for one beat	0·8 s

Each beat is initiated at a small area of special tissue known as the **pacemaker,** which is situated among the heart muscle just beneath

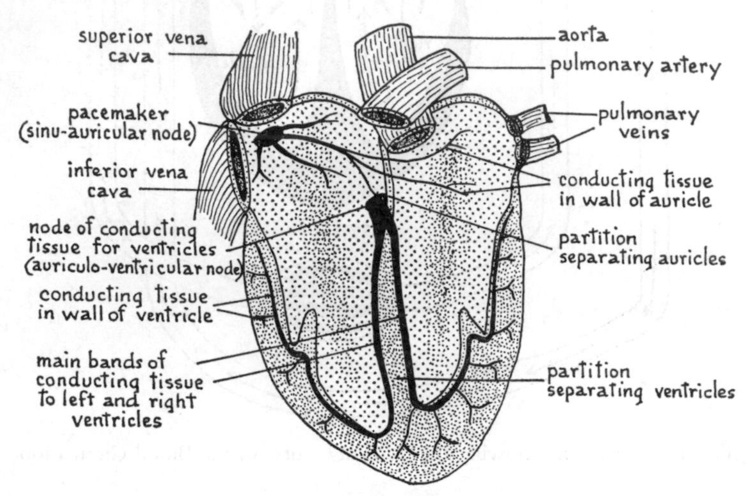

FIG. 114. Diagram of the Heart, with Front Wall Removed, Showing the Pacemaker and the Main Strands of Conducting Tissue

the entrance of the superior vena cava (*see* Fig. 114). From this pacemaker region special conducting tissue spreads out among the muscle of the auricles and excites another region, the **auriculo-ventricular node,** from which further strands of the conducting tissue spread into the muscle of the ventricles. The specialized conducting tissue and both nodes consist of primitive, elongated cardiac muscle cells.

Although the heart will continue to beat rhythmically after all nervous connections have been cut, it is constantly under the influence of nervous impulses that arise from a **cardiac control centre** in the medulla of the brain. There are **parasympathetic nerves,** which convey impulses that slow the heart-beat, and **sympathetic nerves,** which convey impulses that accelerate it. In normal healthy people, the rate is accelerated during exercise, by emotional excitement, during digestion and at high temperatures. Many diseases also cause acceleration of the rate of beat.

The Arterial and Venous Systems

The blood-vessels that supply the major organs of the body are shown and named in Figs. 115 and 116. Certain important points need special mention. The muscle of the heart itself is supplied by right and left **coronary arteries,** which open from the aorta close to the heart (*see* Fig. 109). The venous blood is taken away from the heart by **coronary veins,** which open directly into the right auricle.

The blood supply to and from the liver is different from that of any other organ. Arterial blood enters it from the **hepatic artery** and venous blood enters it from the **hepatic portal vein.** The blood in the artery carries oxygen to enable the liver cells to carry out respiration; the blood in the portal vein carries digested food materials from the gut for treatment in the liver (*see* pp. 151 and 152). The only exit for blood is by the **hepatic veins,** which take the blood into the posterior vena cava. The liver is the only organ in the body that has arterial and venous blood flowing into it.

The Lymphatic System

Apart from the cells that line the heart and the blood-vessels and small channels in the liver, the blood does not come into contact with the cells of the body. Instead, the cells are bathed in **lymph,** which exudes from the capillaries and is usually called **tissue fluid.** Lymph is very similar to plasma except that there is little if any of the blood protein in it. It is the actual medium of exchange between the cells

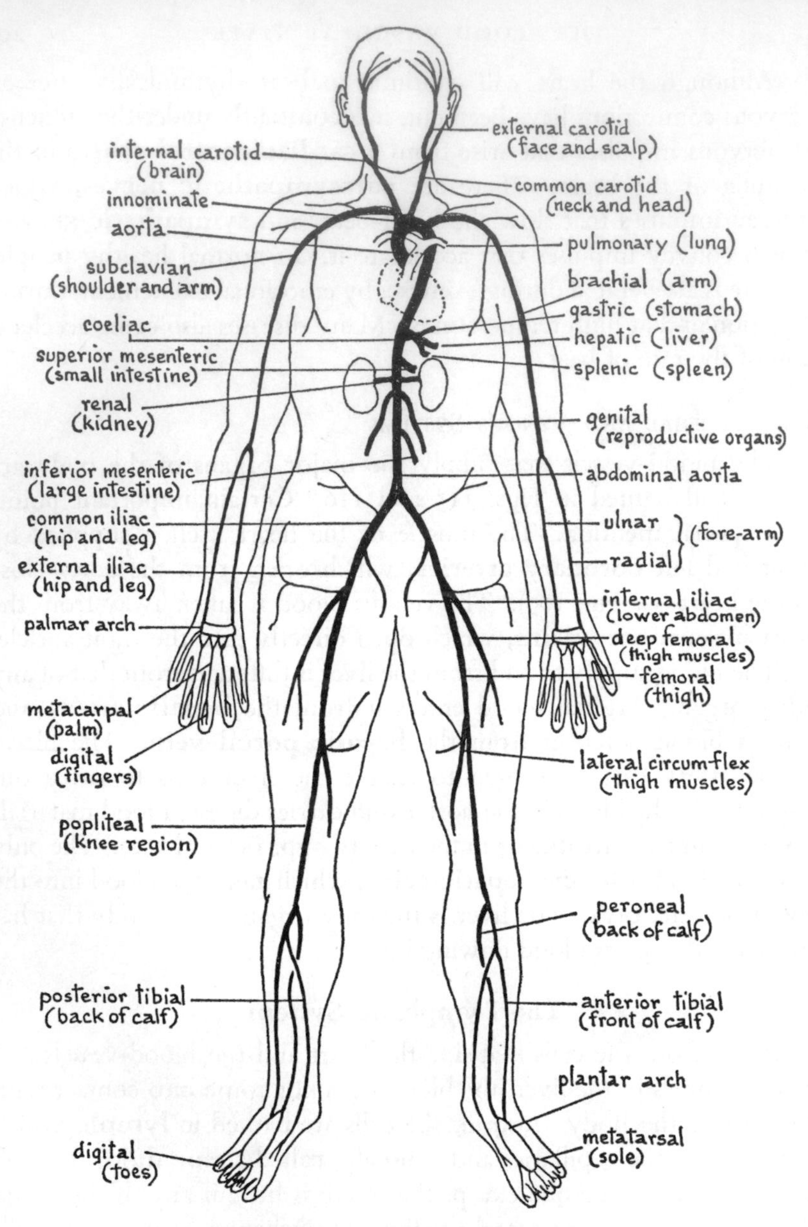

FIG. 115. Main Arteries of the Body

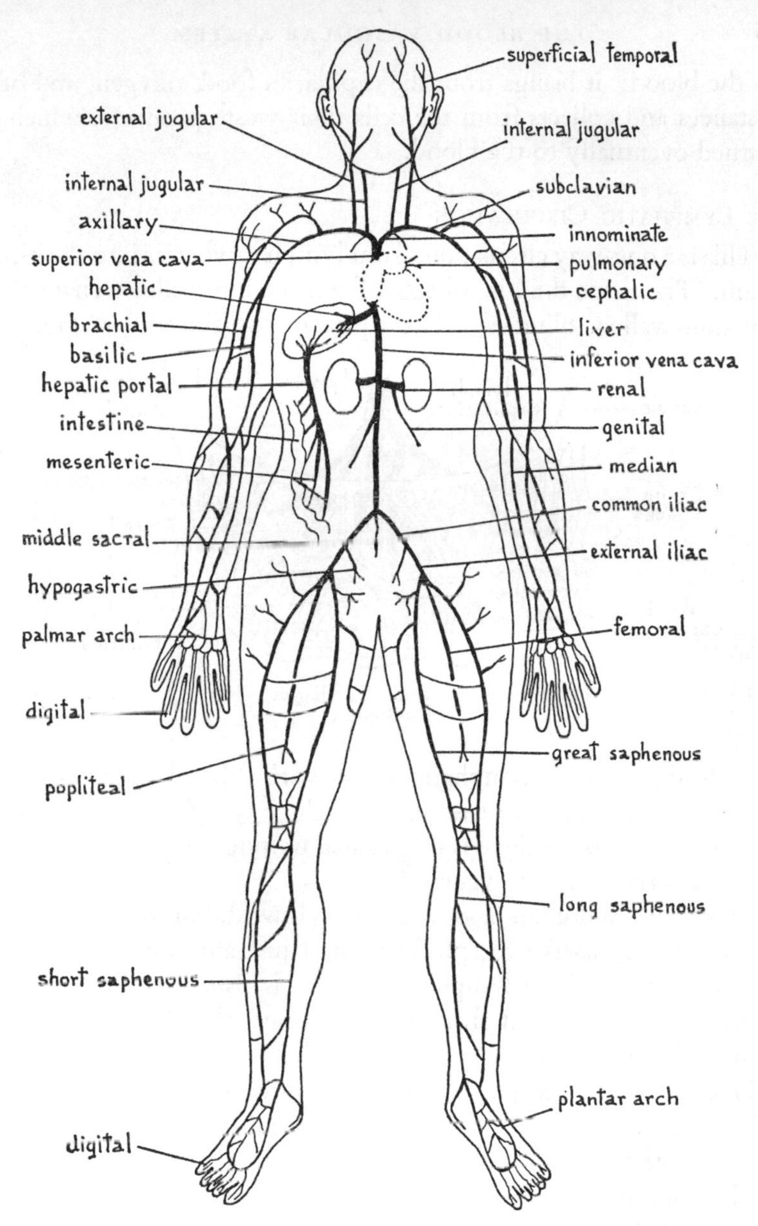

FIG. 116. Main Veins of the Body

and the blood; it brings from the capillaries food, oxygen, and other substances and collects from the cells their waste products, which are returned eventually to the blood.

THE LYMPHATIC CIRCULATION

This is a one-way circulation, from lymph capillaries into the bloodstream. The tissue fluid is collected from the intercellular channels by very thin-walled, blind-ended **lymphatic capillaries** (*see* Fig. 117),

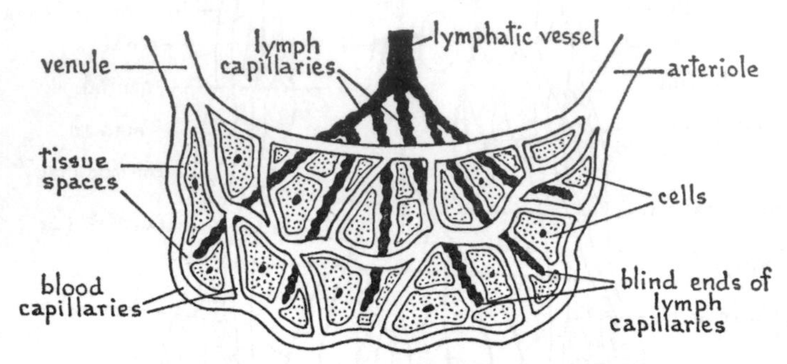

FIG. 117. Relationship between Blood Capillaries, Tissue Spaces and Lymph Capillaries

which lead into larger lymphatic vessels with valves like those of veins (*see* Fig. 107, p. 194). Vast numbers of these lymphatic vessels pursue independent courses; there is no gradual uniting into larger and larger vessels as is the case with veins.

From the abdomen and regions below it, all these lymphatic vessels enter a sac-like receptacle located just anterior to the second lumbar vertebra. The lymph in this sac is usually milky with the droplets of fat absorbed from the intestine; the fluid is known as **chyle** and the sac as the **chyle receptacle.**

From the chyle receptacle, the lymph flows up the **thoracic duct,** which lies just in front of the vertebral column, and is joined by numerous vessels from the thorax. Finally, this duct, joined by more vessels from the left shoulder and arm, opens into the **subclavian vein** just below the base of the neck. The right lymphatic duct collects the fluid from the right arm and shoulder, the right side of the thorax and the right half of the head and then enters the right subclavian

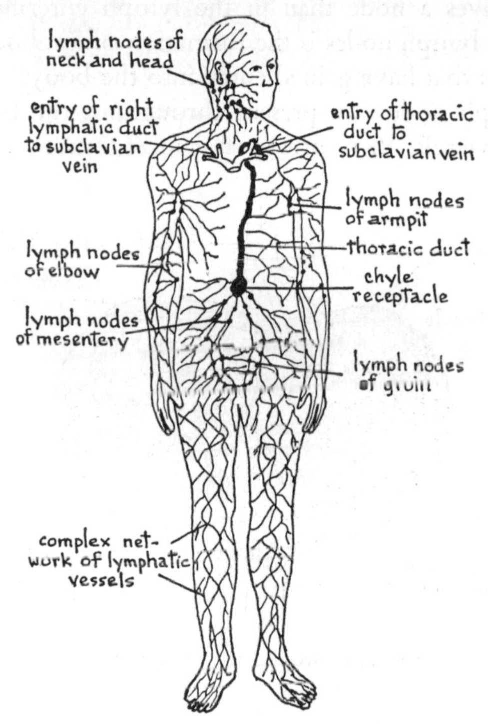

FIG. 118. The Lymphatic System

vein. Thus the lymph is returned to the blood. The main parts of the system are shown in Fig. 118.

LYMPH NODES (LYMPHATIC GLANDS)

Along the course of the lymphatic vessels there are numerous ovoid organs called **lymph nodes,** or lymphatic glands (*see* Fig. 119). Every vessel passes through at least one node. Each node has a framework of connective tissue and a maze of passage-ways, which are lined by phagocytes. These engulf bacteria or any other foreign matter in the lymph stream. Another function of the nodes is the manufacture of lymphocytes; there are always many more of these cells in the lymph that leaves a node than in the lymph entering it. Another function of the lymph nodes is the formation of antibodies, which act against antigens that have gained entry into the body.

While lymph nodes are present throughout the body along the course of the vessels, there are great concentrations of them in the

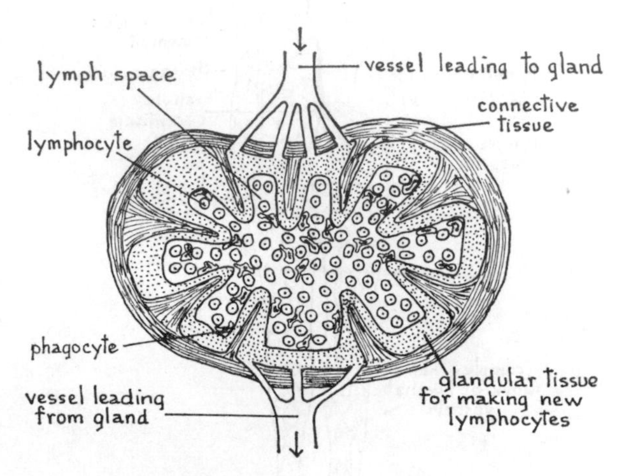

FIG. 119. Section of a Lymph Node

following regions—in the groin, around the armpits, in front of the elbow joint, at the sides of the neck and in the mesenteries supporting the gut. During severe infection the nodes become swollen and inflamed; most parents will be familiar with swollen glands in the neck of a child during certain diseases.

Apart from the lymph nodes, there are other regions of the body that contain a great deal of lymphoid tissue similar to that found in the nodes. The spleen, the thymus gland, the tonsils and the appendix are the most noteworthy. The spleen, besides its lymphoid functions, is also capable of holding a reserve supply of blood and is concerned with the destruction of worn-out blood corpuscles. In the lining of the small intestine, there are numerous small areas of white lymphoid tissue, known as **Peyer's patches.**

The lymph nodes and lymphoid tissues form a second line of defence against germs, the first defence being the action of phagocytes and lymphocytes at the point of entry. For example, in the case of a splinter in a finger, disease-causing germs may be introduced; if they succeed in establishing themselves, the germs are carried in the lymphatic circulation and there is swelling and pain in the lymph nodes of the armpit.

SUGGESTED EXERCISES

1.* Examine, on the highest power of the microscope or microprojector, a prepared slide of mammalian blood. Note how numerous the red corpuscles are in relation to the white, and the very small size of the platelets.

2.* Open the thorax of a freshly killed rat and display its contents. Note that the front of the heart is covered by a mass of soft pink tissue—the thymus gland. Make a large drawing (×4) showing the contents of the thorax.

3.* Carefully remove the thymus gland and locate the pulmonary artery and the aorta. Turn the heart forwards and note the inferior vena cava. Turn it to the left and note the right superior vena cava. Turn it to the right and note the left superior vena cava.

4.* Examine and draw in ventral view the heart of a sheep (this can be obtained from a butcher's shop). Examine in dorsal view and find the entrances of the venae cavae and pulmonary veins.

5.* Dissect the heart of a sheep by slicing away the thick ventral walls of the ventricles to expose the exits into the pulmonary artery and the aorta. Then slice away the ventral walls of the auricles to expose the bicuspid and tricuspid valves, the entrances of the venae cavae and pulmonary veins. This, and any other specimens desired, may be preserved in a five per cent solution of commercial formalin.

6.* Try to obtain from a butcher a piece of a large vein, about five or six inches. Cut it open longitudinally, clean out under the tap, and then pin it out in a wax dish or on a board. Examine the valves.

7.* Try to obtain from a butcher's shop a short length of a vein and an artery of the same external diameter, e.g. the renal artery and vein. Compare the thickness of the walls of these two.

8. Count your own rate of heartbeat (*a*) by feeling the beat of the left ventricle at the middle of the ribs on the left side, (*b*) by feeling with your first finger the pulse of the radial artery in the wrist. Compare the rates after a period of rest and after one minute of "running on the spot."

9. How many pulses can you find on your head and neck? What is the value of knowing the positions of these?

10. Work out the following blood routes in as much detail as you can: heart—kidney—heart; heart—small intestine—heart; heart—hind-limb—heart.

11. Make a permanent preparation of your own blood; see instructions in the appendix.

12. If a haemocytometer is available, make a count of your own red blood cells after diluting with Ringer's solution (*see* pp. 181–2).

CHAPTER 11

EXCRETION

SOME of the chemical changes that take place in the body have already been described. All the living cells are performing respiration with glucose and fat to replenish their stores of energy in ATP. In each case, the waste products are carbon dioxide and water. Cells in the liver deal with excess amino-acids, converting their nitrogenous part into urea, which is the waste product in this case. These, and all other chemical changes that take place in the body, are together called **metabolism,** which merely means change. As a result of metabolism, waste products may be produced that are of no value to the body, and indeed, if they accumulate, they may be very harmful. Therefore, they have to be eliminated, and this function is performed by special parts of the body called **excretory organs.** They are the **kidneys,** the **skin** and the **lungs.** Some excretory products are also passed into the gut, to be expelled from the body in the faeces.

The Urinary System

Urine is passed out of the body by a series of organs that together constitute the **urinary system.** All the materials eliminated by the kidneys together make up the **urine,** which passes from the kidneys down two narrow tubes called the **ureters** into the bladder (*see* Fig. 120). From the bladder, the urine is expelled through a single tube, the **urethra,** which opens at the tip of the **penis** in the male, and into the **vulva** in the female (*see* Figs. 128 and 129, pp. 224 and 227).

THE KIDNEYS

The kidneys are dark red, bean-shaped organs lying against the dorsal (back) wall of the abdomen, embedded in fat and covered by the peritoneum, which separates them from the coelomic fluid of the abdominal cavity. Each kidney is about 8 centimetres long and 5 centimetres wide; the right kidney is usually lower than the left, but

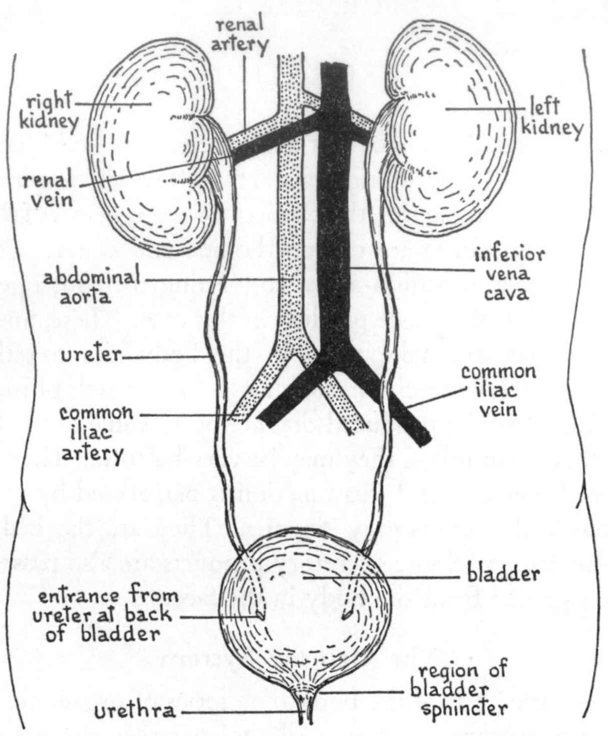

FIG. 120. Kidneys, Ureters and Bladder, Seen from the Front

roughly the centre-points of both are opposite the second lumbar vertebra.

There is a distinct indentation, called the **hilus** or **hilum,** at the middle of the inner border of each kidney; in this region, the renal artery and vein, and the ureter are located (*see* Fig. 121). In the hilus region, the ureter is considerably expanded to form a chamber called the **pelvis.** (Note, this is also the name of the hip-bone.) The renal artery and vein extend into the pelvis and branches of both supply the individual **pyramids.** These somewhat conical structures, of which there are eight to twelve, contain large numbers of **renal tubules,** which open into a central duct that has its outlet in the pelvis. These renal tubules are extensions of the functional units of the kidney, which lie in the outer cortex region. These functional units are known as **nephrons** and there are over one million in each kidney. With a hand-lens they can be seen as minute spots on the cut surface

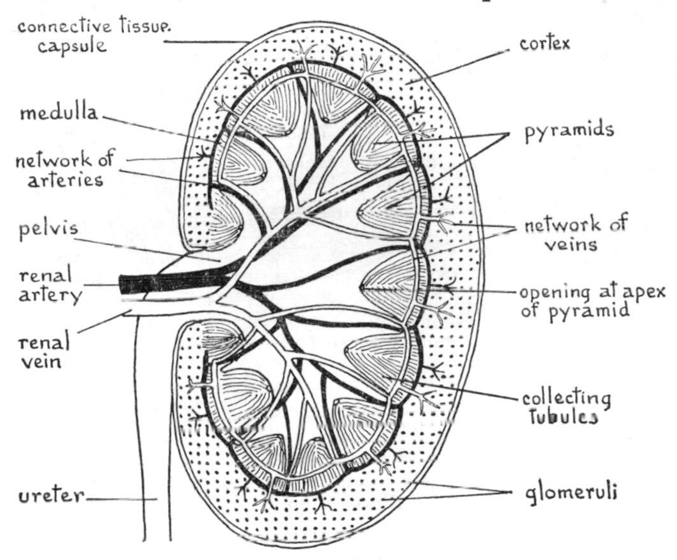

FIG. 121. Longitudinal Section of a Kidney

of a kidney. Enclosing the whole kidney is a tough capsule of fibrous connective tissue.

Each renal tubule (nephron) has a small, cup-like ending called a **Bowman's capsule,** which lies in the cortex. From the capsule, a fine duct, which is somewhat coiled, extends into the medulla and then back to the cortex again, the part in the medulla being known as the **loop of Henlé** (*see* Fig. 122). There is then a second coiled portion, after which the tubule joins a collecting duct, which is also joined by many other tubules. The collecting duct finally opens into the pelvis at the apex of a pyramid.

A small arteriole, which is a fine sub-branch of the renal artery, extends into every Bowman's capsule and leads to a network of capillaries called a **glomerulus** (little knot). Blood from the glomerulus flows out of the Bowman's capsule in another arteriole, which then divides up into a second network of capillaries around the first coiled portion of the renal tubule. Similar networks of capillaries surround the

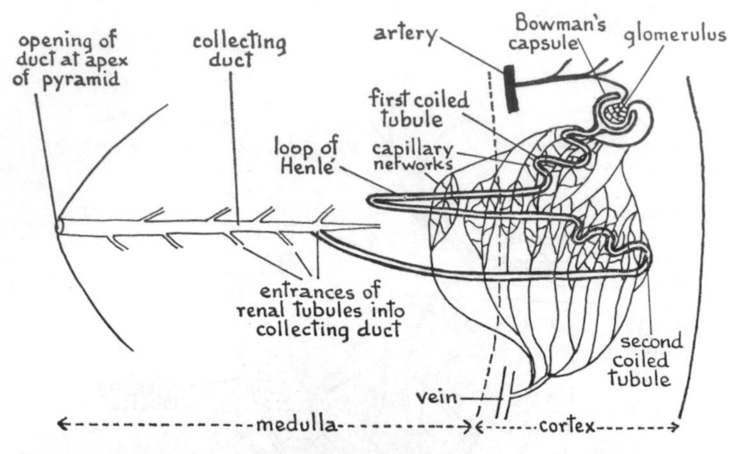

FIG. 122. Diagram Showing a Single Renal Tubule (Nephron) and its Blood Supply

loop of Henlé and the second coiled portion. All the capillaries lead to venules and finally the blood passes out of the kidney in the renal vein.

How a Renal Tubule Functions

Because the arteriole that enters a Bowman's capsule is larger in diameter than the arteriole that leaves it, a high blood pressure is developed in the capillaries of the glomerulus; it is actually three times the normal blood pressure found in capillaries. Under this high pressure, much of the blood plasma, except for the proteins, is forced through the wall of the Bowman's capsule into the cavity of the tubule (*see* Fig. 123). This fluid, called the **kidney filtrate,** then passes along the first coiled portion, the loop of Henlé and the second coiled portion. As it passes along, most of the water, the glucose, and the salts required by the body, are reabsorbed. By the time the fluid reaches the collecting duct, it is urine, consisting of about ninety-six per cent of water, two per cent of salts and two per cent of urea, with traces of other

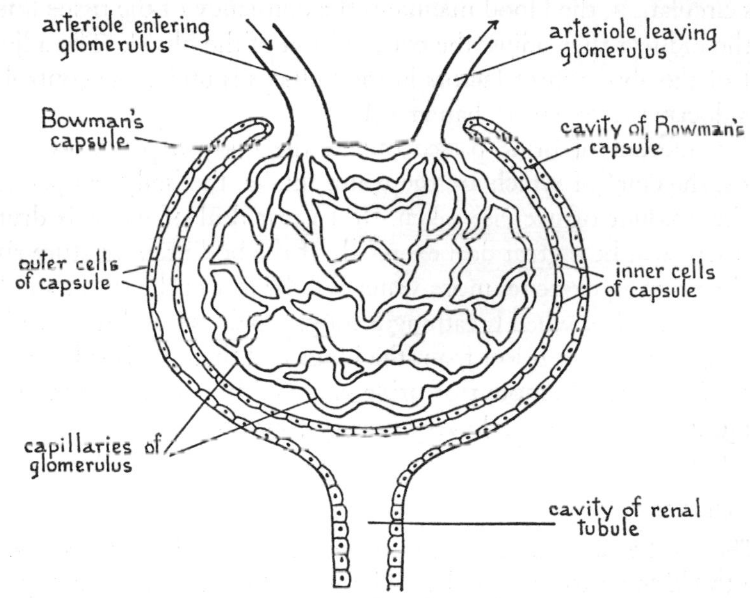

FIG. 123. A Bowman's Capsule and a Glomerulus in Section

nitrogenous substances; its pH is about 6. This urine is constantly passing into the pelvis, down the ureters and into the bladder, which gradually fills with the liquid. The outlet from the bladder is controlled by a sphincter, which opens when the muscular walls of the bladder contract, and thus, at intervals, the urine is forced from the bladder, along the urethra and so out of the body.

Functions of the Kidneys

The kidneys perform two very important functions, those of **excretion** and of **regulation of the composition of the blood.**

The excretory function consists of the removal from the blood of urea and certain other waste substances, all dissolved in water. Of equal importance is the function of regulation, which consists of maintaining these properties of the blood in a constant state—1. The quantity of water. 2. The pH. 3. The numbers and kinds of ions. In its circulation, the blood maintains the constancy of the tissue fluids, and the kidneys then adjust the composition of the blood. The adjustment of the above three factors in the kidneys is under the control of the endocrine system (*see* Chapter 15).

The amount of urine passed out of the body depends on many factors, the chief of which are the water intake, the body temperature and the amount of exercise taken. If a great deal of water is drunk, then there will be a great deal excreted. If the body temperature rises, as it does during exercise, more water will be lost in the sweat and less in the urine. Also, when breathing is more rapid, as it is during violent exercise, more water is lost from the lungs and there will be less in the urine. The normal amount of urine excreted by the average man, if resting, is about 1·1 to 1·4 l every twenty-four hours.

The Ureters

These tubes are about 20 to 30 centimetres in length, extending from the kidneys to the bladder. The wall of each ureter is composed of three coats: an outer connective tissue layer, a muscular coat and an inner mucous membrane. The muscular layer squeezes the urine along by peristaltic movements and it enters the back of the bladder by two slit-like apertures. These close when the bladder contracts and thus reflux of the urine up the ureters is prevented.

The Bladder

The bladder is a strong muscular sac lying in the cavity of the pelvic girdle. It has an outer connective tissue coat, then three layers of unstriated muscle, outer longitudinal, middle circular, inner longitudinal, and finally soft mucous membrane. Its capacity in an average male is about 500 cm^3 ($\frac{7}{8}$ of a pint) when fully distended. The outlet is normally closed by a sphincter, and as the bladder becomes distended with urine, slight warning contractions occur and finally there is complete contraction, the sphincter opens and the urine is expelled. **Urination** or **micturition** is a reflex act in babies but they learn voluntary control, normally by the age of three. Inability to establish such control is called **enuresis**; the nocturnal variety is by far the more common. **Excretion by the skin** is described on p. 54.

Excretion via the Gut

A number of excretory substances are passed out from the body by way of the gut. The most important of these substances are contained in the bile, which contains waste substances such as the bile pigments. They are derived from the breakdown of red corpuscles in the liver. The enzymes and the mucus that are produced along the course of the gut are also passed out with the faeces. The roughage of the food provides the bulk that keeps these substances moving towards the anus, where they finally leave the body in the faeces. Indeed, one of the evils of **constipation,** which is often due to lack of roughage in the diet, is that certain excretory products present in the gut are reabsorbed, with consequent ill-effects on the body.

SUGGESTED EXERCISES

1.* Dissect a rat to display the kidneys, renal arteries and veins, ureters, bladder, urethra, and penis. With forceps, clear away any fat that may obscure the kidneys or the blood-vessels. Carefully dissect out the ureters to the points where they enter the base of the bladder on its dorsal surface. Cut through the pelvic girdle in the mid-line. Press the knees outwards and downwards to widen the cut and pin them out. Then the urethra can be traced out to its exit at the end of the penis. Draw the dissection the full size of your page.

2.* Remove a kidney with a short length of ureter, renal artery and vein. Using a razor or a sharp scalpel, cut the kidney and, if possible, the other structures, into halves. Examine the cut surface with a hand-lens. Draw with a magnification of about six.

3.* Examine on a microscope or microprojector a prepared slide of a kidney section. Note, in the cortical region, the Bowman's capsules with their glomeruli, and in the medulla, the parallel rows of tubules, which are mainly loops of Henlé.

4. Kidneys of larger animals can usually be purchased from butchers, and they are very suitable for examination and dissection. Those of the cow, sheep and pig are all of suitable size.

CHAPTER 12

THE REPRODUCTIVE SYSTEM

THERE are two kinds of reproduction in living creatures, **asexual** and **sexual.** Asexual reproduction is common among plants and among lower animals. It consists of the detachment of a portion of the body

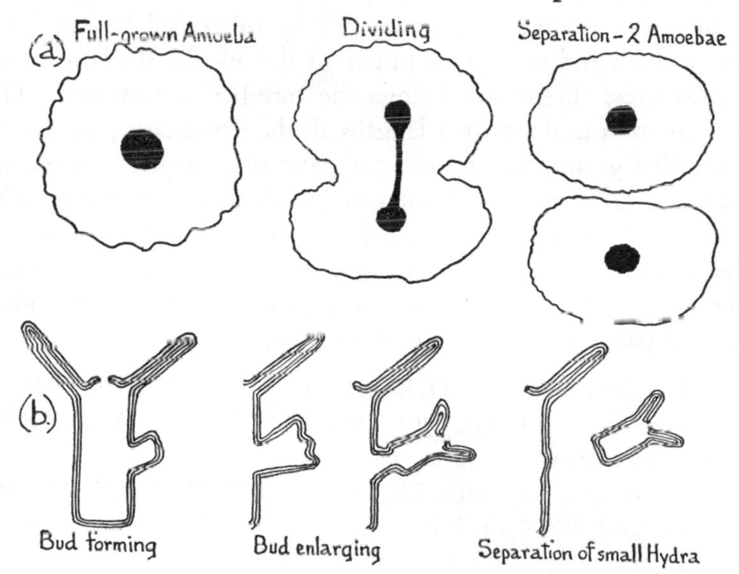

FIG. 124. Asexual Reproduction in (*a*) *Amoeba* (*b*) *Hydra*

of one organism, this portion giving rise to another creature of the same species. **Amoeba** and **Hydra** reproduce by this method (*see* Fig. 124). In *An Introduction to Living Things*, in this series, many examples of asexual reproduction from the plant and animal kingdoms are described.

All vertebrate animals reproduce by the sexual method and there is no asexual reproduction. There are two sexes, male and female, and each plays a part in the reproductive process. The sexes differ in their reproductive organs, and in the case of mammals, external differences usually make sex recognition comparatively easy. Each sex produces special reproductive cells called **gametes,** in organs called **gonads.** The male produces **spermatozoa** (sperm) in the **testes** and the female produces **ova** (eggs) in the **ovaries.** Testes and ovaries are familiar to most people as the soft roes (testes) and hard roes (ovaries) of herrings.

These gametes, sperm and eggs, differ from all other cells in the body in one very important way. Every other cell has within its nucleus a fixed and constant number of thread-like structures called **chromosomes**; in human beings the number is forty-six. These chromosomes bear along their lengths all the hereditary potentialities, usually called **genes,** which are aggregates of deoxyribonucleic acid. The chromosomes exist in similar pairs, so that there are twenty-three pairs. Thus if we call one particular chromosome A, then there is another A of the same shape, thickness and length. Males differ from females in the fact that they have twenty-two matched pairs and one unmatched pair. The arrangement is set out below—

Female: AA, BB, CC, DD, EE, FF, GG, HH, II, JJ, KK, LL, MM, NN, OO, PP, QQ, RR, SS, TT, UU, VV, XX (sex chromosomes).

Male: AA, BB, CC, DD, EE, FF, GG, HH, II, JJ, KK, LL, MM, NN, OO, PP, QQ, RR, SS, TT, UU, VV, XY (sex chromosomes).

Sex is determined by the possession of the XX pair (female) or the XY pair (male) (*see* later).

The special reproductive cells, gametes, have only one member of each pair in their chromosome complement, i.e. they have twenty-three chromosomes instead of forty-six, the number having been halved in a special reduction division that led to their formation in the testes and ovaries respectively. Thus male gametes (sperm) can have A, B, C, D, E, F, G, H, I, J, K, L, M, N, O, P, Q, R, S, T, U, V, X, or A, B, C, D, E, F, G, H, I, J, K, L, M, N, O, P, Q, R, S, T, U, V, Y and equal numbers of X-carrying, and Y-carrying sperm are produced. Female gametes can have only A, B, C, D, E, F, G, H, I, J, K, L, M, N, O, P, Q, R, S, T, U, V, X chromosomes.

When a sperm and an ovum unite, the process is called **fertilization** and the fertilized egg is known as a **zygote.** The twenty-three chromosomes of a sperm join with the twenty-three chromosomes of an ovum to give the full human number forty-six again in the zygote. If the sperm carried an X-chromosome, the zygote will develop into a female, but if the sperm carried a Y-chromosome, the zygote will develop into a male. The chances of a boy or a girl being born are theoretically equal (*see* p. 41).

If all goes well, the zygote will develop inside the mother's uterus until the embryo is sufficiently mature to be born. The new-born baby is thus partly a product of its father's sperm and its mother's egg; it has a mixture of paternal and maternal characteristics.

The Female Reproductive System

The female reproductive organs consist of the ovaries, in which the ova are formed, the **oviducts** along which they pass, the **uterus** in which the embryos develop, the passage (**vagina**) through which they pass at birth, and the outlet (**vestibule** and **vulva**) through which they finally emerge.

The Ovaries

The ovaries are solid flattened bodies each about the size and shape of a large almond; their length is between 25 and 40 millimetres their width between 19 and 21 millimetres and their thickness about 7 to 8 millimetres. They lie one on each side of the abdominal cavity, laterally placed just beneath the upper rim of the pelvic girdle. Each ovary is strongly supported by mesenteries (*see* Fig. 125).

An ovary is enclosed in a connective tissue layer, beneath which is a layer of cubical cells called the **germinal epithelium.** The cells of this layer are capable of successive division, like those of the skin (*see* p. 49), and the products of their division are passed inwards to become either actual ova, or cells that surround the developing ova (**follicle cells**). It has recently been shown that probably all the potential ova are already formed from the germinal epithelium before birth. During the age of **puberty** (thirteen to fifteen years), the ova

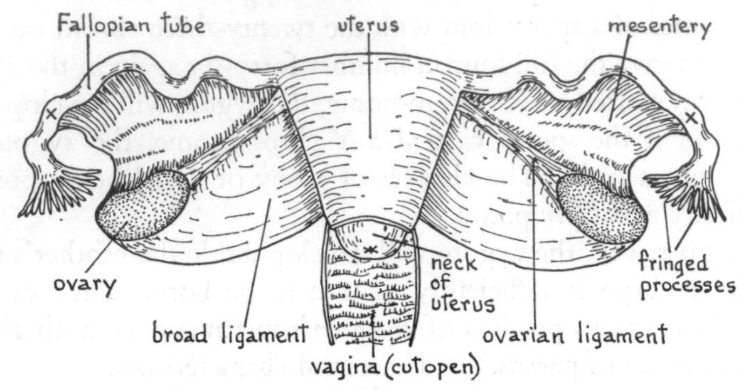

FIG. 125. Ovaries, Fallopian Tubes and Uterus
X marks the approximate position of fertilization.

begin to develop, having been stimulated by a hormone produced in the pituitary gland (*see* Chapter 15).

Each actual egg-cell becomes surrounded by follicle cells, and next, a split that appears in the follicle becomes filled with fluid. The whole structure, consisting of egg-cell, follicle cells and fluid, is known as a **Graafian follicle** (*see* Fig. 126). These are produced only by mammals.

By the time the follicle is mature, it bulges against the connective tissue on the outside of the ovary. The increase of fluid causes great pressure, which eventually bursts the connective tissue, and the egg-cell, surrounded by follicle cells, passes into the funnel of the oviduct. This process of releasing eggs from the ovary into the oviduct is known as **ovulation** (*see* page 223). The actual egg-cell is about 0·1 mm in diameter, but the whole structure, with the follicle cells, is nearly 1 cm in diameter (*see* Fig. 127).

The Oviducts

These are normally called **Fallopian tubes,** after their discoverer, Fallopius (Italian), 1523–63. They lie in a horizontal plane above the

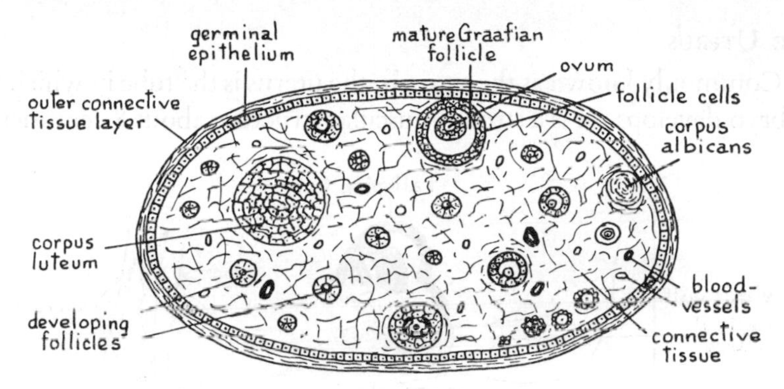

Fig. 126. Longitudinal Section of an Ovary

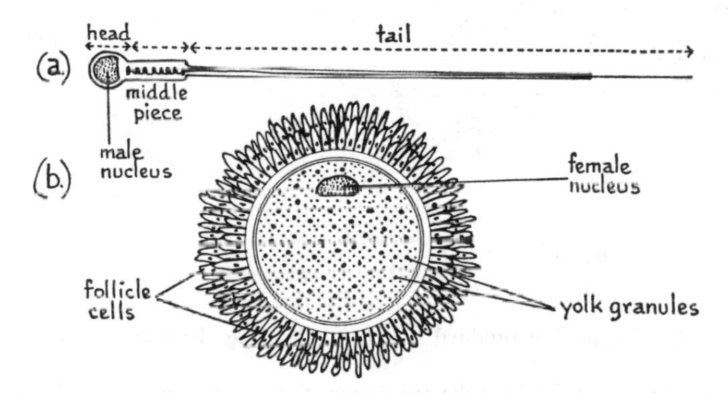

Fig. 127. (*a*) Human Sperm, × 1250 (*b*) Human Ovum with Follicle Cells, × 200

ovaries. Their open ends bear fringed processes, which lie close to the ovaries, and at the time of ovulation, these processes are erected to form a fringe that captures the eggs. If sperm are present, fertilization occurs well up the oviduct (*see* X in Fig. 125, p. 222). Normally only one egg is ovulated per month and probably the two sides alternate. An egg is moved along the Fallopian tube partly by the cilia that line its cavity and partly by peristalsis brought about by the muscular wall.

The Uterus

Commonly known as the **womb,** the uterus is the tube in which the embryo develops. It is a strong muscular structure about 8 centimetres

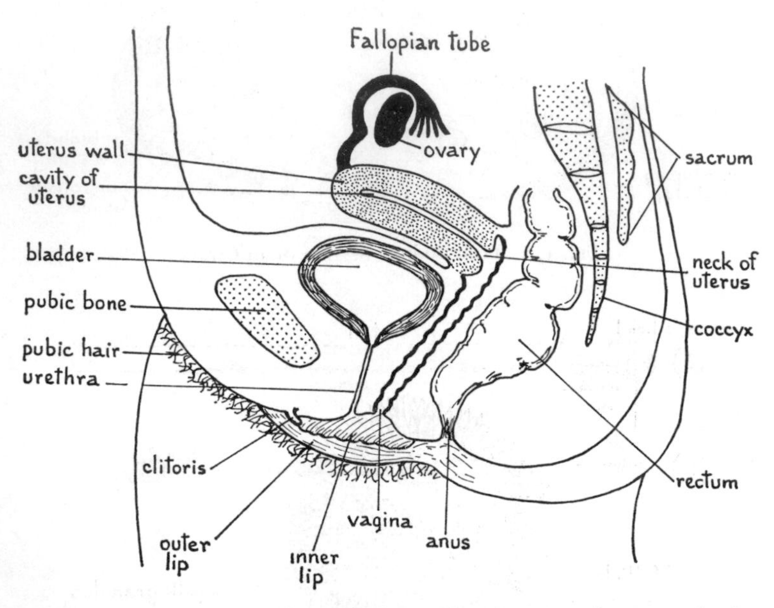

Fig. 128. Median Longitudinal Section, Showing Female Reproductive System
(One ovary and Fallopian tube are shown, though they are not in the plane of the section.)

long, 5 centimetres wide and 2·5 centimetres thick. After the first pregnancy it is somewhat enlarged. The uterus is capable of great enlargement during pregnancy when it extends high into the abdominal cavity. Normally it lies almost horizontally over the urinary bladder (*see* Fig. 128).

The Vagina

This is the passage extending from the neck of the uterus to the external aperture, a length of about 8 centimetres. Again, it is a muscular canal lined with mucous membrane and capable of considerable dilatation. It receives the penis of the male during sexual intercourse, the sperm being ejaculated near the neck of the uterus. During childbirth, it becomes greatly distended as the baby passes from the uterus to the exterior. In a virgin, the aperture from the vagina into the vestibule is partly closed by a thin fold called the **hymen.**

The Vestibule and Vulva

The external opening of the female genital tract is a long slit, the vulva, lying between two fleshy folds covered with hair, which is continuous with the pubic hair on the lower abdomen. Within these outer folds are two thinner folds devoid of hair; at their forward end, they close round a tiny protuberance, the **clitoris,** which is a vestigial penis. Finally, within these inner lips, the vagina and the urinary aperture open into a vestibule.

The Menstrual Cycle

After the age of puberty, females undergo a regular sequence of events in the ovaries and uterus, known as the **menstrual cycle** (Latin, mensis = a month). The month referred to is a lunar month of twenty-eight days. The sequence is set out below and explained later.

Days	Occurrence
1 to 4	Period of loss of blood
5 to 12	Development of follicle in ovary
	Growth of uterus lining
13 to 15	Ovulation
16 to 20	Breakdown of ovum
	Development of corpus luteum
21 to 28	Deterioration of uterus lining
	Regression of corpus luteum

Thirteen to fifteen days after the beginning of the loss of blood, a mature ovum leaves an ovary on the right or left side (rarely both). If the ovum is fertilized, the cycle will not recur until some months after birth. If the ovum is not fertilized, it will disintegrate in the next four or five days. The cavity left in the ovary after ovulation becomes filled with follicle cells and is called a **corpus luteum** (yellow body). This produces a hormone, which prepares the uterus to receive and implant the fertilized ovum. In the absence of fertilization, the cells of the corpus luteum deteriorate and the structure becomes filled with white connective tissue, being then called a **corpus albicans** (white body). The corpora albicantes (plural) remain as white spots in the ovary throughout life. During the last week of the cycle, the uterus lining deteriorates; cells become loosened and ready to be cast off and capillary blood vessels become very distended. The lining of the uterus is finally cast off gradually and with it portions of the capillaries, whose broken walls give rise to the bleeding. Within three or four days, the whole structure heals up; another follicle is developing ready for ovulation and the uterus lining is restored to its intact condition.

The menstrual cycle is controlled by various hormones produced by the pituitary gland, by the ovaries and by the corpus luteum. These are further explained in Chapter 15.

The Male Reproductive System

The male reproductive organs consist of the testes, in which the male gametes (sperm) are developed, a duct from each testis called a **vas deferens,** which expands into an **ejaculatory duct,** and the **urethra,** which extends through the organ of copulation called the **penis** (*see* Fig. 129). Along the passages from the testes to the outlet in the penis, there are several glands whose secretions contribute towards the **seminal fluid.**

The Testes

The testes are almost ovoid in shape, being slightly flattened laterally. Each is approximately thirty-eight millimetres long, twenty-five wide and nineteen millimetres thick. They are developed in the embryo in the abdomen, just beneath the kidneys, but during the later part of embryonic life they migrate slowly into the **scrotal sacs** behind the penis. As they migrate, their ducts and blood-vessels are pulled down with them, forming the **spermatic cord.** Descent into the scrotal sacs is normally completed before birth; failure of the

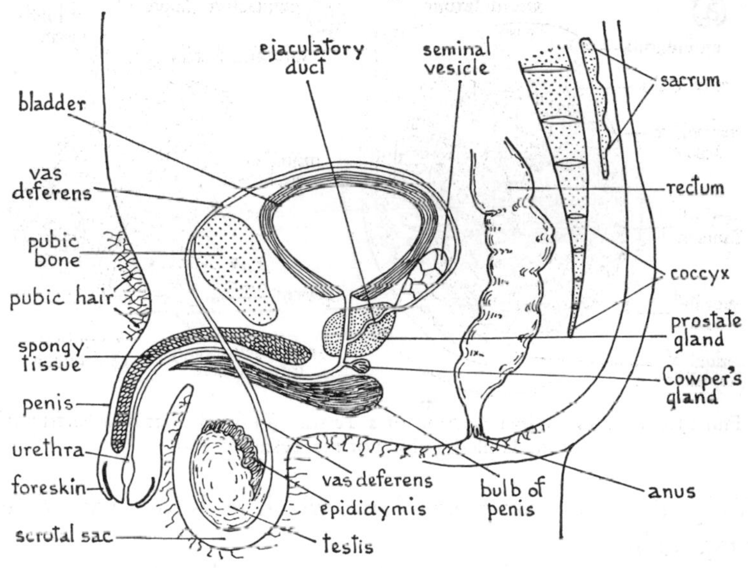

FIG. 129. Median Longitudinal Section, Showing Male Reproductive System

testes to descend results in inability to produce sperm that are capable of fertilizing the ova. It is considered that the sperm need a lower temperature for the completion of their development than that found in the abdomen. The temperature in the scrotal sacs is several degrees lower than that in the middle of the abdomen.

Each testis consists of compartments containing minute coiled tubules in which the sperm are developed from the germinal epithelium (*see* Fig. 130). Beginning at some point in the period of puberty (thirteen to fifteen years) the cells of the germinal epithelium continually divide. The inner cell, after each division, migrates towards the centre of the tubule, undergoing further divisions and changes, which result in the production of sperm. At this stage, the sperm are not motile, are not fully mature, and are not capable of effecting fertilization. They mature in a network of fine tubules called the **rete testis,** located in the upper region of the testis, and in a very coiled tube, the

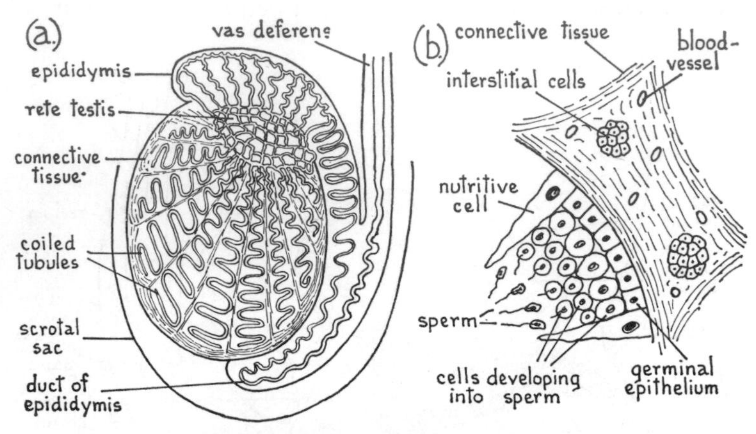

FIG. 130. (*a*) Longitudinal Section of a Testis, (*b*) Small Part of a Section of One Coiled Tubule of the Testis, × 150

epididymis, almost six metres long, which leads to the vas deferens (*see* Fig. 130).

Among the connective tissue that separates the tubules of the testis from one another, there are small groups of cells that produce the male hormones. These hormones stimulate the development of the male secondary sexual characteristics during puberty: the deepening voice, the male form of body and the growth of hair in various parts (*see* Chapter 15).

The Vas Deferens

Each vas deferens is a white muscular tube extending from a testis back into the abdomen, and down behind the bladder to end in a dilated muscular ejaculatory duct. The sperm lie in a clear, lymph-like fluid, which is squeezed along the vas deferens by peristalsis. Above the ejaculatory duct, on each side, there is a glandular **seminal vesicle,** which secretes a fluid that is thought to make the sperm capable of fertilization. During sexual intercourse, the muscular wall of the ejaculatory duct contracts, ejaculating (throwing out) the seminal fluid into the urethra. The right and left ejaculatory ducts open into the urethra among the substance of the **prostate gland.** The secretion of the prostate gland passes through twenty or thirty very small ducts that lead into the urethra; the fluid stimulates the sperm so that they begin to swim actively. Two small yellow glands about the size of

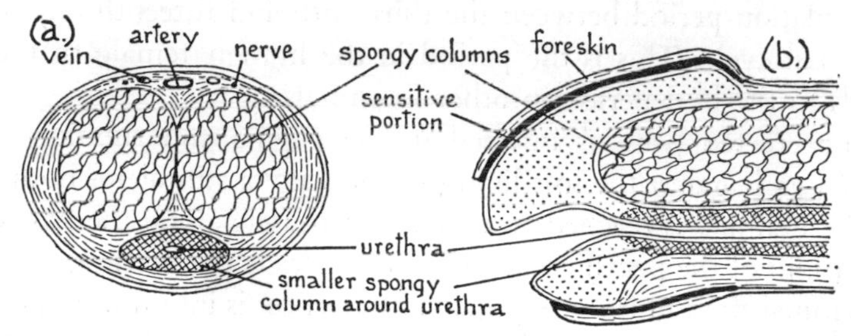

FIG. 131. (*a*) Transverse Section of the Penis in the Middle Region (*b*) Longitudinal Section of the End Portion of the Penis

peas, located near the base of the penis, also add their secretion to the seminal fluid. The secretion of these **Cowper's glands** partly precedes the ejaculation of the seminal fluid and serves to neutralize the acidity of the vagina, which would otherwise kill the sperm.

The Penis

This is the male copulatory organ, covered with skin continuous with that of the scrotal sacs. Internally it consists essentially of three longitudinal **spongy columns** of tissue, through one of which the urethra passes (*see* Fig. 131). Normally the penis is soft and hangs downwards, but during sexual excitement, these spongy cavities are dilated with blood and it becomes hard and erect. It is in this condition that it is inserted into the female vagina and its increased length results in the seminal fluid being ejaculated near the neck of the uterus. After ejaculation, blood is drained out of the spongy cavities into veins and the penis becomes soft again.

The sensitive portion of the penis is the smooth bulb at the end. This is covered by loose skin called the **prepuce** or **foreskin.**

Fertilization

Fertilization can occur within only a short time of six to twelve hours after ovulation. This means that sexual intercourse can normally be effective in the production of offspring only if it takes place within the ovulation period between the thirteenth and fifteenth days of the menstrual cycle. This is the period in the human female that corresponds to the heat period in other mammals.

The erected penis is inserted in the vagina and rubbing of the swollen end against the soft lining of the vagina stimulates both sexes. Finally a climax is reached and a tract of muscle along the base of the penis (bulb of the penis) contracts and constricts the urethra, squeezing the seminal fluid into the vagina. The ejection is intermittent, coming in several waves; it is followed by relaxation of the reproductive organs. A single emission of seminal fluid has a volume of about 3 cm^3 and it contains normally about three hundred million sperm; this seems to be a gross waste of substance when only one can fertilize the ovum. The sperm may remain active and capable of fertilization for about two days.

Males are sexually active almost throughout life, although there is some waning of sexual power in old age. Females cease to produce ova at about the age of forty-seven years, a time of life known as the **menopause.** Thus a woman's sexual life has an average span of about thirty-two years.

Venereal Diseases

Several serious diseases are acquired by sexual contact; they are known as **venereal** diseases. The most important of these are **syphilis** and **gonorrhoea.**

Syphilis is responsible for a great deal of human unhappiness; it has caused many people to become lifelong invalids and has been responsible for many premature deaths. The organism that causes the disease is a spirally coiled microbe, about 12 μm in length, known as *Treponema pallidum*. It can enter the body through a soft, moist surface, which is usually the surface of the penis in the male or the vagina or vestibule in the female. After a period of about three weeks, an ulcer develops at the point of entry. The ulcer may not be noticed if it is within the vagina. These ulcers contain millions of the spirochaetes and are highly infectious at this stage. After about six weeks, the ulcers disappear but the disease has not yet run its course. The spirochaetes enter the blood-vessels and spread all over the body. During the next six to twelve weeks, a syphilitic rash usually appears on the skin and many organs in the body may become inflamed and painful. These symptoms subside and the disease may become latent for a long period of time, often several years. The third stage of the disease is localized and one or more particular parts of the body become seriously damaged. The most severe cases occur when the brain, the heart, the spinal cord, or the skeleton are affected. In the case of the brain, the usual result is insanity, sooner or later followed by death. In the heart, there is serious interference with its action and the invalid is bedridden and unable to undergo even slight exertion. A common result of attack on the spinal cord is paralysis. These are only a few of the very serious consequences that may result. If a pregnant mother is infected, it is very likely that the baby will be born with **congenital syphilis,** which may not be evident for some years.

Syphilis can be cured if treatment is begun during the early stages of disease; the antibiotic **penicillin** has proved to be of the greatest value.

Gonorrhoea is caused by another type of bacterium, *Neisseria gonorrhoeae*. This is again acquired by contact and the organisms invade the genital tract of the female or the urethra of the male, causing inflammation of mucous membranes. Small ducts may be closed by the inflammation, resulting in sterility. Later, the bacteria spread throughout the body in the blood-stream and localized attacks begin, especially in the joints. The eyes can be affected from the hands or from a contaminated towel; this is a serious condition, which often results in blindness. Cases of new-born babies acquiring gonorrhoea of the eyes during the passage through the vagina were formerly common.

Venereal disease is on the increase and the cause is undoubtedly promiscuous sexual intercourse. There are cases on record of one infected woman transmitting syphilis to dozens of men and of one man infecting several women. It is important that people should be aware of these great dangers and that any person who suspects infection should seek medical advice without delay.

SUGGESTED EXERCISES

1. Examine, with microscope or microprojector, sections of a mammalian ovary and testis.

2.* Dissect male and female rats to display the reproductive system. To locate the outlets through the pelvic girdle, cut through the girdle in the mid-line and press the knees downwards until they lie flat on the dissection board. Pin down firmly and then cut away about a centimetre of bone on each side of the mid-line. Both male and female systems should be drawn.

Chapter 13

DEVELOPMENT

After fertilization in the Fallopian tube, the zygote begins to divide. At first, it is a single cell, but by the end of the first day after fertilization, several divisions will have taken place. The first division results in two cells, each of which then divides, forming an embryo of four cells. During the second day there are further divisions so that first eight and then sixteen cells are formed. These cells are progressively smaller, since no food material supplies the embryo in its passage through the Fallopian tube. It is subsisting on the minute granules of yolk that are scattered evenly throughout the egg (*see* Fig. 127, p. 223). These early divisions are known as **cleavage stages**; they are illustrated in Fig. 132.

The Blastocyst

The embryo absorbs fluid from the Fallopian tube and soon a cavity appears, separating an outer layer of cells from an inner mass. The outer layer is concerned with future nutrition of the embryo and is called the **trophoblast.** The inner mass will form the future baby and is called the **embryonic knot.** At this stage, the whole structure is known as a **blastocyst**; it is found only in the mammals (*see* Fig. 133). The blastocyst reaches the cavity of the uterus on the fourth or fifth day after ovulation and at this stage it consists of more than one hundred cells. Stimulated by the hormone from the corpus luteum in the ovary, the uterine lining gradually becomes thicker and softer. Glands in this lining secrete a milky fluid, which is absorbed through the cells of the trophoblast. This "**uterine milk**" nourishes the embryo.

Implantation

This term is used to denote the fixation of the embryo to the uterine wall. The process begins five or six days after fertilization and

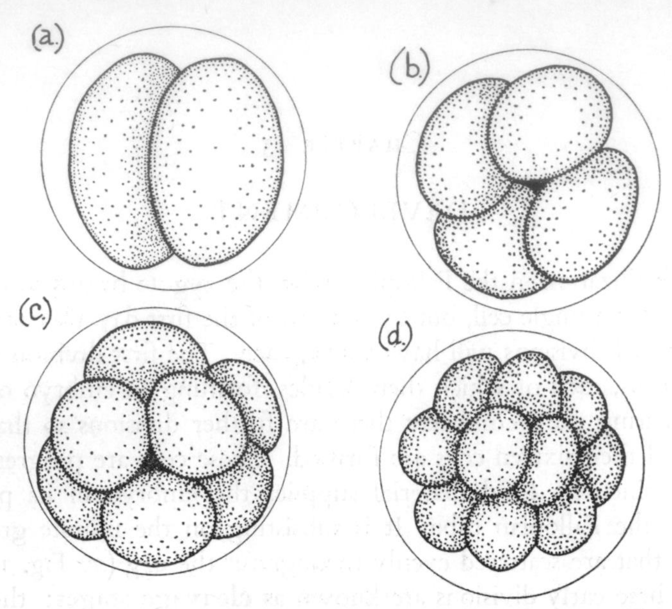

FIG. 132. Cleavage Stages of the Fertilized Egg (*a*) 2-cell Stage (*b*) 4-cell Stage (*c*) 8-cell Stage (*d*) 16-cell Stage

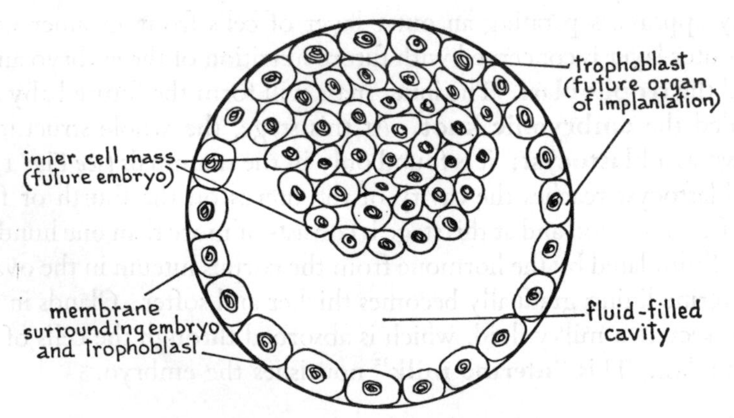

FIG. 133. Vertical Section of an Early Blastocyst

is rapid in human beings. It occurs partly as a result of burrowing activity by the embryo, probably loosening the cells of the uterus by enzyme action, and partly by folding of the uterus wall, which gradually covers the embryo.

The trophoblast becomes several cells thick and the outer cells secrete enzymes that destroy and liquefy uterine cells. The material formed is absorbed as food by the embryo. By the twelfth day after

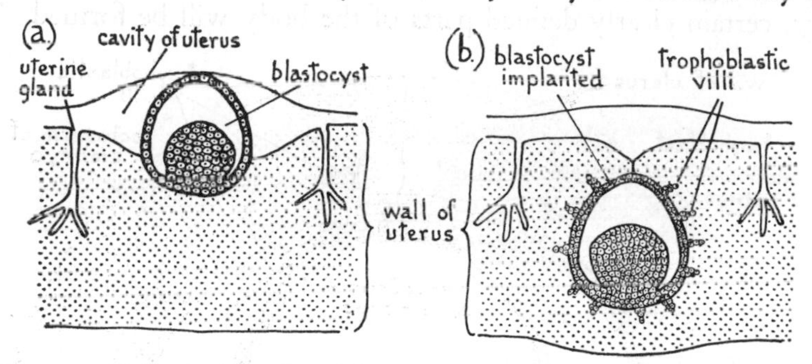

FIG. 134. (*a*) Beginning of Implantation (*b*) Blastocyst Implanted and Villi Growing into Uterus Wall

fertilization, the blastocyst is firmly implanted. Small projections like the villi of the intestine develop on the outside of the trophoblast; they serve to increase the area available for absorption of nutritious material and later some of these develop into part of the **placenta,** which will form the permanent attachment of the embryo until birth. The appearance of the blastocyst at this stage is illustrated in Fig. 134.

FORMATION OF THE GERM LAYERS

By the twelfth day after fertilization the embryonic knot becomes flattened by amoeboid movements of the cells. Then there is a clear separation into an upper and a lower layer of cells. The upper layer has been called the **ectoderm** (outer layer) and the lower layer the **endoderm** (inner layer). It is important to remember that cell divisions continue and that the number of cells is increasing all the time. The endoderm grows rapidly down into the cavity of the blastocyst to form at first a hemisphere and finally a closed vesicle called the **yolk sac** (*see* p. 240). All round the embryo between ectoderm and endoderm, a third layer of cells appears, probably derived by migration of cells from the other two layers. This third layer is known as the **mesoderm** (*see* Fig. 135).

These three layers, ectoderm, endoderm and mesoderm have been called the **germ layers,** meaning that from these germs or beginnings, certain clearly defined parts of the body will be formed. The

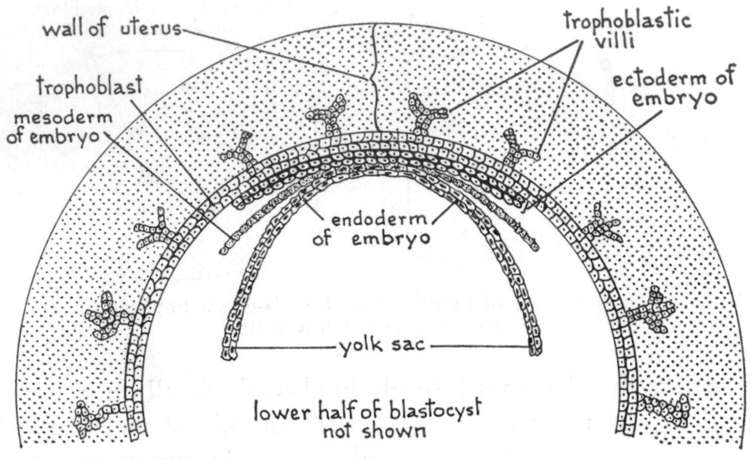

FIG. 135. Vertical Section of Blastocyst, Showing the Three Germ Layers Forming

ectoderm will give rise to the epidermis of the skin and to the central nervous system (brain and spinal cord); the endoderm will give rise to the inner layer of the alimentary canal and to the glands associated with it. All the rest of the body between the epidermis and the gut lining will be formed from mesoderm, i.e. all the skeleton, the muscles, the blood system, the excretory system, the reproductive system, etc. It should be evident that by far the greater proportion of the body is derived from mesoderm.

The embryo can now be visualized from above as a flattened pear-shaped region of ectoderm, which spreads out beyond the actual embryonic area (*see* Fig. 136(*a*)). Beneath this is a layer of endoderm growing outwards and downwards to form the yolk sac. The mesoderm lies between ectoderm and endoderm except in the central

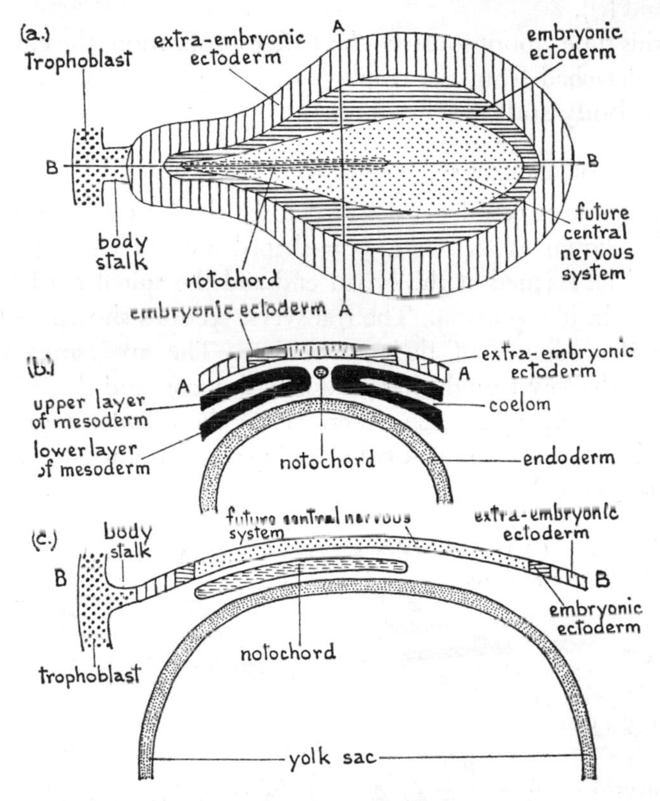

FIG. 136. Embryo, 12 to 14 Days after Fertilization (*a*) View from Above (*b*) Transverse Section at AA (*c*) Longitudinal Section at BB

region (*see* Fig. 136(*b*)). The mesoderm splits into upper and lower layers; the upper layer becomes applied to the ectoderm and the lower layer to the endoderm. The cavity between these two layers of mesoderm is the **coelomic cavity**; it will give rise in the future to the abdominal cavity, the pericardial cavity and the pleural cavities.

Beneath the area that will form the central nervous system (*see* Fig. 136(*a*)), a slender rod of cells, the **notochord,** is produced. Around this, the future vertebral column will be formed (*see* Fig. 136(*b*) and (*c*)).

By this time, about fourteen days after fertilization, the embryo has become detached from the trophoblast except for a posterior region called the **body stalk** (*see* Fig. 136(*c*)).

Formation of the Central Nervous System

The region of ectoderm destined to be the future central nervous system is shown as an inner, pear-shaped area in Fig. 136(*a*). The brain will be formed at the broad end and the spinal cord from the narrow, stalk-like portion. The transverse sections shown in Fig. 137 indicate the manner of their formation. The area sinks inwards, forming a shallow trough that gradually deepens, and the sides of this gutter eventually meet each other. Then the ectoderm grows over the tube thus formed (**neural tube**) and in this way the central nervous system is enclosed.

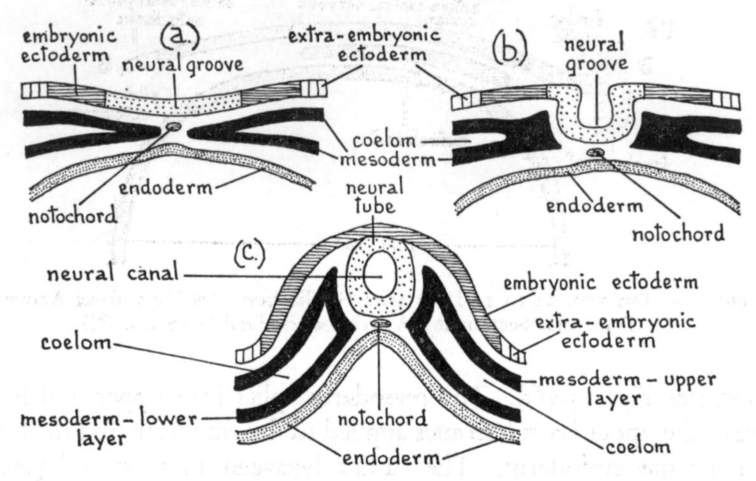

Fig. 137. Transverse Sections of Embryo, Showing Formation of the Central Nervous System (*a*) Early Neural Groove Stage (*b*) Later Neural Groove Stage (*c*) Neural Tube Formed and Enclosed in the Body

The Embryonic Membranes

It has already been indicated that the germ layers, ectoderm, mesoderm and endoderm, grow out into the cavity of the trophoblast beyond the confines of the embryo proper. These **extra-embryonic** (outside the embryo) **sheets** will form structures of which some are of great importance in future development. They are called **embryonic membranes** and are four in number: the **yolk sac,** the **amnion,** the **chorion** and the **allantois.**

The Yolk Sac

The yolk sac is formed at first of endoderm growing beneath the embryo and eventually forming a closed bag. Later, mesoderm covers the endoderm. In birds and reptiles, there are large, yolky eggs and the function of the yolk sac is to enclose and digest the yolk. At a later stage, blood vessels arise in the yolk sac and convey the digested food to the embryo proper. In most mammals, including human beings, the yolk sac remains very small and does not appear to perform any useful function (*see* Fig. 138(*a*), (*b*) and (*c*)).

The Amnion

The amnion consists at first of a fold of ectoderm, which arises all round the embryo, extends over the top of it and finally closes. Later, mesoderm becomes applied to the outside of the amnion. When fully formed it encloses the embryo in a clear watery fluid, which affords protection against injury (*see* Fig. 138(*a*) and (*b*)).

The Allantois

In reptiles and birds, the allantois is a very important structure. It grows out as a small sac from the hind gut, close to the yolk sac, and serves at first to contain the solid excreta (**uric acid**) of the embryo. In the later stages of development, it becomes very large and is pressed against the shell. Many blood-vessels develop in it and it serves as a

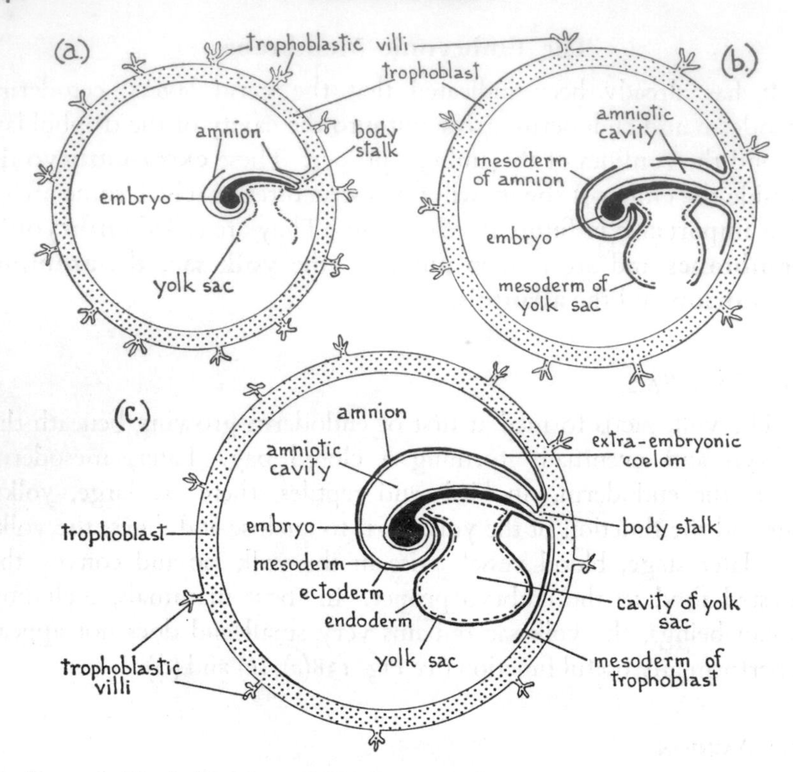

FIG. 138. Vertical Sections of the Blastocyst, Showing Development of the Amnion and Yolk Sac (*a*) 14 Days (*b*) 16 Days (*c*) 18 Days

respiratory organ in the exchange of gases with the atmosphere outside the porous shell. In the mammals, it is vestigial (*see* Fig. 139), growing out as a small sac into the body stalk.

THE CHORION

At an early stage in development, a layer of mesoderm is added to the inner surface of the trophoblast. This mesoderm penetrates the trophoblastic villi, which now become known as **chorionic villi.** It is usual to call the whole wall, i.e. trophoblast plus mesoderm, the chorion (*see* Fig. 139).

By the twenty-eighth day the chorionic villi have disappeared except in the region of the body stalk. The amnion has become much larger and its mesoderm has fused with the chorionic mesoderm. The important region is now the body-stalk region. Here the villi penetrate deeply into the uterine wall to form the placenta (*see* p. 242)

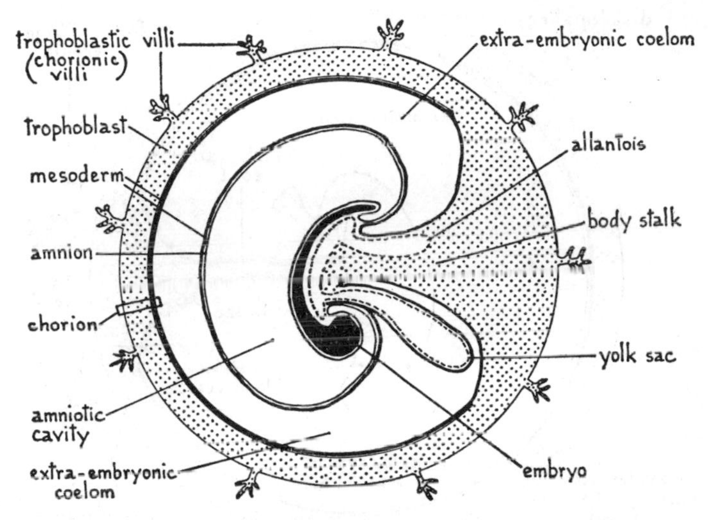

FIG. 139. Vertical Section of Blastocyst, Showing Allantois Growing into Body Stalk

and the body stalk is now called the **umbilical cord** (*see* Fig. 140). Through the placenta and the umbilical cord, all future exchanges between the uterus and the embryo take place. From this stage onwards, the embryo is known as a **foetus** and the membranes as the **foetal membranes.** It should be pointed out here that the early stages of human development are not known with great certainty, much of the information having been obtained from studies of closely related Primates, especially monkeys. It is fairly certain that human development does not differ greatly from that of the monkeys. From the beginning of the foetal stage, future development has been very thoroughly investigated.

The Placenta

The placenta arises from two sources, maternal tissue and foetal tissue. When mature, it is a disc about 20 centimetres in diameter and 2·5 centimetres thick. At twenty-eight days it is less than a millimetre

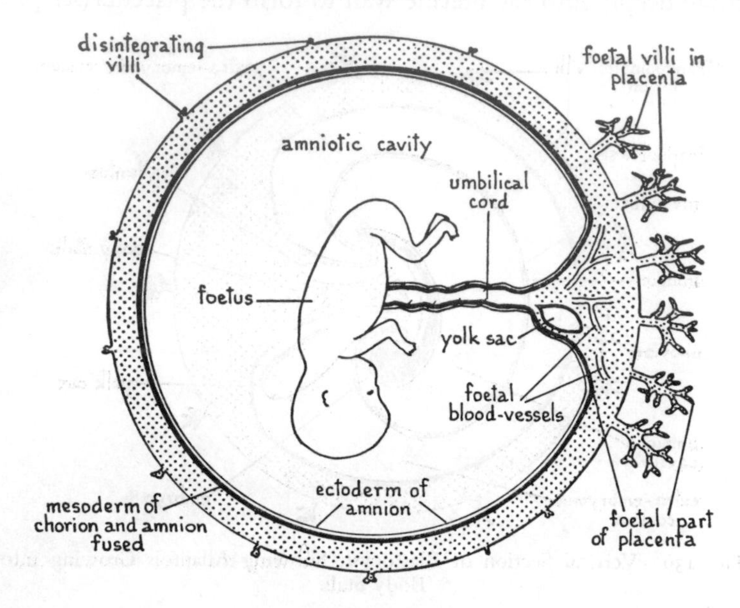

FIG. 140. Foetus of About 80 Days: Placenta and Umbilical Cord Well-developed

across. The uterine wall becomes greatly thickened and large blood spaces appear in it. Chorionic villi penetrate into these blood spaces, enlarge and become branched like trees. As the foetal blood system develops, capillaries are formed in the villi connected with two umbilical arteries and one umbilical vein. The blood of the foetus thus circulates to the placenta in the umbilical arteries, through the capillaries of the villi and back to the foetus in the umbilical vein (*see* Fig. 141).

Through the delicate walls of the villi and the walls of the foetal capillaries, exchanges take place between the blood of the mother and the blood of the foetus. It is important to note that the two bloodstreams are never in direct contact; the high maternal blood-pressure would burst the foetal blood-vessels. Thus the foetus obtains from the mother's blood digested food and oxygen. The excretory products of the foetus, mainly urea and carbon dioxide, pass into the mother's blood and are eliminated by the maternal excretory organs.

The membranes of the villi and the capillary walls also provide an effective barrier against bacteria, which may be present in the

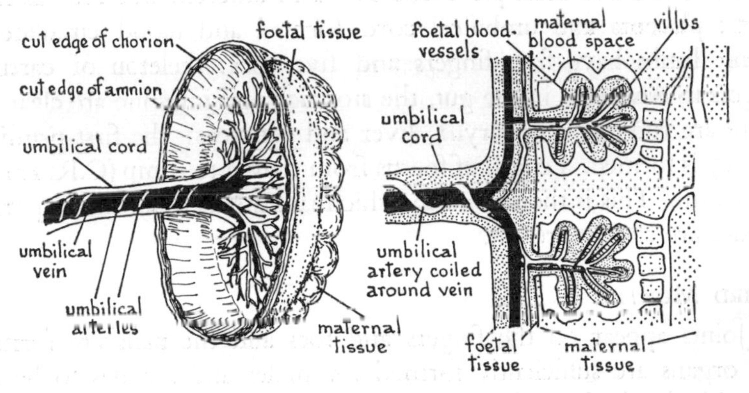

FIG. 141. (*a*) Placenta Just Before Birth, × 1/5 (*b*) Diagrammatic Section of Placenta

mother's blood. However, viruses such as that causing German measles, and certain microbes such as the spirochaetes of syphilis, can cross the barrier and affect the foetus seriously if not fatally.

Shortly after the baby is born, the placenta becomes detached from the uterus and forms the larger part of the **afterbirth** (*see* p. 248).

General Development

As the nervous system and the embryonic membranes are developing, other parts of the body are also progressing. It is not possible in a book of this scope to describe details of development but the main outlines are summarized below in the form of a monthly timetable.

First Month

Cleavage of the egg: formation of the blastocyst: implantation: notochord, amnion and yolk sac formed: neural groove closed: head and tail ends distinct: visceral clefts formed: heart consists of four chambers and blood circulation has begun: limb buds formed: length of embryo 2·5 mm.

Second Month

Eyes, ears and nasal pits formed: brain differentiated into its main parts: placenta and umbilical cord formed and blood circulates in them: limbs have five fingers and five toes: skeleton of cartilage begins to develop: in the gut, the stomach and intestine are clear and there are rudiments of larynx, liver and pancreas: the first primitive kidney is formed. Length of foetus from crown to rump (C.R. length) is 25 mm. This is the month in which rudiments of all the important organs are clearly present.

Third Month

Joints appear on the fingers and toes and the nails are formed: sex organs are sufficiently formed for males and females to be distinguished: the lungs begin to form as empty sacs at the base of the trachea: muscular tissue becomes well developed: bone development begins in the skull. C.R. length is 100 mm. Note that the greatest increase in length takes place in the second month.

Fourth Month

The skin is quite firm though thin: fine hair develops: sweat glands are forming: eyelids are fused together over the eyes: movements of the foetus are now sufficiently strong for the mother to feel them: this is known as the "**quickening.**" C.R. length is 145 mm.

Fifth Month

Much more hair is developed all over the body except on the head: the sebaceous glands are active and they secrete a white waxy substance (**sebum**) on the surface of the skin: for the first time, the legs almost equal the arms in length. C.R. length 190 mm.

SIXTH MONTH

The face assumes a more human expression: the skin becomes wrinkled and brownish-red in colour: the hair becomes strong and black: quite a thick layer of sebum all over the body. C.R. length 230 mm.

SEVENTH MONTH

The foetus now becomes quite plump due to the deposition of fat in the dermis: the eyelids are open: sweat glands are functional: growth of hair on the head: a foetus born prematurely at this time, or even a little earlier, may survive with careful nursing. C.R. length 265 mm.

EIGHTH MONTH

The skin becomes bright red and is covered with the waxy sebum: bone formation is well advanced: movements are very vigorous: the adult type of kidney is well developed. C.R. length 300 mm.

NINTH MONTH

Hair gradually lost from the trunk and limbs but remains on the head: the skin becomes paler in colour: the body is still plumper: if male, the testes descend into the scrotal sacs. C.R. length 350 mm.

Birth

Just before birth, the foetus has a chubby appearance with head relatively large compared with the rest of the body. The face is broad, the nose flattened and the cheeks bulge. The size of the face is quite small, the large head consisting mainly of the cranium. The jaws are very small and usually there are no teeth.

The neck is short and the thorax rounded, and the abdomen is quite distended, mainly because of the large size of the liver, which occupies almost half of the abdominal cavity. There is usually fine, dark, primary hair on the head but little on the rest of the body; this primary hair is shed within a few weeks after birth and replaced by secondary hair, which may not be the same colour as the primary hair.

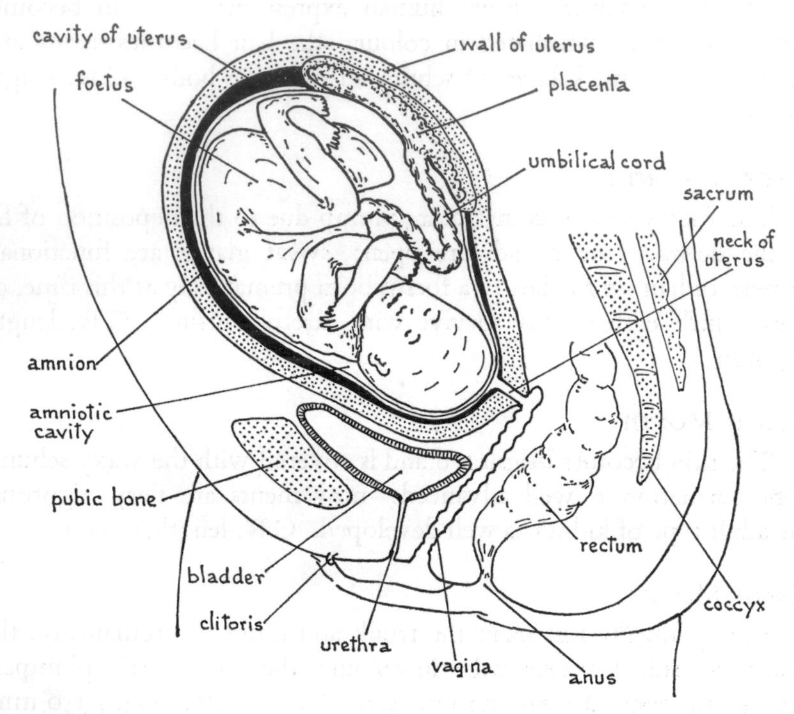

FIG. 142. Position of Foetus Just Before Birth

The normal position of the full-term foetus in the uterus is shown in Fig. 142. It lies with the head downwards near the neck of the uterus.

As the birth process begins, the neck of the uterus gradually widens and the amnion bursts, causing a rush of fluid down the vagina and out of the vulva. This is known as the "**breaking of the waters.**" The widening of the neck of the uterus is a slow process and as it occurs the foetus is slowly propelled through the neck by rhythmical involuntary contractions of the uterine muscular wall assisted by voluntary contractions of the abdominal muscles. When the contractions are in progress, the mother suffers the **"labour pains,"** which are due mainly to the stretching of the neck of the uterus as the head of the foetus is being forced through it. After the passage of the

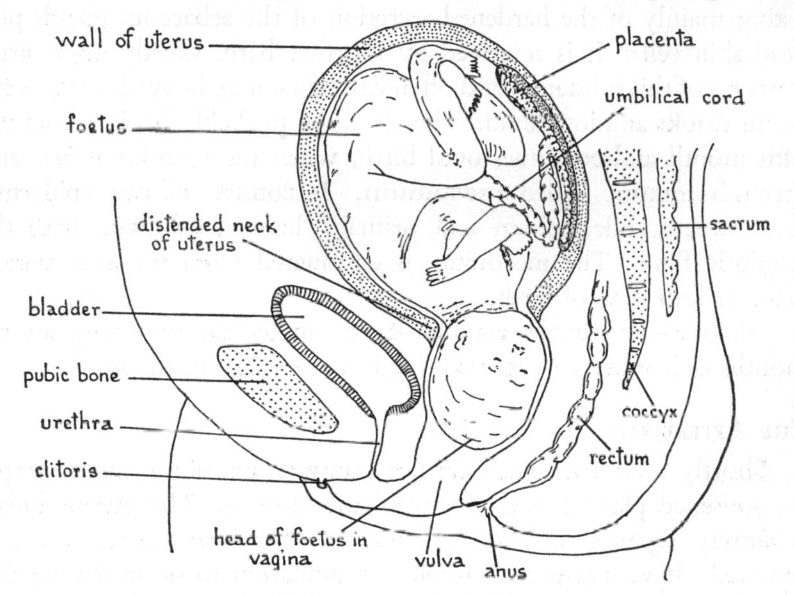

FIG. 143. The Birth Process: Head of Foetus in Vagina

head, the remainder of the body follows with less difficulty to the mother. The position as the head is in the vagina is shown in Fig. 143.

When the head is in view in the dilated vagina, it can be grasped by the nurse or doctor and the emergence of the foetus assisted by gentle pulling. Finally the foetus emerges, still attached to the placenta by the umbilical cord. This is tied tightly in two places near the foetal abdomen and cut between the two; the small portion left is bound against the abdomen by a bandage; it soon shrivels up to form the **navel** or **umbilicus.**

The emergence of the infant involves rapid changes in its mode of life. It has to breathe for itself and the first breath, if not taken quickly, is assisted by a slap on the back. The heart has to make rapid adjustment in view of the added work of the pulmonary circulation. Soon the infant must take its food by the mouth.

The new-born baby is covered with a white waxy material consisting mainly of the hardened secretion of the sebaceous glands plus dead skin cells; it is removed by the first bath, shortly after birth. Portions of this substance, and often cast hairs, may be swallowed as the foetus drinks amniotic fluid. This drinking probably begins about the fifth month and continues until birth, when the intestine is full of a greenish mixture called **meconium.** It consists of cast epidermal cells, mucus, bile, sebum and primary hairs, swallowed with the amniotic fluid. The meconium is eliminated from the anus within three or four days of birth.

All new-born babies are blue-eyed but by the time they are six months old, other colours may begin to develop in the iris.

The Afterbirth

Shortly after the birth, further contractions of the uterus expel the loosened placenta and the foetal membranes. The uterine lining is slowly repaired and about three months later menstruation is resumed. It is, however, possible for ovulation to occur during this period and pregnancy can thus be started again. In mammals other than human beings, the afterbirth is eaten by the mother; it is probable that the hormones of the placenta stimulate milk production.

Suckling

The mammary glands consist essentially of large numbers of **milk-secreting tubules,** though much of their size before pregnancy is due to accumulation of fat. Secretion of milk does not begin until the end of pregnancy, the process being stimulated by a hormone (prolactin) from the pituitary glands (*see* Chapter 15). The baby is adapted for sucking by the shape of its mouth and by an extra pad of fat in the cheeks and the strongly developed muscles of lips and cheeks. The milk, under suction, emerges from fifteen to twenty fine pores in each teat. The first fluid secreted is not milk but a watery yellowish fluid called **colostrum.** It has less fat, more salts than milk and at least twice as much protein, especially globulin. About three days after birth, ordinary milk begins to appear and gradually its composition approaches that of mature milk (*see* p. 124). Secretion of milk normally continues for several months; in some cases it may be continued for more than a year, and, provided the mother is healthy, it is undoubtedly the best food for the baby in its early stages. Soon, however, it will need food that is more concentrated, and it will need roughage. Many excellent cereal and vegetable foods are now readily available and diet charts are obtainable for the asking at the appropriate clinics or from any doctor. The gradual change-over from breast milk to other materials is known as **weaning.**

SUGGESTED EXERCISES

1. A specimen of a mammalian foetus attached to the placenta can be obtained from a biological dealer. Such a specimen should be examined with a hand-lens and illustrated. The specimens usually supplied are from a rabbit or a cat.

2. Various stages in development of the foetus can be examined if pregnant female rats are dissected.

Chapter 14

THE NERVOUS SYSTEM

THE conditions which exist outside the body constitute the **external environment,** which includes physical factors such as temperature, pressure, light, sound, humidity, air movements, chemical factors such as the amount of oxygen or carbon dioxide or other gases in the atmosphere, and biotic factors such as the effects of other living creatures. The conditions that exist within the body constitute the **internal environment,** which consists of the body fluids. Here again there are many factors concerned: temperature, pressure, pH, the quantity of water, the amounts and kinds of the various substances in solution, etc. Within each living cell, there is a third environment, the **intracellular,** which differs in different kinds of cells and even in different parts of the same cell.

We may consider that the body is engaged in an incessant struggle to overcome unsuitable conditions of the external environment and to maintain constancy in the internal environment. To do these things, it is necessary that there should be appreciation of changes in the environment. Any change that is big enough to excite the body into action is called a **stimulus.** Some stimuli, such as a rise or fall in temperature of 10°C, are readily perceived and we are evidently conscious of them; others, such as a slight rise in blood temperature, are also perceived, and action is taken by the body, although we are not conscious of this. The word "**perceive**" used in this sense means the ability to detect the change. For continued survival, any living creature must be able to perceive changes in its environment and it must be able to respond suitably. There are two distinct mechanisms in higher animals that accomplish these processes. They are the nervous system and the more primitive endocrine system (*see* Chapter 15).

The nervous system consists of three main parts, the **sense organs,** the **nerves** and the **central nervous system** (brain and spinal cord). The sense organs perceive changes (stimuli) and send impulses along nerves to the central nervous system. In the central nervous system, action is determined and impulses are sent out along other nerves to the organs that will make the required response.

The sense organs are **receptors** that are sensitive to certain kinds of stimuli; the organs that make responses are called **effectors**; they are muscles or glands.

The Sense Organs

There are five major sense organs with which we are able to perceive external stimuli. They are the **eyes,** the **nose,** the **ears,** the **tongue** and the **skin.** In actual fact, the perceptive area is only a small part of each of the organs. The senses associated with these five organs are those of **sight, smell, hearing, taste** and **touch.** Some of these can be subdivided; the sense of sight includes the perception not only of the intensity or brightness of the light, but also of the shapes of objects, their distances and their colour. In the skin, besides the sense of touch, there are senses of pressure, pain, heat and cold (*see* p. 53). In the ears, there are not only special organs for perceiving sound, but also organs known as **balance receptors,** which perceive movements of the head, and **position receptors,** which perceive changes in the position of the head. Besides these five major sense organs, there are countless numbers of minute receptors in the muscles (*see* p. 84), in the tendons, ligaments and joints, in the mesenteries, in all parts of the gut, in the walls of blood-vessels, and in fact, everywhere in the body. All these tiny organs are sensitive to changes in the internal environment. Only the major sense organs will be described here.

The Eye

The eye includes not only the actual eyeball, but the muscles that move it, the socket in which it lies, and the structures that protect it in front, the eyelids and eyelashes.

The eyeball is composed of parts of two spheres; a large part of a larger sphere at the back and a small part of a smaller sphere in front (*see* Fig. 144). It is encased in three layers, none of which is complete. The outer layer, called the **sclerotic,** consists of strong fibrous connective tissue, which is white and opaque except in front, where it

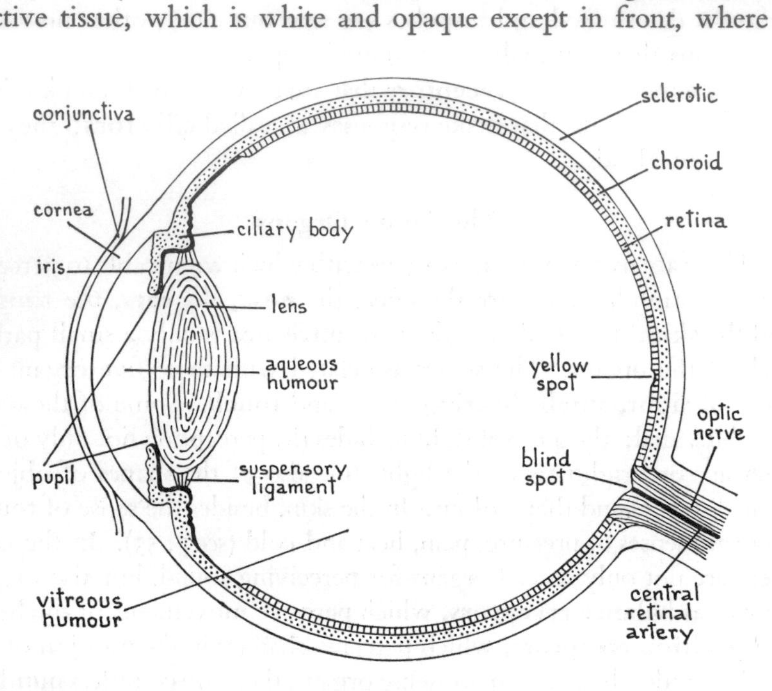

FIG. 144. Horizontal Section of the Eyeball

forms the transparent **cornea.** The sclerotic layer is incomplete where the optic nerve joins the eye.

The middle coat, the **choroid,** which is dark brown to black in colour, prevents internal reflection and contains many blood-vessels, which supply food and oxygen to the inner layer, the **retina.** In front, the choroid is thickened to form a ring of muscle called the **ciliary body,** and then continues forwards to form the coloured **iris,** which is pierced by a round hole, the **pupil.** Through the pupil, light enters the eye. From the ciliary body, a circular **suspensory ligament** encloses the **lens** and holds it in its correct position.

The inner layer, the retina, which ends behind the ciliary body, is the layer that is sensitive to light. From all parts of the retina, small nerve fibres pass towards the optic nerve and then extend to the brain. Where the optic nerve passes through the retina, there is no sensitivity to light; the area is known as the **blind spot.**

The suspensory ligament, holding the lens, divides the eyeball into two chambers. The chamber in front of the lens contains watery fluid called the **aqueous humour**; the posterior chamber contains transparent jelly called the **vitreous humour.** (The word "humour" was once used as a name for any fluid in the body.) The two humours keep the eyeball in its correct shape.

The eyeball is well protected from damage; it lies in a bony socket, the **orbit,** which is padded with fatty tissue. Over the cornea, there is a thin transparent skin, continuous with that which covers the inner surface of the eyelids. This skin, the **conjunctiva,** is a protective structure, shielding the cornea from damage. It contains blood vessels and many nerve-endings, and is very sensitive to irritation. Inflammation of the conjunctiva is known as **conjunctivitis**; the blood-vessels become dilated and plainly visible.

Above the eyeball in front there is a protective bony ridge; with the eyebrows, this wards off objects that might fall into the eye from above, it directs perspiration so that it drips away at the sides of the eyes, and shields the eyeballs from the direct rays of the sun. The eyelids can be closed to form a protective covering over the front of the eyeball; during blinking, the surface of the conjunctiva is washed by fluid from the **tear gland.** The eyelashes are stiff, protruding and protective hairs. Sebaceous glands at the bases of the eyelashes secrete a fluid that prevents the eyelids sticking together. Infection of one of these glands leads to a painful swelling called a **stye.** At the inner angle of each eye there is a fold of the conjunctiva called the **nictitating membrane** (*see* Fig. 145); in reptiles, birds and some mammals, it can be moved across the eye to act as another protective covering.

Above the outer angle of each eye, just beneath the eyebrow, there is a soft mass of tissue known as a **lachrymal,** or tear gland. The fluid secreted, known as the **tears,** contains small quantities of salts and mucin in solution. Although it is slightly antiseptic, its main function is to keep the surface of the conjunctiva moist. The fluid is passed through several ducts, and most of it is lost by evaporation, the remainder being drained off through two small ducts into the nasal

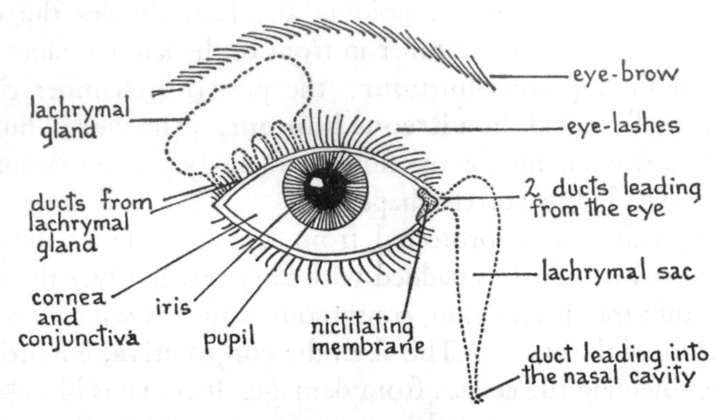

FIG. 145. Front View of Right Eye
The lachrymal structures, actually beneath the skin, are shown in broken lines.

passage (*see* Fig. 145). Irritation of the conjunctiva, and certain emotional states, increase the flow of tears. A young baby's eyes should be protected against strong light, winds and dust, because the glands are rarely functional at birth.

Muscles of the Eyeball

There are six small bands of muscle inserted in the sclerotic. Four of these are called **rectus muscles—superior, inferior, lateral** (external) and **medial** (internal), and the other two are the **superior** and **inferior oblique muscles.** The inferior oblique muscle has its origin anteriorly in the bone of the orbit near the lachrymal gland. It extends across the floor of the orbit and is inserted in the sclerotic underneath the lateral rectus muscle (*see* Fig. 146). The other five muscles have their origin in the bone at the back of the orbit, uniting to form a **ring tendon** that surrounds the optic nerve. The superior oblique muscle is inserted in the upper surface of the eyeball; it then extends inwards to pass through a ring of cartilage and then backwards to the ring tendon.

When all the muscles are relaxed, the eyeball is facing forwards. Contraction of one of the rectus muscles will cause it to face slightly

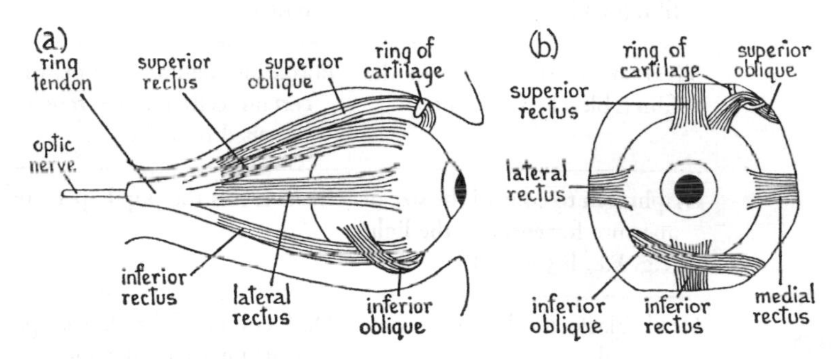

FIG. 146. Two Views of the Muscles of the Right Eyeball (*a*) From the Right Side (*b*) From the Front

upwards, downwards, right or left. The oblique muscles act in combination with one or more of the rectus muscles, causing the eyeball to roll. By suitable combinations of these six muscles, the front of the eyeball can be directed at any point in the circle of vision.

Formation of Images on the Retina

The way in which images are formed on the retina is very similar to the way in which photographs are taken with a camera (*see* Fig. 147), except that light is allowed to enter the camera for short periods and that only single photographs are taken. The similarities and the differences are listed in the table on p. 256.

When rays of light pass from one material to another, e.g. from air to water, they are bent; the bending is called **refraction.** The light that enters the eye is refracted five times before it reaches the retina, having passed through a number of different materials—the conjunctiva, cornea, aqueous humour, lens, and vitreous humour. The greatest amount of refraction takes place through the cornea, and the lens gives fine adjustment, so that in a normal eye, the rays are focused

	Simple Camera	The Eyeball
Similarities	Lens for focusing light on the film (or plate)	Lens for focusing light on the retina
	Silver salt in the gelatin of the film is bleached by light.	Purple pigment in protoplasm of certain cells of the retina is bleached by light.
	Diaphragm to control the size of aperture for entry of the light (e.g. F.8, F.5·6 etc.)	Iris controls the pupil aperture.
	Dead black inside to prevent internal reflection	Dark colour of the choroid prevents internal reflection.
	The image produced on the film is inverted.	The image produced on the retina is inverted.
Differences	Focusing by moving the lens in and out	Focusing by changing the shape of the lens
	Single pictures (Cine camera can take a series.)	Series of pictures

on the exact centre of the retina, where there is a slight depression called the **yellow spot.** The sensitive cells of the retina are called **rods** and **cones** (*see* Fig. 148). They are situated near the back of the retina and the light that reaches them has passed through all the refracting materials and through several layers of nerve cells and fibres. The rods contain a purple pigment, and the cones, a violet pigment, both of which are bleached by light. The pigment in the rods is extremely sensitive to light and is bleached very quickly when the light is strong.

Hence the rods are of little use in bright light but are of great value in poor light. For example, on a dark night, it takes at least half an hour

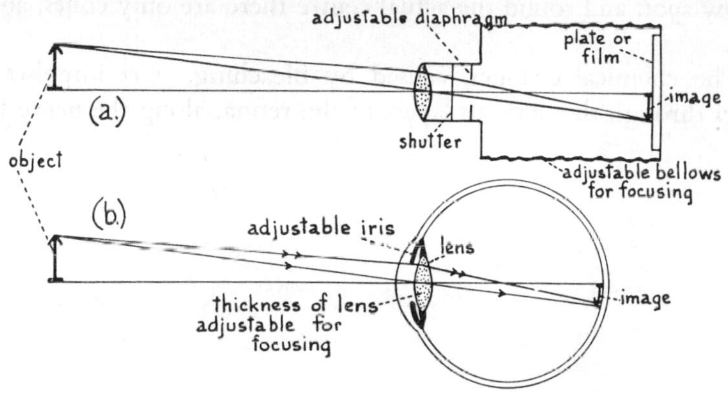

FIG. 147. Comparison between (*a*) A Camera (*b*) The Eye

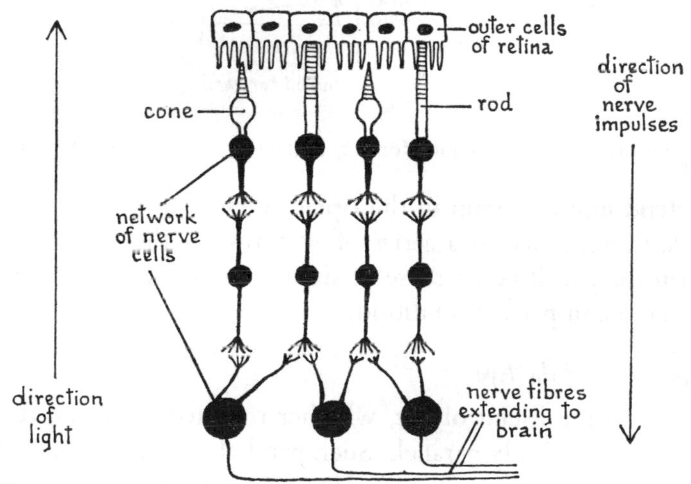

FIG. 148. Vertical Section Showing the Structure of the Retina, × 250

for the rods to reach maximum sensitivity, i.e. for the eye to become **dark-adapted.** The pigment in the cones is less sensitive and is bleached much more slowly. Hence for daylight vision and colour vision, the cones are used. It is significant that the number of cones compared with rods increases gradually towards the middle of the yellow spot, and round the actual centre there are only cones, and no rods.

The chemical changes, caused by bleaching, start impulses that travel through the nervous layers of the retina, along the nerve fibres

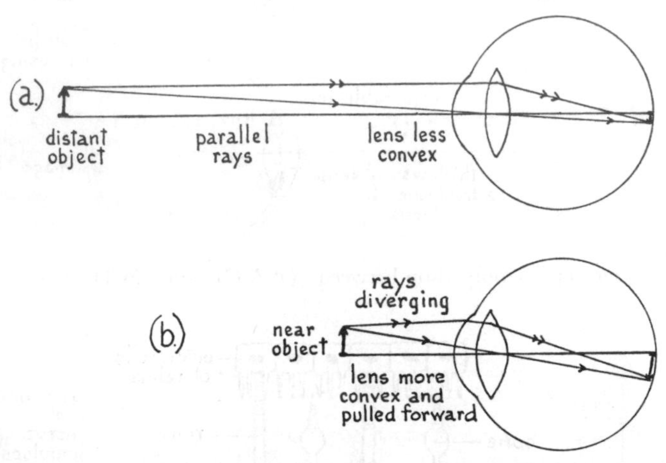

FIG. 149. Accommodation for (*a*) Distant Vision (*b*) Near Vision

that extend into the brain in the **optic nerve.** In the brain, the impulses are interpreted as a series of pictures. If the optic nerves are cut, then there will be no sense of sight, even though the eyes themselves may be in perfect condition.

Accommodation of the Eye

Light from a distant object, whether reflected or direct, travels in rays that are practically parallel. Such parallel rays can be brought to a focus on the retina by a convex lens with relatively little curvature (*see* Fig. 149). But when an object is near to the eye, the rays diverge and a more convex lens is necessary to focus them on the retina. This change in convexity of the lens is accomplished by the muscles of the ciliary body. When the ciliary muscles are relaxed, the suspensory ligament is tightened and the tension causes the lens to become less convex, as when it adapts for seeing distant objects. When the ciliary muscles contract, the whole ciliary body and choroid are pulled forward towards the cornea, the suspensory ligament is slackened and the lens will become as convex as its elasticity will allow, thus adapting for seeing near objects. It is this muscular tension that causes the eyes to become tired with long periods of close work. Looking at distant objects will relieve this tension, since the ciliary muscles then relax.

An object may be brought so close to the eyes that we cannot focus it sharply. The closest point at which an object can be clearly seen is called the **near point.** In children, the near point is very close to the eye; for children of ten to eleven years, its distance from the eye is nearly nine centimetres. As the age increases, the distance of the near point from the eyes becomes greater, and between the ages of forty and fifty there is a marked change; the near point may be fifty centimetres or more from the eye. When they are reading, most old people have to hold the books quite a long distance from the eye. The condition is known as **presbyopia**, or long sight of the aged. It is due to the gradual loss of elasticity of the lens, so that although the ciliary muscles are still able to contract strongly, the necessary accommodation cannot be effected. Correction can be made by using suitable convex lenses in spectacles—"reading glasses."

The **far point** is the distance beyond which we can focus without any contraction of the ciliary muscles. It is normally about six metres, i.e. from six metres to infinity, we can see without tension of the ciliary muscles. Acuteness of vision (**visual acuity**) is tested at six metres distance; the individual has to read rows of letters, each row smaller than the last one. At the end of each row is a number indicating the distance at which a normal (average) eye can read that particular size of type. Two examples are given below (not the actual size used on test cards).

(1) a e i d t p u v w 12

(2) l h b m n t d a c 6

A normal eye could read line (1) at twelve metres and line (2) at six metres and is graded 6/6. But an individual who can read line (1) but not line (2) would have less acute vision and would be graded 6/12.

Binocular Vision

When an object is viewed from a distance of more than six metres, there is very little difference between the images formed by the two eyes. But as the object is approached, and still kept in view, the eyes gradually converge, i.e. turn inwards, so that the two images formed are slightly different; the right image has more of the right side of the object, and the left image more of the left side. Both images are interpreted as one in the brain, the slightly differing images contributing towards a sense of distance, solidity and reality. Other factors that help in judgement of distance are the relative sizes of objects, colour, and shadows.

Common Defects of the Eye

Among the commoner defects of the eye are **myopia** (short sight), **hypermetropia** (long sight, but not that normal with increasing age—presbyopia), **astigmatism** and **cataract. Colour-blindness** is discussed briefly in Chapters 16 and 17.

In myopia, either the refractive power of the eye is too great, or the eyeball is too long. In either case, rays of light are brought to a focus in front of the retina and the image is blurred. Wearing spectacles with concave lenses will correct the condition, since the lenses will cause the rays of light to diverge slightly before they reach the eye (*see* Fig. 150(*a*)). In hypermetropia, either the refractive power of the eye is too little or the eyeball is too short; in either case, rays of light are brought to a focus behind the retina and the image is consequently blurred. The spectacles for this condition need to have convex lenses so that the rays are converged slightly before they reach the eye (*see* Fig. 150(*b*)).

Astigmatism is due to non-uniform curvature of the lens or cornea. Light rays passing through a greater curvature will focus at a shorter distance than rays passing through a lesser curvature, and again a blurred image will be produced. The condition is alleviated by using lenses that are parts of hollow cylinders, in which the greater curvature of the cylinder compensates for the lesser curvature of the cornea and vice versa.

Cataract is a more serious condition. In this disease, the lens, or a part of it, becomes milky-white and opaque. In bad cases, there is

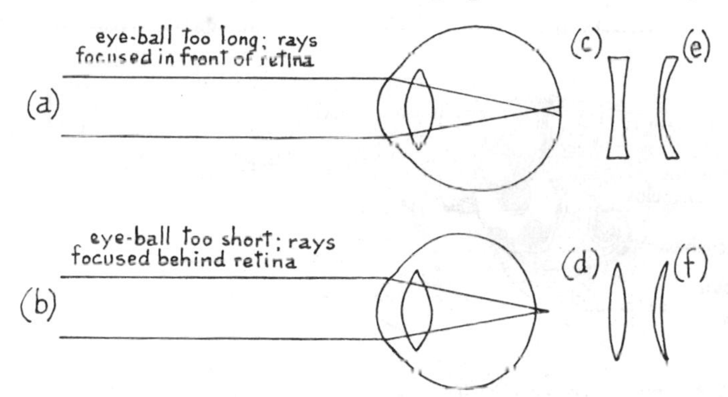

FIG. 150. (*a*) Myopia (*b*) Hypermetropia (*c*) Type of Lens Used to Correct Myopia (*d*) Type of Lens Used to Correct Hypermetropia (*e*) and (*f*) Modern Types of These Lenses

total blindness and the remedy is to have the lens removed. Subsequently there is, of course, no power of accommodation, so strong, bifocal spectacles are worn. The wearer is able to read by finding the best near point. Implantation of artificial lenses has been practised to some extent, with varying success.

THE EAR

There are three major parts of the ear: outer, middle and inner In fish, there is only the inner ear; in amphibians, the middle ear is added; in reptiles and birds, there is a trace of the outer ear in the form of a short tube leading to the ear-drum. In mammals, the outer ear is completed by an external flap called a **pinna.**

The Outer Ear

The pinna in man is of a peculiar shape and has lost much of the original function, which was to collect sound waves and reflect them down the external canal. This faculty is seen at its best in animals such as dogs and cats when they "cock the ears." The **external canal,** leading to the ear-drum, is about 2·5 cm long and curved slightly upwards in its middle region (*see* Fig. 151). Close to the external

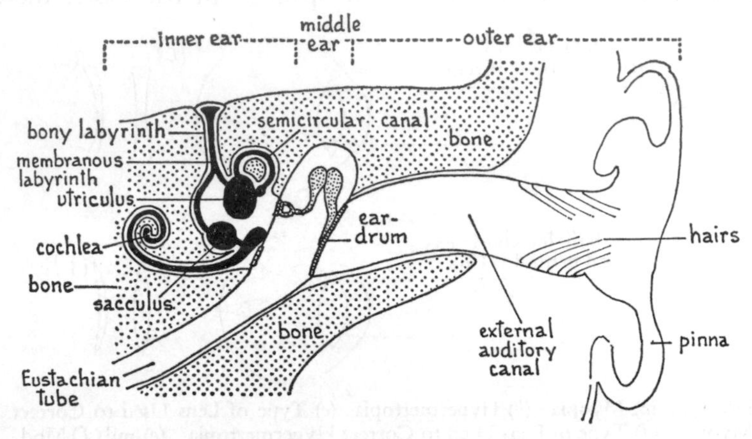

FIG. 151. Diagrammatic Vertical Section of the Left Ear

opening, there are stiff protective hairs, and glands that secrete **ear-wax.** The hairs tend to prevent the entry of foreign objects, and the wax protects the delicate thin skin that lines the inner part of the canal and the **ear-drum.**

The Middle Ear

This is an expanded chamber connected to the pharynx by the narrow **Eustachian tube.** The chamber contains air, the pressure of which is regulated by opening the valve at the junction with the pharynx. On the outer surface of this tympanic chamber is the ear-drum (**tympanic membrane**) and on the inner surface are two smaller membranes, which because of their transparency during life, are known as the **oval window** and the **round window.** Connecting the ear-drum with the oval window, across the chamber, is a chain of three small bones called the **hammer,** the **anvil** and the **stirrup** (*see* Fig. 152). The hammer is firmly attached to the ear-drum and is jointed to the anvil by a synovial joint; these two bones move as one, being very tightly bound together by ligaments. Between the anvil and the stirrup there is another synovial joint, which is freely

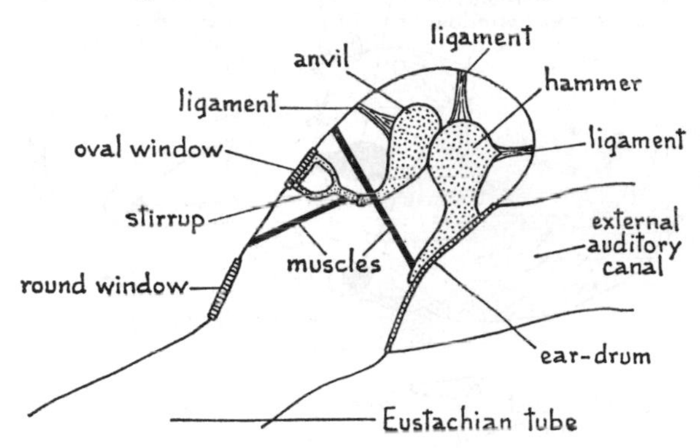

FIG. 152. The Middle Ear, in Vertical Section

movable. In addition, the bones are bound to the upper end of the tympanic chamber by three threadlike ligaments. Two small muscles, one attached to the hammer and one to the stirrup, adjust the tension in the chain of bones. The function of the chain is to transmit vibrations across the tympanic chamber from the ear-drum to the oval window.

The Inner Ear

This is a very complicated structure consisting essentially of a membrane filled with fluid lying within a bony tunnel also filled with fluid. Because of the winding pathways in both structures, they are known as the **membranous labyrinth** and the **bony labyrinth** (*see* Fig. 153). There are five distinct regions in the membranous labyrinth: three **semicircular canals** that lead into a chamber called the **utriculus,** and an ovoid chamber, the **sacculus,** connected to the utriculus by a narrow canal. From the sacculus the fluid-filled membrane extends round two and a half turns of the **cochlea,** which is

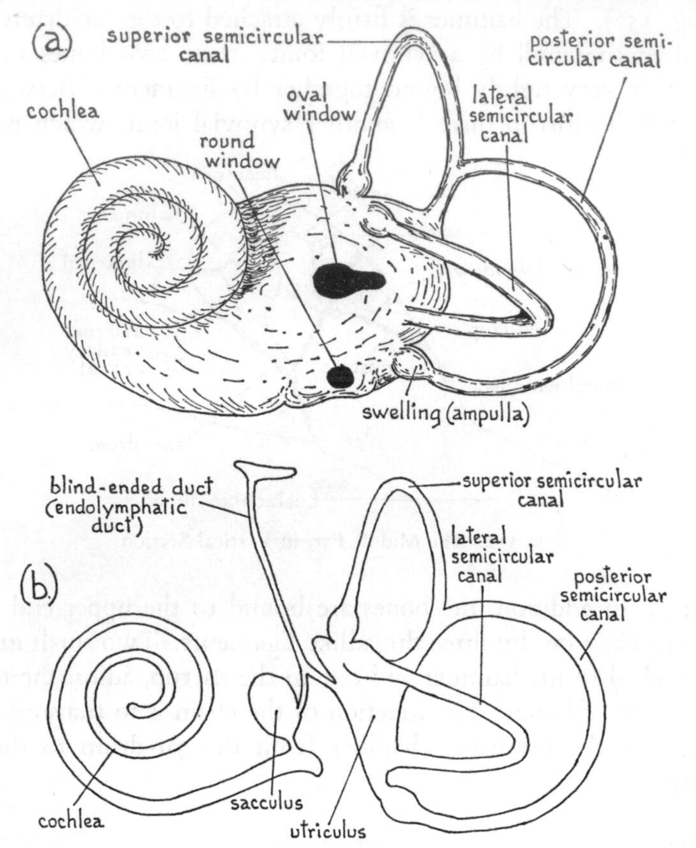

FIG. 153. (*a*) Left Bony Labyrinth (*b*) Left Membranous Labyrinth (*both as seen from the left side*)

somewhat like a small snail's shell. From the junction of the sacculus and cochlea the labyrinth extends outwards to the oval window. From the junction of the utriculus and sacculus a small duct extends inwards, to end blindly in the bone of the skull. This is a vestige of a duct that opened on the top of the head in our remote marine ancestors; it is still present in some fishes, for example, the dogfish and shark.

The bony labyrinth follows the outline of the semicircular canals and the cochlea, but between these two it forms an expanded chamber. Each semicircular canal has a swollen end where it joins the utriculus and in these swellings there are sense organs, which are sensitive to the movements of the head. Impulses from these sense organs are interpreted by the brain in terms of balance. In the utriculus and sacculus there are sense organs concerned with the perception of body position and acceleration. Finally, in the cochlea, there is a long, coiled series of sense organs concerned with the perception of sound.

Sound Perception

Sound waves set the ear-drum vibrating and the vibrations are transmitted by the chain of bones to the oval window. Vibrations of this window travel through the fluid **perilymph** of the bony labyrinth and proceed round the cochlea to the round window. This is a kind of safety device that bulges outwards when the oval window is pushed inwards, and moves inwards when the oval window bulges outwards. In the cochlea, vibrations in the perilymph set the membranous labyrinth vibrating, and these vibrations pass into the inner fluid, the **endolymph.** Finally, the vibrations in the endolymph affect the **organs of Corti** (*see* Fig. 154), which are arranged in a long series like a coiled piano keyboard with strings. According to the frequency of the sound waves, certain particular regions of the cochlea are affected and they set up impulses that are conveyed along nerve fibres of the **auditory nerve** to the brain, where they are interpreted as various kinds of sound.

We are able to interpret a number of properties concerned with sound—

1. The **pitch**: this property is related to the frequency of sound waves, which ranges from slow vibrations of sixteen per second to very rapid vibrations of twenty thousand per second. There are

frequencies above and below these, but they are inaudible to human beings; for example, the so-called "silent whistle" for dogs has a

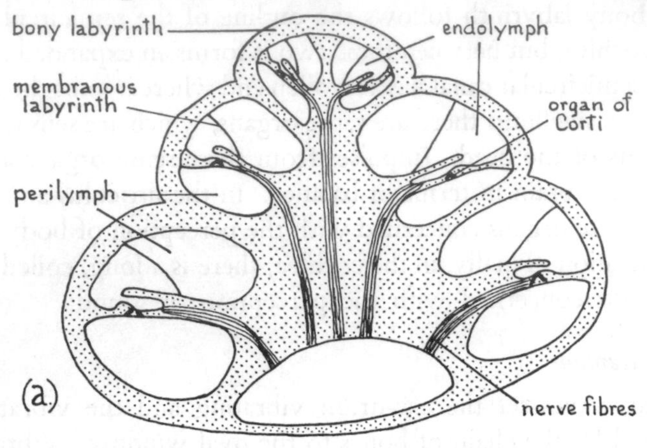

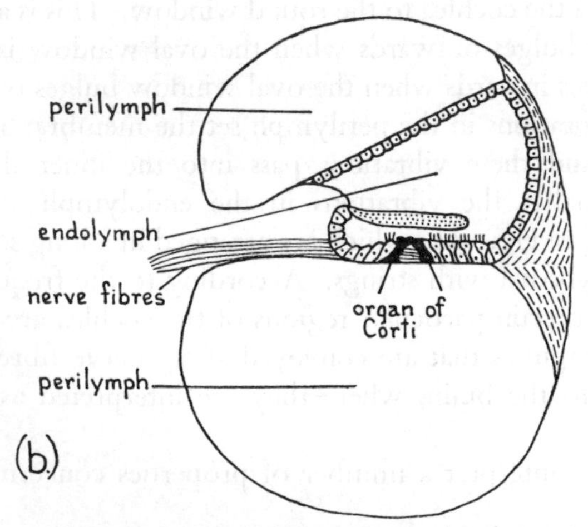

FIG. 154. (*a*) Vertical Section through the Cochlea (*b*) Enlargement of Small Portion, to Show an Organ of Corti

frequency of about twenty-five thousand per second. We interpret rapid frequencies as high, piercing sounds and slow frequencies as low, bass sounds. Middle C on a piano has a frequency of two hundred and fifty-six.

2. **Intensity**: this is produced by the physical energy of the sound transmitted through the atmosphere; we can distinguish between grades ranging from very loud to very soft.

3. **Quality**: the musical term for the quality of a sound is known as **timbre**; it results from the production of various overtones. Thus we can distinguish between the sounds produced by a piano, a violin and a trumpet, even though a note of the same pitch is played on each.

SENSE ORGANS OF THE NOSE, TONGUE AND SKIN

These have been described and illustrated on pp. 158, 133 and 53, respectively.

INTERNAL SENSE ORGANS

In all the organs and tissues of the body there are very tiny sense organs. Irritation, movement, pressure and pain in most organs can be experienced, but many sense organs are stimulated without our being aware of this. The presence of food in the stomach starts a rhythmic contraction and expansion; the presence of food in the small intestine is the stimulus that results in peristalsis. A constant stream of impulses from internal organs is arriving in the central nervous system and the appropriate adjustments are brought about.

Nerve Cells and Fibres

The central nervous system is made up of many millions of cells called **neurons**; in the adult human brain, there are at least six thousand million of these. Every neuron has a **cell body,** within which is the nucleus, and from the cell body there are outgrowths of two kinds. Those that convey impulses into the cell are called **dendrites**; they usually have brush-like endings (*see* Fig. 155). Only one fibre conveys impulses out of the cell; it is called an **axon** and it may be very long. Every neuron must have one axon and at least one dendrite. Most of the neurons in the brain have many dendrites, but in the retina of the eye, for example, many of the neurons have one dendrite (*see* Fig. 148, p. 257).

The neurons are not actually joined together to form a network; the junctions between them are of a peculiar type called **synapses.**

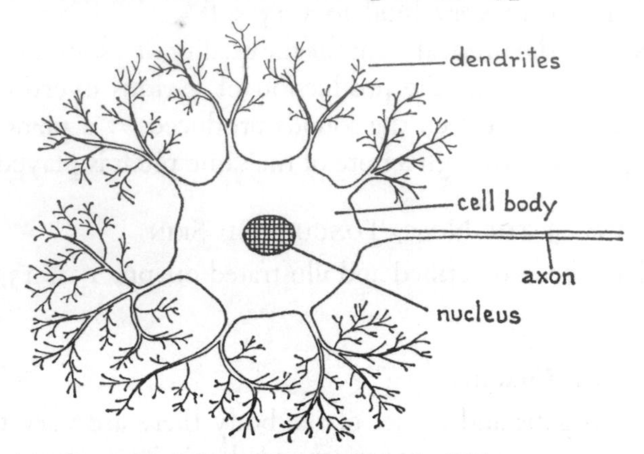

FIG. 155. A Single Neuron from the Brain, × 150

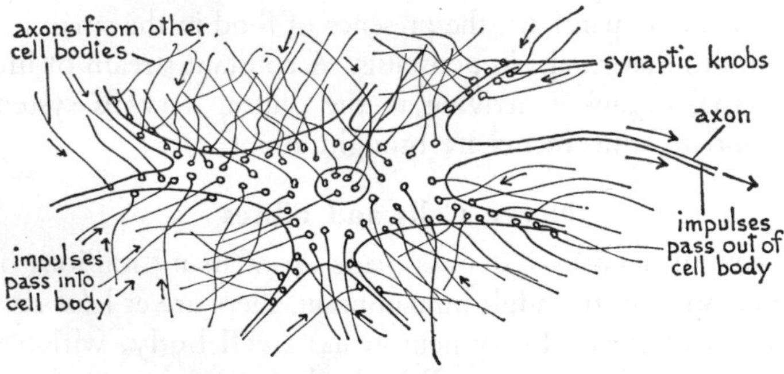

FIG. 156. Synaptic Knobs Ending on the Surface of a Neuron, × 200

In the brain and spinal cord there are numerous axons leading towards each neuron. Each axon, which itself extends from another neuron, ends in a small knob that nearly touches the cell membrane either on the cell body or on a dendrite, but never on the axon (*see* Fig. 156).

If a synaptic knob is sufficiently stimulated by an impulse, it will secrete the hormone acetylcholine, which stimulates the adjacent neuron (*see* Chapter 15). Two definite facts are known about this system of communication. Firstly, impulses have to reach a certain strength to pass the synapse; the "all or none" rule again applies here. Secondly, there must be impulses from a sufficient number of synaptic knobs before the receiving cell will send out impulses along its own axon.

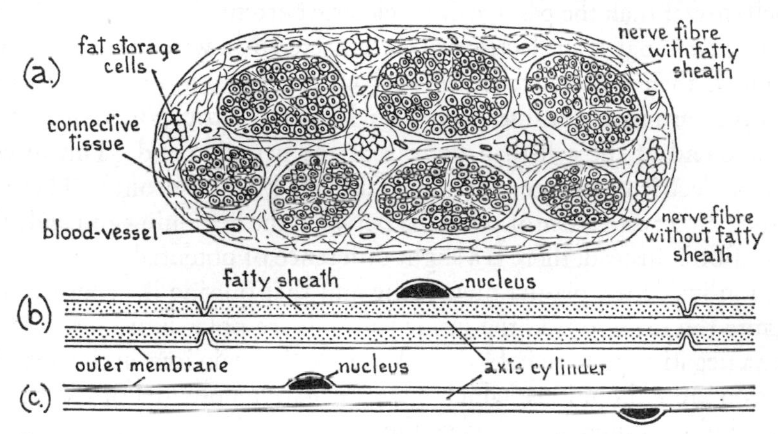

FIG. 157. (*a*) Transverse Section of a Whole Nerve, × 40 (*b*) A Single Nerve Fibre with Fatty Sheath, × 200 (*c*) A Single Nerve Fibre without Fatty Sheath, × 200

NERVES

A nerve consists of a number of separate fibres bound up with connective tissue, which contains blood-vessels and often fat storage cells (*see* Fig. 157(*a*)). Each fibre contains a central cylindrical strand of cytoplasm with a plasma membrane; this strand is known as the **axis cylinder** and the actual nerve impulses pass along it. Some axis cylinders are surrounded by fatty material, while others are not; both kinds are enclosed in a delicate membrane with nuclei at fairly wide intervals (*see* Fig. 157(*b*) and (*c*)). Each axis cylinder conducts impulses in one direction, but not in both. Those that conduct impulses into the central nervous system are called **sensory fibres,** because they carry impulses initiated in sense organs. Those that convey impulses out of the central nervous system are called **motor fibres,** since the majority lead to muscles, which bring about movement as a response. A nerve that carries only sensory fibres is called a **sensory nerve,** while a nerve that carries only motor fibres is called a **motor nerve.** However, most nerves are **mixed,** carrying both sensory and motor fibres.

The conduction of impulses along nerves is electrical, although it is much slower than the passage of an electric current along a wire. If the positive terminal of an electric cell is connected by a wire to the negative terminal, then a current will flow from the positive terminal to the negative terminal, because there is a higher electric potential at the positive end of the cell than there is at the negative end. This direct flow of electricity is extremely rapid (300 000 km per second). There is no such direct difference of potential between the beginning and end of a nerve fibre. Instead, there is a slight difference of potential between the solution inside the plasma membrane and that outside it. This occurs because the main ions in the inside solution are potassium ions, which have a negative potential of —2·92 V, while the main ions in the outside solution are sodium ions, which have a negative potential of —2·71 V. If we find the difference between these—

$$\text{i.e. } (-2{\cdot}92) - (-2{\cdot}71)$$

$$\text{we have } -2{\cdot}92 + 2{\cdot}71 = -0{\cdot}21 \text{ V.}$$

The potentials given above refer to mole quantities. Since the ions are sparsely scattered in nerve fibres, the actual difference in potential is smaller than this and has been shown to be 0·05 volt. Therefore there is a slight negative potential on the inside of the axis cylinder with respect to the outside (*see* Fig. 158).

When an impulse passes along the axis cylinder, there is a very rapid exchange of sodium and potassium ions across the plasma membrane, so that the potential inside becomes slightly positive, but it quickly recovers and becomes negative again. It is this change of potential passing like a wave along the axis cylinder that causes the actual transmission of nerve impulses. The speed of conduction varies in different nerve fibres; the maximum is about 160 metres per second and the minimum in the neighbourhood of 2 metres per second.

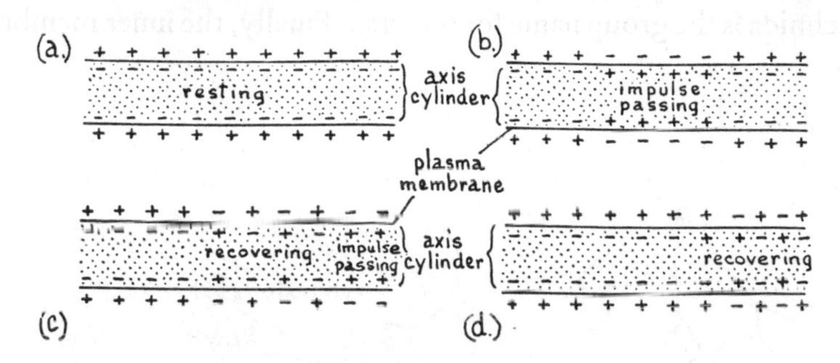

FIG. 158. (*a*) Distribution of Charges in a Resting Nerve Fibre (*b*) Distribution When Impulse Passes (*c*) Recovering to Resting State (*d*) Centre Region in the Resting State Again

The Central Nervous System

This consists of the brain and spinal cord. The brain is protected by the cranium and the spinal cord by the vertebral column. Both consist essentially of three types of material—

1. Cell bodies, which constitute the **grey matter.**
2. Nerve fibres, which constitute the **white matter.**
3. Branched jelly-like **neuroglia** cells, which perform the functions of connective tissue, of which there is very little in the central nervous system (*see* Fig. 159).

Both the brain and the spinal cord are enclosed in three membranes called **meninges**; any inflammation or infection of these leads to the serious disease, **meningitis.** The outer membrane, the **dura mater,** is made of tough fibrous connective tissue, which is partly fused to the bone outside it. Beneath the dura mater there is a second delicate membrane consisting of widely spaced, fine fibres. The spaces are filled with **cerebro-spinal fluid,** which is almost exactly like lymph. This fluid acts as a cushion to protect the brain and spinal cord from mechanical damage. Because the fibres of this second membrane form a network like a spider's web, it is called the **arachnoid membrane** (Arachnida is the group name for spiders). Finally, the inner membrane,

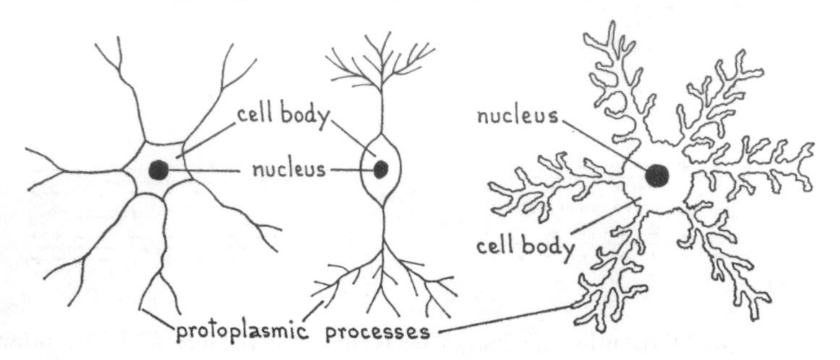

FIG. 159. Three Types of Branched Neuroglia Cells from the Brain

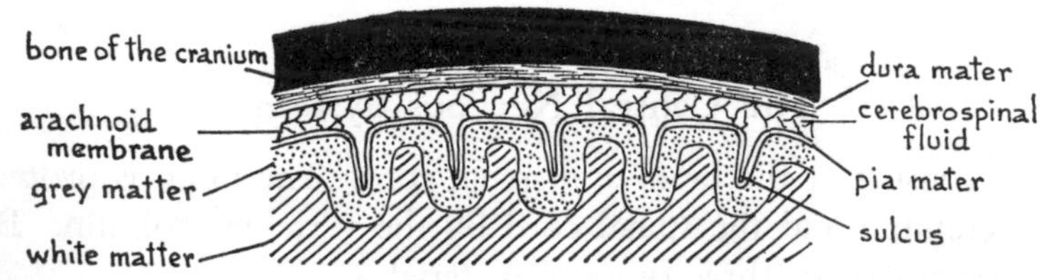

FIG. 160. Vertical Section of the Cranium and Upper Part of the Brain, Showing the Meninges

the **pia mater,** contains a dense network of blood-vessels and is applied closely to the whole surface of the brain and spinal cord (*see* Fig. 160).

Although both the brain and spinal cord have essentially the same structure and the same origin (*see* p. 238), there are important differences, especially in function, so they are here described separately.

The Brain

The average weight of the adult human brain is approximately 1370 grams, though there is considerable variation in this respect. It consists of five main parts, which are shown in Fig. 161 and described below.

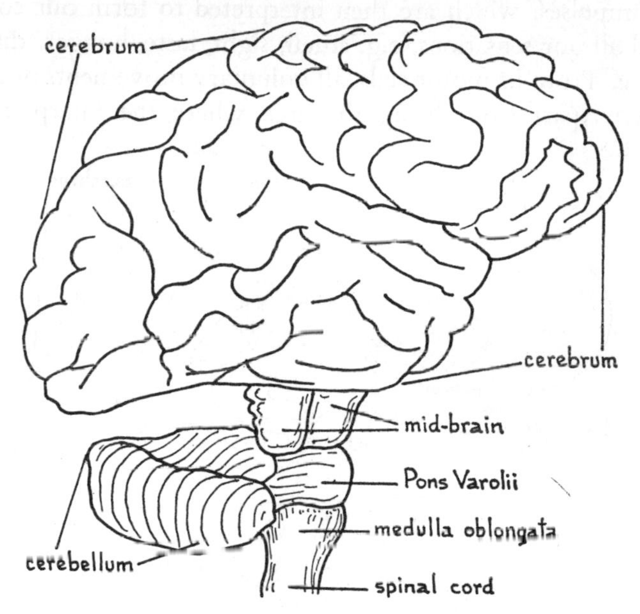

FIG. 161. Side View of the Brain with the Lower Parts Slightly Detached to Show their Shape and Size

The Cerebrum

This consists of two large **hemispheres,** partly separated by a deep cleft called the **longitudinal fissure.** Each hemisphere is composed of a layer of grey matter (*see* p. 271), ranging from 1·5 to 3·0 mm in thickness; beneath this there is a much thicker region of white nerve fibres. The grey matter has numerous folds called **convolutions,** separated by clefts called **sulci** (singular—sulcus). In each cerebral hemisphere there is a fluid-filled cavity known as a **lateral ventricle** containing cerebro-spinal fluid; these ventricles both communicate by a narrow passage with the third ventricle in the mid-brain (*see* Fig. 162).

The **cortex** (grey matter) of the cerebrum is really an enormously enlarged collection of sensory and motor neurons. It receives all the sensory impulses, which are then interpreted to form our consciousness, and all our sensations, e.g. smell, sight, taste, hunger, thirst, etc., arise there. From its motor cells, all voluntary movements of the body originate. Apart from being the area where the interpretation of

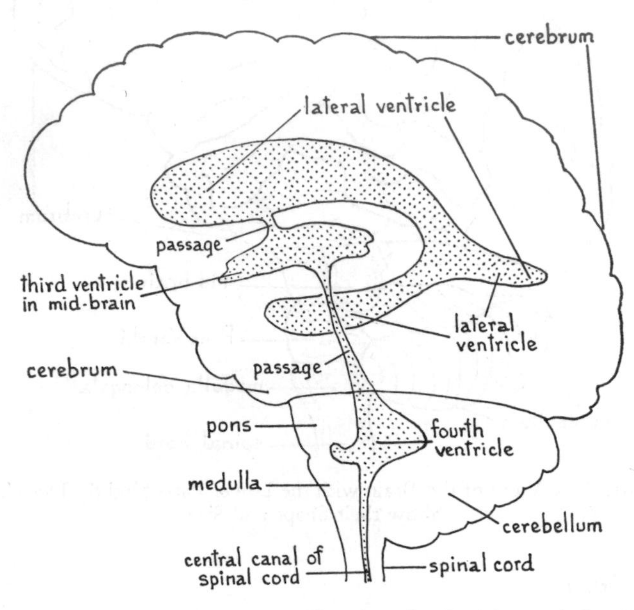

FIG. 162. Side View of the Brain, Showing the Positions of the Cavities

sensory impulses and the despatch of motor impulses take place, it is also the region where memory, intelligence and the various emotions such as joy, sorrow, fear, anger, etc., are located.

The cerebral cortex has been gradually mapped out so that we know the regions where sensory reception and motor control, affecting many areas of the body, are located (*see* Fig. 163). Undoubtedly one of the principal reasons for man's great success is the enormous surface area of grey matter in the cerebral cortex. Throughout human evolution, there has been not only an increase in the size of this region, but also an increase in the number of convolutions and in the depth of the sulci.

The Cerebellum

This lies below the posterior portion of the cerebrum and it consists of a cortex of grey matter and a central region of white matter.

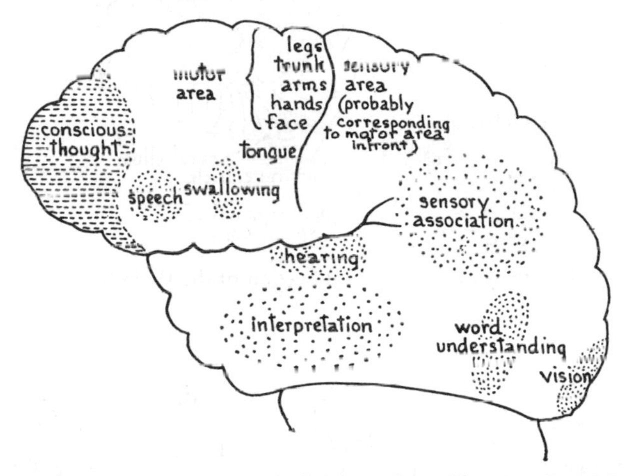

FIG. 163. Side View of Left Cerebral Hemisphere, Showing the Main Areas that Have Been Mapped

The grey matter penetrates into the white matter, forming a peculiar branched shape when seen in vertical section (*see* Fig. 164). This arrangement has been called the **tree of life** (arbor vitae).

The chief function of the cerebellum appears to be co-ordination of movement; motor impulses originating in the cerebrum, first pass into the cerebellum, and these impulses, concerned with group action of muscles, are co-ordinated in the cerebellum and then relayed to the muscles. It is concerned also with balance and receives fibres from the semicircular canals of the ears. No sensations are interpreted in the cerebellum; all its activity is below the conscious level.

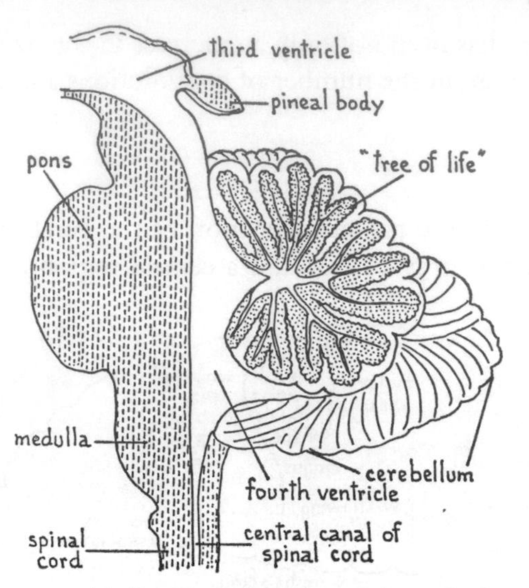

FIG. 164. Median Vertical Section of the Brain Stem

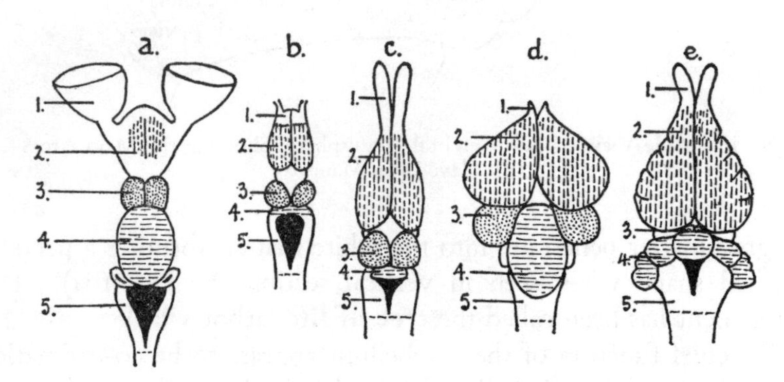

FIG. 165. Dorsal Views of Vertebrate Brains (*a*) Dogfish (*b*) Frog (*c*) Lizard (*d*) Pigeon (*e*) Rabbit

Similar structures are shaded in the same way. 1. Olfactory Lobes 2. Cerebrum 3. Optic Lobes 4. Cerebellum 5. Medulla

The Mid-brain

This is a very small portion of the brain lying between the pons in front and the cerebellum at the back, with the medulla below it. It consists mainly of large bundles of fibres connecting the cerebrum with the cerebellum, medulla and spinal cord. In the other classes of vertebrates, apart from the mammals, the mid-brain is characterized by two large optic lobes concerned with the reception of impulses from the eyes (*see* Fig. 165), but in the mammals, and particularly in man, these lobes have become very small and are completely covered by the cerebrum. They have also become divided into an anterior and posterior pair. In some mammals the anterior pair are still concerned with visual stimuli, but in man these are now referred to the back of the cerebrum (the visual area in Fig. 163, p. 275). The posterior pair are concerned with reflex response (see p. 284) to auditory stimuli. In front of the mid-brain is the **pituitary gland,** which is described and illustrated on p. 292.

The Pons

"Pons" is the Latin word for bridge and it consists mainly of numerous fibres connecting the right and left sides of the cerebellum, and fibres connecting the cerebrum with the spinal cord.

The Medulla

This part of the brain resembles the spinal cord in structure, since the grey matter is no longer near the surface. Numerous nerve fibres pass through it and it is interesting to note that motor fibres from the cerebellum to the muscles cross in the medulla from left to right and vice-versa. This has the consequence that a damaged motor area in the left cerebral hemisphere, for example, will cause paralysis in some part of the right side of the body.

The most important function of the medulla is the control of certain vital involuntary or **reflex actions.** For example, the vasomotor centre, the respiratory centre, and the cardiac inhibitory centre all lie in the medulla. It also controls many of the digestive processes and a number of common reflex actions such as laughing, sneezing, coughing, hiccoughing, etc. As in the cerebellum, all the activity of the medulla is below the level of consciousness.

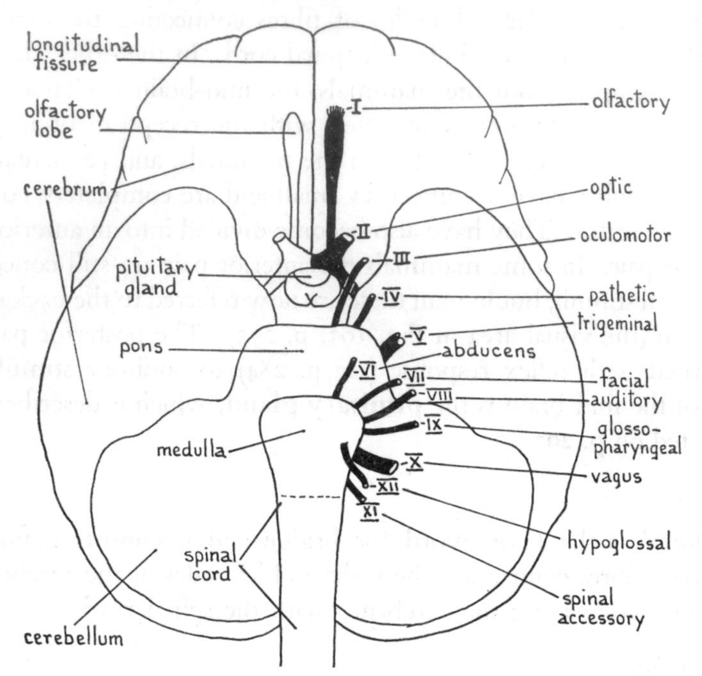

FIG. 166. The Brain Seen from Below
Cranial nerves are shown in black on one side.

Cranial Nerves

Nerves that are directly connected to the brain are called cranial nerves. There are twelve pairs; some carry only sensory fibres, some carry only motor fibres, and some are mixed nerves. (*See* Fig. 166 and the table on p. 279.)

THE SPINAL CORD

The spinal cord extends from the medulla oblongata to the posterior border of the first lumbar vertebra. It is about 45 centimetres long

NUMBER	NAME	KIND	REGIONS SUPPLIED
I	Olfactory	Sensory	From sense organs in the nose
II	Optic	Sensory	From the retina of the eye
III	Oculomotor	Motor	To four eye muscles
IV	Pathetic	Motor	To one eye muscle (superior oblique)
V	Trigeminal	Mixed	To and from face, teeth, jaws
VI	Abducens	Motor	To one eye muscle (external rectus)
VII	Facial	Mixed	To and from tongue, face, lips, eyelids
VIII	Auditory	Sensory	From sense organs in the inner ear
IX	Glossopharyngeal	Mixed	From back of tongue, tonsils, palate To salivary glands and some muscles of the throat
X	Vagus	Mixed	To and from larynx, lungs, heart, gut, pancreas, liver, muscles of trunk and limbs
XI	Spinal accessory	Motor	To muscles of larynx, pharynx, and soft palate
XII	Hypoglossal	Motor	To muscles of the tongue

and approximately the thickness of the little finger, tapering to a fine point at its extremity. It is enclosed in the same three membranes as the brain—the dura mater, arachnoid membrane and pia mater, but the dura mater does not adhere to the vertebrae. As in the brain, there are external and internal cavities filled with cerebro-spinal fluid: externally, the arachnoid membrane and internally the central canal.

The grey matter occupies an H-shaped region and the white matter is outermost (*see* Fig. 167). The white matter contains ascending (sensory) bundles of nerve fibres and descending (motor) bundles. The essential function of the spinal cord is to act as a pathway for impulses between the body and the brain. Damaged or severed fibres are not regenerated; if they are ascending fibres, then there is a loss of sensation; if they are descending fibres then there is muscular paralysis. Complete severance of the spinal cord results in paralysis of the body below that level.

Spinal Nerves

There are thirty-one pairs of spinal nerves: eight cervical, twelve thoracic, five lumbar, five sacral and one coccygeal. Each spinal nerve has both **dorsal sensory** and **ventral motor roots** (*see* Fig. 167), which unite to form a single **mixed nerve.** Sensory impulses travelling along fibres of the mixed nerve proceed via the dorsal root into

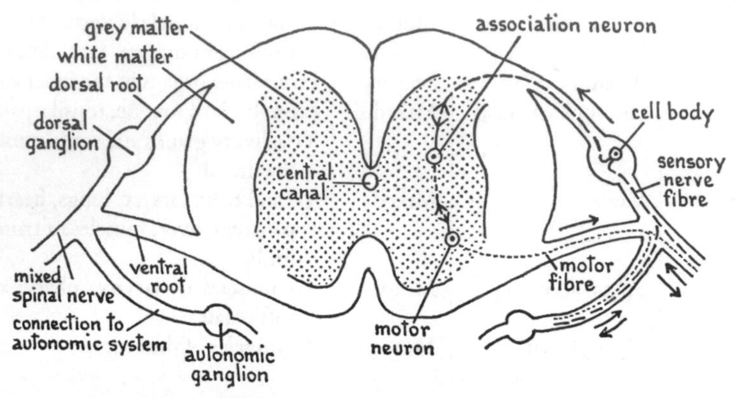

Fig. 167. Transverse Section of the Spinal Cord
The main nervous pathways are shown on the right.

the **dorsal ganglion,** where their cell bodies are located; from these cell bodies, axons pass into the cord. The fibres along which motor impulses travel extend from cell bodies in the cord, out through the ventral root.

Certain of the spinal nerves unite to form interlacing networks known as **plexuses** (*see* Fig. 168). The most prominent of these are the **cervical plexus,** which supplies nerve fibres to the muscles of the neck, shoulder and diaphragm, the **brachial plexus,** which supplies nerves to the arm, and the **lumbo-sacral plexus,** which supplies nerves to the leg and hip. All these plexuses are, of course, paired.

The Cerebrospinal Fluid

This fluid, which exudes from capillaries of the pia mater, is almost exactly like lymph except that it contains few white corpuscles. The

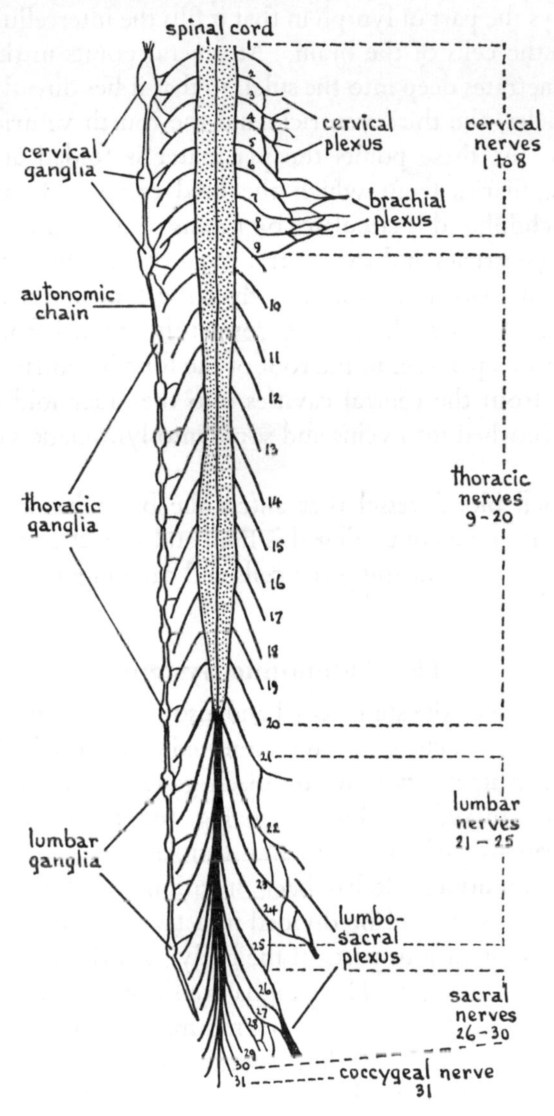

FIG. 168. The Spinal Cord and Spinal Nerves
The autonomic chain is shown on one side.

fluid also plays the part of lymph in that it fills the intercellular channels surrounding the cells of the brain. At several points in the brain the pia mater penetrates deep into the sulci, so that it lies directly above the lateral ventricles, the third ventricle and the fourth ventricle (*see* Fig. 162, p. 274). At these points the pia mater is folded and possesses numerous capillaries from which the fluid escapes. In the embryo and in early childhood it circulates by means of the beating of the cilia that line the ventricles and the central canal of the spinal cord. These cilia are lost at maturity, and, after that, the fluid circulates because of the pressure of more fluid being constantly exuded from the capillaries. There are apertures in the roof of the fourth ventricle where the fluid escapes from the central cavities into the arachnoid membrane, some to be absorbed into veins and some into lymphatic vessels of the scalp.

Every small blood-vessel that enters the brain has a covering of arachnoid membrane containing this fluid and thus it comes to fill the intercellular spaces, nourishing the cells and carrying away their waste products.

The Autonomic System

On either side of the spinal cord and also on the course of some of the cranial nerves is a chain of interconnected ganglia, which constitute the **autonomic nervous system.** Each ganglion is connected by a short nerve to a mixed spinal nerve (*see* Fig. 167, p. 280). Nerves from the ganglia pass to various organs and stimulate automatic adjustment to changing conditions. It has been mentioned before (p. 85) that this autonomic system can be divided into two portions, which exert opposite effects on various parts of the body. All the activity is below the level of consciousness. The ganglia are shown on one side in Fig. 168, p. 281 and Fig. 169 gives a diagrammatic representation of the **sympathetic** and **parasympathetic** portions of the system. The whole autonomic system has the dual functions of regulating the internal environment, and of controlling the automatic adjustment of the internal organs to the changing conditions of the external environment.

Much of the system is concerned with relaying impulses that originate in the brain or spinal cord.

Among the important regulatory activities of the autonomic system are the adjustment of the iris of the eye, the control of the amount

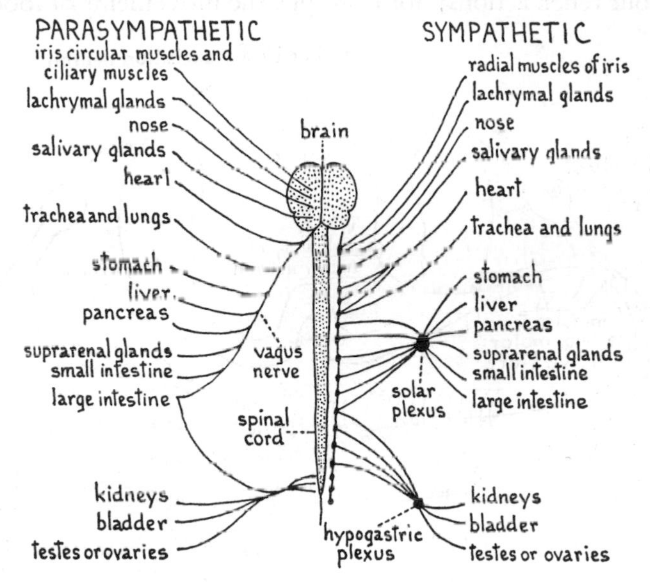

FIG. 169. The Main Nerves of the Autonomic System
The sympathetic and parasympathetic portions are each shown on one side only.

of perspiration, the production of saliva, the acceleration of heart-beat, and control of the muscle of blood-vessels to the viscera, etc.

Behaviour

Behaviour may be described as the total of all the responses an individual makes to the stimuli he perceives. Behaviour is the result of activity of the nervous system and, in some cases, of the endocrine system (*see* Chapter 15). There are different levels of complexity in behaviour, the simplest of these being the reflex action.

Reflex Actions

A true **reflex action** is an automatic or involuntary response over which the individual has normally no control; the same stimulus always produces the same response. We are quite unconscious of most of our reflex actions; for example, the movements of food in the

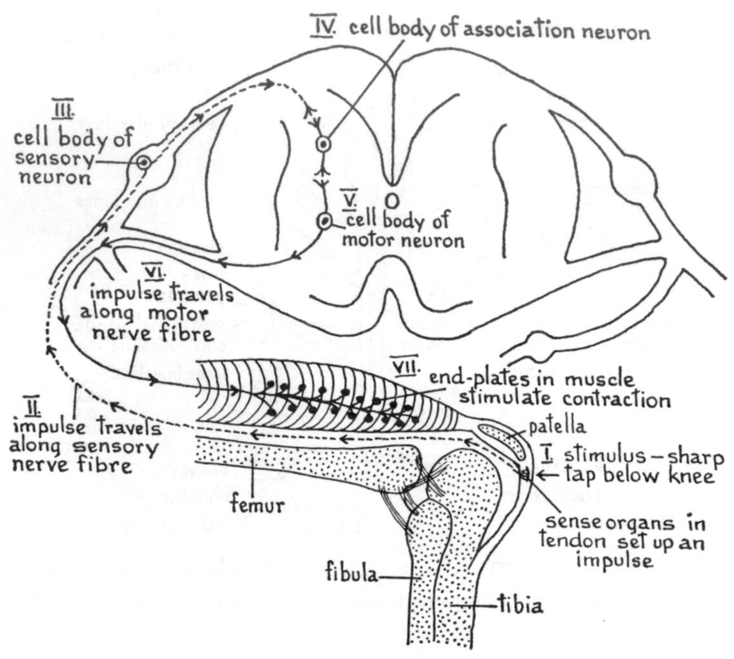

Fig. 170. Diagram Showing the Nervous Connections Involved in the Knee-jerk Reflex
I to VII shows the sequence along the nervous path.

intestines, variations in diameter of the small blood-vessels, slight variations in heart rate and in respiration. We are, however, aware of many other reflexes and, in some cases, we can exercise some measure of control over them; for example, the quick closure of the eyelids when some object touches them, the recovery of balance after a slip, the sudden removal of the hand when it is painfully stimulated. We are conscious of some reflexes that we cannot control at all; for example, the change in diameter of the pupil under different conditions of light, the knee-jerk reflex and the action of the ciliary muscles of the eye when adjusting the shape of the lens.

The type of nervous circuit involved in a reflex action is called a **reflex arc** (*see* Fig. 170). Many of these circuits do not involve the brain but are dealt with in the spinal cord; they are known as **spinal reflexes.** Others, which involve the brain, are called **cranial reflexes.** There are three essential parts in a reflex arc—

1. An impulse travels into the central nervous system as the result of a stimulus.
2. An **association neurone** in the central nervous system is stimulated by the incoming impulse, and this impulse is transmitted to a motor neurone.
3. The motor neurone transmits an impulse to an effector organ, which makes the response.

No nervous circuit is as simple as this, which involves only three neurones. There are always many association neurones concerned and each motor impulse must go out along many fibres.

The Conditioned Reflex

The work of Pavlov, the Russian physiologist, showed that some reflex actions could take place in response to stimuli other than the normal ones. For example, a dog salivates at the sight or smell of its food. If a bell is rung every time the food is brought, then eventually the dog will salivate when the bell rings, whether the food appears or not. Such an altered reflex action is known as a **conditioned reflex.** Urination in young children is controlled by a spinal reflex; as they grow, they learn to control the act by imposing cerebral control over the simple reflex. Conditioned reflexes play a large part in habit formation. It is sometimes considered that all the processes we learn that eventually become automatic should also be called conditioned reflexes. For example, we learn to walk, but when we have learnt, walking becomes a reflex response. Nevertheless we can always impose cerebral control over such learned reflexes, as, for example, when one makes a conscious effort to walk in a particular manner.

When a person's behaviour is recorded for a period of, say, one day, it is found that much of it is what we call **habitual behaviour.** The sequence in which one washes, dresses and prepares to go out, consists of a long series of individual actions, each of which had to be learnt at one time. Nevertheless any unusual stimulus, such as a different bathroom or a different house in a different locality, can alter the whole sequence and a new series of actions is learnt until they in their turn become habitual. A young child has considerable difficulty in learning to play a new scale on a pianoforte; a competent pianist can play very complex compositions while appreciating not only the actual sequences of notes but also their interpretation in terms of speed, intensity of sound, etc.

Instinct

An **instinct** is an inherited series of reflex responses, each one providing the stimulus for the next one. Birds build nests, perform complicated courtship activities and migrate, all by instinct. Social insects, such as bees and ants, are controlled almost entirely by instinct. Human beings learn, as they mature, to impose cerebral control over much of their instinctive behaviour.

Insight

Intelligent behaviour, or **insight,** is the highest type of nervous response. A situation consisting of a number of separate factors arises, each of the separate factors having been encountered before. The intelligent person can survey the factors; choose the relevant ones and find the solution without trial-and-error. Many animals exhibit intelligent behaviour as this is defined above. A fox, approaching a farm, may smell the poultry, hear the dog barking and see the farmer. The fox must consider the situation and decide which of several courses to adopt.

In human affairs, intelligence normally means the ability to consider a situation both in the light of past experiences and of future effects. Undoubtedly, there are two components in intelligence: it is partly inheritance and partly experience. It is exceedingly difficult to determine what fraction of intelligence is due to each factor. While we cannot improve the inherited component in any particular person, the environmental component can be improved by all-round improvement in the conditions under which people live, and the opportunities that they have for education.

SUGGESTED EXERCISES

1.* Bulls' eyes (or sheep's or pigs') can sometimes be obtained from butchers, and always from biological dealers. Examine the external appearance, noting the six eye muscles, the sclerotic, cornea, iris, pupil, and the optic nerve. With sharp scissors, the eye can be cut from the cornea in two planes at right-angles to each other and the four quarters opened out. The humours will be obvious during the cutting. The lens can be easily removed from the suspensory ligament and examined. The thickened ring of the ciliary body can be seen, and the blackened posterior chamber.

2. The iris reflex can be observed by holding a small electric torch close to someone's eye and then switching on the light. Note the size of the pupil changing quickly when the light is switched on.

3. To demonstrate the existence of the "blind spot," mark a small circle and a cross on your exercise book, 5 cm apart—thus ●—5 cm—×. Close the right eye and with the left, look straight at the cross. Bring the book slowly towards your eye and the circle will disappear from view as its image crosses the blind spot.

4. Examine microscope slides showing nerve cells and nerve fibres. These can be purchased from any biological dealer or can be made in the laboratory.

5. Examine a prepared slide of a vertical section of the cochlea.

6. Examine a prepared slide of a transverse section of the spinal cord.

7.* The brain of a rat can be extracted fairly easily if it is first hardened by the following method. Cut off the heads of any rats that have been used for dissection. Skin the top of the head and with a scalpel point prise off a small piece of bone, so that the hardening agent (ninety per cent alcohol) can penetrate. The heads can be kept in this fluid until required. The bones on top of the skull can be prised up with a scalpel and then pulled off with a strong pair of forceps. At the sides of the skull, the bones are best removed by breaking off small pieces with a strong forceps or by cutting off small pieces with a bone forceps. (A curved pair of nail clippers will do the job very well on the rat's skull.) Great care is needed when lifting the brain from the floor of the cranium as the pituitary body tends to become detached rather easily.

It is best to lift the brain with a flattened scalpel, first from the front, then from the back, having cut the spinal cord, and then from the sides. Look underneath while lifting, so that the pituitary body can be eased up with the scalpel.

Draw dorsal and ventral views of the brain and then cut transverse blocks at various levels. Using another brain, try to cut a horizontal longitudinal section, and with yet another, cut a median vertical section. Examination of these will give some idea of the interior.

8. Try to recall any examples of intelligent behaviour in animals. During your observations of live rats, or any other animals, watch for any signs of intelligence and record these in your exercise book.

CHAPTER 15

THE ENDOCRINE SYSTEM

THE control and co-ordination of the body by the nervous system has been outlined in Chapter 14. Since nerve-endings do not actually extend into cells, chemical regulation within the cells is stimulated, controlled and co-ordinated by substances called **hormones.** Sometimes called "chemical messengers," they are produced in special glands or groups of cells, and are transported in the blood, to exert their effects elsewhere in the body. Because these secretory structures have no ducts, but pass their products directly into the blood, they are known as **endocrine or ductless glands.** Other types of glands which have ducts are called **exocrine glands,** e.g. the salivary glands, the mammary glands and the sweat glands. The hormones, which are of enormous importance, are all soluble in water, and are either proteins or fat-like substances called **steroids.**

Endocrine glands

There are a number of these glands; their nature and functions are described below; their locations in the body are illustrated in Fig. 171.

THE HYPOTHALAMUS

This is a part of the brain which exercises overall control of most of the endocrine system. It is situated immediately above the pituitary gland, almost at the middle of the base of the brain (*see* Figs. 171 and 172). The hypothalamus has nervous connections from all parts of the brain, but apart from its nervous functions, it has specialized groups of neurons which secrete hormones; hence they are called **neurosecretory cells.** Some have long axons which terminate in the pituitary; droplets of hormone in solution, produced by these neurosecretory cells, pass

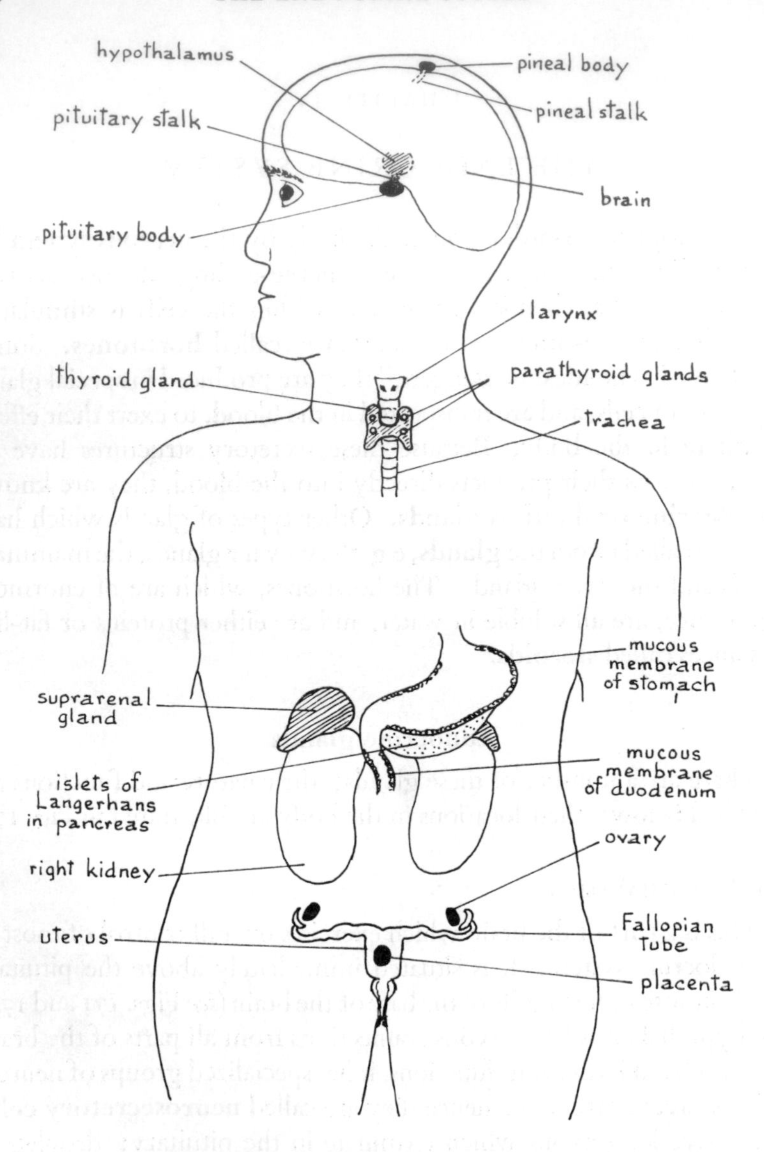

FIG. 171. Location of the Endocrine Glands, Shown in the Female

along the axons into the pituitary, where they are liberated, absorbed into the blood, and distributed throughout the body. Such are the hormones **oxytocin** and **vasopressin,** each of which has two important functions. Oxytocin causes contraction of the uterine muscle during birth and also stimulates the mammary glands to release milk during suckling. Vasopressin causes a rise in blood-pressure by inducing contraction of the unstriated muscle in the walls of blood-vessels; it also affects osmoregulation by the kidney tubules. Deficiency of vasopressin results in excessive loss of water from the kidneys, leading to constant thirst. The condition is known as **diabetes insipidus**; it can be relieved by suitable doses of the hormone.

Some groups of neurosecretory cells produce hormones called **releasing factors.** They are liberated into the hypothalamus itself, absorbed into blood capillaries and carried to the pituitary in portal veins. These veins split up into capillaries and the hormones are released from the blood to be absorbed by certain groups of cells in the pituitary. Each releasing factor stimulates the secretion of a particular hormone.

Until recent years, it was believed that the pituitary controlled most of the other endocrine glands, but it is now recognized that the pituitary itself is controlled by the hypothalamus.

THE PITUITARY GLAND

This is a small spherical gland, about 1 cm in diameter, suspended from the hypothalamus by the pituitary stalk. It lies in a depression of the floor of the cranium known as the Turkish saddle. The gland is partly divided internally into three regions separated by connective tissue partitions (*see* Fig. 172). In the anterior part, six hormones are secreted, each in response to a particular releasing factor from the hypothalamus; in the intermediate part, two hormones are produced; the posterior part receives the hormones oxytocin and vasopressin from the hypothalamus. It is important to notice that several of the pituitary hormones have the sole function of stimulating other endocrine glands into action. The various pituitary secretions are summarized below.

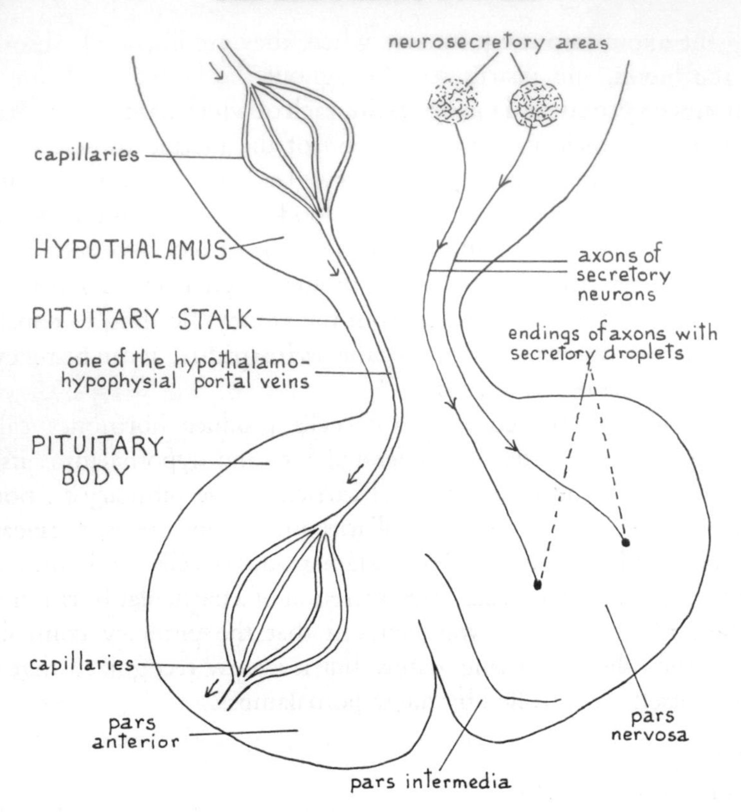

FIG. 172. Diagrammatic Representation of the Relationships between the Hypothalamus and the Pituitary Body

The pars anterior secretions are:

1. **Somatotropin** stimulates growth; excess causes **pituitary gigantism**; deficiency before puberty causes **pituitary dwarfism**. Somatotropin is also concerned in the control of blood sugar (*see* p. 296).

2. **Thyrotropic hormone** induces secretion of **thyroxin** from the thyroid gland.

3. **Adrenocorticotropic hormone** excites activity of the adrenal cortex.

4. **Follicle-stimulating hormone** induces development of Graafian follicles in females and maturation of sperm in males.

5. **Luteinizing hormone** promotes ovulation and subsequent development of corpora lutea; a similar hormone in males stimulates production of hormones in the testes.

6. **Prolactin** induces production of milk in the mammary glands.

The pars intermedia secretes two hormones, both of which are concerned with the production and maintenance of the pigment cells in the skin. Each is called a **melanocyte-stimulating hormone.**

The Thyroid Gland

This is situated in the neck, and it consists of two small pink lateral masses of tissue, one on each side of the top of the trachea; these two masses are joined by a narrow connection across the front of the trachea. The gland secretes **thyroxin,** a hormone containing iodine; the main function of the hormone is the control of the basal rate of metabolism (see p. 112). In the young, deficiency of thyroxin lowers this rate and, in extreme cases, will cause retardation of physical and mental development, eventually producing a type of idiocy called **cretinism.** In adults, the results of thyroxin deficiency are not so drastic; there is sluggish metabolism, atrophy of the thyroid, and swelling of the tissues below the epidermis; the condition is known as **myxoedema.** Nowadays, it is usually possible to treat both of these diseases by suitable doses of thyroxin. Excessive production of thyroxin leads to **exophthalmic goitre,** which is characterized by a high metabolic rate, restless and feverish activity, rapid heart-beat and often wasting of the body. The characteristic protrusion of the eye-balls and swelling of the thyroid give the disease its name. Removal of portions of the thyroid will usually alleviate the condition. **Simple goitre** is due to lack of iodides in the diet; it is treated by addition of potassium iodide to suitable articles of food; some brands of table salt already have this iodide added.

Thyroxin is also important in temperature regulation. At low temperatures, there is increased production of the hormone and hence a rise in the basal rate of metabolism, with the consequent release of more energy as heat.

The Parathyroid Glands

These are two pairs of minute brown glands embedded in the posterior surface of the thyroid. They secrete **parathormone** which affects calcium metabolism. The glands are sensitive to the level of calcium ions in the blood; when the level is below normal, secretion is stimulated; when the level is normal, secretion ceases. The effects of parathormone are soon apparent; there is increased absorption of calcium from the intestine, withdrawal of calcium salts from bones, and retention of calcium ions by the renal tubules. Deficiency of the hormone is comparatively rare; it is characterized by extreme irritability of the nervous system and by muscular spasm. Excess of parathormone is somewhat more common; in this case, the bones become decalcified and then break very easily; calcium ions are excreted copiously, and calcium phosphate stones are found in the kidneys. The disease is known as **osteitis fibrosa.** Complete removal of the parathyroids which may occur in thyroid surgery, is fatal in a few days.

The Adrenal (Suprarenal) Glands

These are paired yellow spherical structures, one above each kidney. Each consists of an outer cortex and an inner medulla, the cortex consisting of three concentric zones (*see* Fig. 173). The medulla secretes **adrenalin** and **nor-adrenalin,** both of which produce dramatic changes in the body in response to fear, shock or stress. The hormones are secreted after the gland is stimulated through nerve-fibres of the sympathetic system. They cause rapid preparation of the body for swift action and have been justly called the hormones of "fright, fight and flight." Adrenalin increases rate of heart-beat and heart output; nor-adrenalin causes constriction of small bloodvessels in the skin and abdominal viscera, so that more blood circulates to the voluntary muscles. The combined action of these hormones causes a rise in blood-pressure and the heart thumps; the pupils are dilated and the eyelids open widely; there is also increased release of glucose from the glycogen of the liver. The general effect is that the whole body is ready for vigorous action which may be necessary in an emergency. This state of tension could not be tolerated for long periods and it is fortunate that the hormones are rapidly destroyed in the tissues.

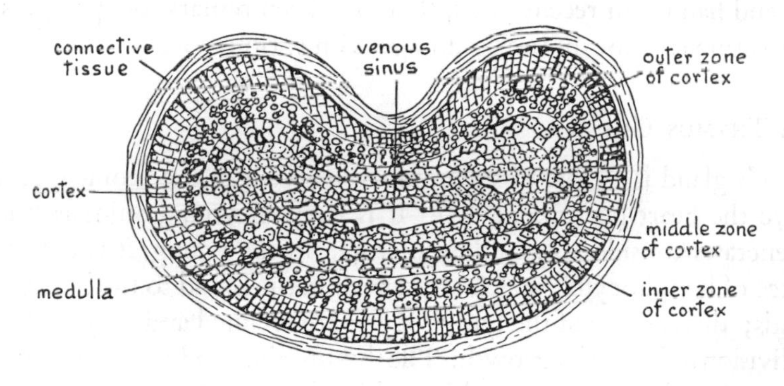

FIG. 173. Section of the Suprarenal Gland, × 1·5

All the hormones secreted by the cortex are known as **cortico-steroids;** they form three distinct groups. The outer zone secretes **mineralocorticoids**, which play an important part in osmoregulation (*see* p. 215), and are also important in connection with the repair of damaged tissue. The middle zone secretes **glucocorticoids**, which affect carbohydrate metabolism and especially the conversion of protein and fat into carbohydrate. They also affect tissue repair, having a somewhat inhibitory effect and hence the rate of healing is dependent upon the balance of these two sets of hormones. The inner zone secretes hormones which, in conjunction with secretions from the sex glands (*see* p. 298), induce the appearance of the male and female secondary sexual characteristics. It is noteworthy that in both sexes, the male hormones (**androgens**), as well as the female hormones (**oestrogens**), are produced. In each case, male or female, it is the balance of these hormones which is important. Excess of androgens in females leads to varying degrees of masculinity, while excess of oestrogens in males leads to some degree of feminity.

Destruction or removal of the adrenal cortex is fatal, while damage or insufficient secretion will cause **Addison's disease**; this is characterized by general wasting of the body, muscular weakness, inability to withstand injury or fatigue, low blood pressure, and pigmentation of the face and hands. In recent years, there has been remarkable progress in the treatment of the disease by dosage with various corticosteroids.

The Thymus Gland

This gland is a mass of soft pink tissue lying in the thoracic cavity above the heart (*see* pp. 44, 45). It is large in young animals but it degenerates considerably as they mature. In the young, it is the main source of lymphocytes, which are carried in the blood to the lymphatic glands; there, some of the lymphocytes are retained and they multiply by division followed by growth. The thymus also produces the hormone **thymosin** which stimulates this multiplication of lymphocytes in the lymphatic glands, hence ensuring an adequate supply in the blood and lymph. To a considerable extent, the thymus gland is the source of antibodies which neutralize the toxins which are produced by bacteria (*see* p. 184).

The Islands of Langerhans

These are small groups of cells scattered throughout the pancreas, hence the name "islands" or "islets" of Langerhans who discovered them. It must be noted that the endocrine function of the pancreas is quite distinct from its other function, the production of the pancreatic juice (*see* p. 149). Each islet consists of two kinds of cells distinguished by size and also by certain chemical tests. The large cells produce the hormone **insulin** which promotes the conversion of hexose sugars, present in the blood, into glycogen, stored in the liver. These cells are directly sensitive to the blood sugar level; if it is high, then insulin is secreted, favouring the glucose→glycogen conversion until the level is normal, when insulin secretion ceases.

The smaller cells produce the hormone **glucagon** in response to stimulation by somatotropin from the pituitary. Glucagon promotes the breakdown of glycogen in the liver to glucose, which is released into the blood. Hence, the balance of these two hormones, insulin and glucagon, is very important in maintaining the level of blood sugar.

Deficiency of insulin leads to the disease **diabetes mellitus,** which is essentially due to the inability to convert hexose sugars in the blood into glycogen stored in the liver. Hence, after a meal, blood sugar rises to such a high level that most of it is excreted in the urine; the small amount left in the blood is rapidly used by the tissues and there is little or no glycogen reserve in the liver. Consequently protein and fat are utilized for energy release and this leads to constant loss of weight. As a result of excessive conversion of fat into carbohydrate, poisonous **ketones** are formed, giving rise to the disease **ketosis**; some of these ketones are excreted in the urine, a condition known as **ketonuria.**

Until 1922, when insulin was discovered by Banting and Best in Toronto, diabetes was always fatal. Since that time, there has been rapid progress in the treatment of diabetics; nowadays they are given graded doses of insulin, and many patients learn to inject themselves with correct and regular doses. The insulin is derived from the pancreas of cattle, sheep and pigs; the material, sold by butchers as "sweet-breads," is collected from slaughterhouses. This may soon be out-dated because insulin can now be synthesized in laboratories.

The Pineal Gland

The pineal gland is a small ovoid organ at the end of a short stalk; it lies almost in the centre of the top of the brain between the cerebral hemispheres; in the embryo it is formed as an outgrowth of the brain. It has had a strange evolutionary history; in the primitive vertebrates, it was a third eye on top of the head, complete with optic stalk and retina. This condition is still found in lamprey larvae and in the tuatara lizard of New Zealand. In all other modern vertebrates, it functions as an endocrine gland, producing the hormone **melatonin**, which affects maturation of the gonads. It seems to act in conjunction with day-length in affecting the onset of sexual cycles in many mammals. Whether this is so in human beings is not known with certainty.

The Ovaries

Apart from their primary function of producing ova, the ovaries secrete hormones both from the cells of the Graafian follicles and later from the corpora lutea. The **menstrual cycle** (*see* p. 225) can be divided into two distinct phases, **follicular** and **luteal.** It must be noted that all the Graafian follicles, though immature, are present at birth; there are probably several hundred thousand, but by the age of puberty there are about ten thousand. Of these, about four hundred ova will actually be produced. The follicular phase is activated by follicle-stimulating hormone from the pituitary; this induces development of Graafian follicles and promotes secretion of **oestrogens** which stimulate the development of the mammary glands, the uterus and the secondary sexual characteristics; the latter include the development of the breasts, the growth of pubic hair and deposition of fat under the skin to give the female figure. This follicular phase, every 28 days, is completed when the egg is ovulated.

The luteal phase is stimulated by luteinizing hormone from the pituitary. This induces growth of the corpus luteum which becomes filled with yellow cells. These secrete the hormone **progesterone** which prevents further ovulation and stimulates the uterus to secrete a milky fluid to nourish the developing embryo (if fertilization has taken place). The uterus is also prepared for implantation of the embryo and the mammary glands are stimulated to further growth.

The production of oestrogens and progesterone reaches its maximum level when the embryo is about three months old. Thereafter, this production declines and the same hormones are then secreted by the placenta. **Relaxin,** from the corpus luteum, inhibits uterine contraction until birth is near, and then it causes dilation of the birth passage. Uterine contraction during the birth process is stimulated by pituitary **oxytocin.**

It must be remembered that if fertilization has not occurred, the ovum is discharged and the menstrual cycle continues its normal sequence every 28 days.

The Placenta

Progesterone and at least one **oestrogen** are produced by the placenta after secretion of these hormones by the ovaries has declined. Together they influence the further development of the uterus and mammary glands, and of great importance is the suppression of further ovulation by progesterone.

The Testes

Apart from their major function of producing sperm, the testes also secrete hormones from small groups of interstitial cells (*see* p. 228). The counterpart of the female luteinizing hormone stimulates the production of **androgens** which induce the development of sperm and control the activity of the seminal vesicles and prostate glands. At the age of puberty, these hormones also promote the development of the male secondary sexual characteristics; there is marked growth of hair on the face, in the armpits and in the pubic region; the voice deepens and the typical male muscular figure develops. In both sexes, this age of puberty is also characterized by some degree of mental and emotional stress and is often marked by instability.

The Kidneys

A short distance before an arteriole enters a Bowman's capsule, it is in close contact with a uriniferous tubule. At this point there is a small cluster of cells packed around both the arteriole and the tubule. These clusters constitute the endocrine portions of the kidneys. The cells are sensitive to the concentration of sodium ions in the blood; when this level falls below a critical value, the cells secrete the hormone **renin**. This acts on a globulin in the blood called **hypertensinogen**, converting it into the active **hypertensin**. In its turn, this acts as a hormone, causing constriction of blood-vessels in the kidneys, and stimulating secretion of mineralocorticoids from the adrenal cortex. These have the effect of increasing the resorption of sodium ions from the tubules back into the blood.

It should now be realized that the function of osmo-regulation in the kidneys is a very complex process controlled by the combined effects of a number of hormones. They include somatotropin from the pituitary, mineralocorticoids, renin from the kidneys, hypertensin in the blood and parathormone from the parathyroids. The net result, in a healthy body, is that the water and salt content of the blood is adjusted to the correct level; throughout the circulation the water and salt content varies; in the kidneys, it is adjusted.

The Alimentary Canal

The wall of the alimentary canal, in several regions, secretes hormones which control the production of some of the digestive juices. From the stomach wall, there is secretion into the blood of the hormone **gastrin** which induces the continued flow of the gastric juice (*see* p. 148); **enterocrinin**, also from the stomach wall, initiates production of digestive juice from the intestinal wall; **gastric secretin** promotes the liberation of part of the pancreatic juice (salt solution). In the wall of the duodenum, four hormones are secreted; **secretin** stimulates continued flow of salt solution from the pancreas; **pancreomysin** induces secretion of the pancreatic enzymes; **cholecystokinin** causes contraction of the gall bladder, thus forcing bile into the duodenum; **enterogastrone** inhibits acid secretion in the stomach and also prevents peristalsis of the stomach wall, so that expulsion of chyme into the duodenum is periodic, not continuous.

The whole system of alimentary hormones is closely integrated so that correct conditions exist at the correct times for the digestion of the food.

Nerve Endings

It has been previously mentioned (*see* p. 269) that hormones are the actual agents which transfer impulses from nerve-endings into cells.

Sympathetic nerve-endings secrete **sympathin**, of which there are two closely allied types. One has effects identical with those of adrenalin, while the other has exactly opposite effects. The nerve-endings which stimulate voluntary muscle (end-plates) secrete **acetylcholine,** while interconnection between the vast network of cells in the brain also depends on the secretion of this hormone from the synaptic knobs. Parasympathetic nerve-endings also secrete acetylcholine.

Tissue Hormones

Two hormones, **histamine** and **acetylcholine** are probably present in all tissues in inactive form. When cells are damaged, these hormones are liberated; they cause inflammation by dilation of blood capillaries, freer release of lymph, and they attract lymphocytes to the area. There is what one might call a local concentration of the body's defences in the damaged area, causing the red colour and the swelling. When histamine is applied to raw flesh, there is acute pain; it is possible that the substance may be the actual mediator in skin pain sensations. There is some reason to believe that these tissue hormones, in conjunction with mineralocorticoids and glucocorticoids, are the main agents in promoting repair of damaged tissues.

Organizers

The early processes of development and the correct sequence of these processes are induced by hormones called **organizers** which are produced in various groups of cells in the embryo. It is probable that there are many different organizers, each responsible for producing a particular structure in the correct place and at the correct time.

Importance of Hormones

All the substances produced in the cells of the body are due to the activity of enzymes (*see* pp. 146–151). Some hormones stimulate enzyme activity and others inhibit it. There is a delicate balance maintained between all the various hormones to ensure that all the vital processes are carried out harmoniously.

It is scarcely possible to separate the nervous and endocrine systems. The centre of control of the endocrine system is the hypothalamus, which also controls the autonomic nervous system. The hypothalamus is itself stimulated by all other parts of the brain. All the nerves eventually exert their effects through the medium of the hormones sympathin and acetylcholine. It is best to regard these two systems as an interconnected and integrated mechanism for the control and co-ordination of all the activities of the body.

Pheromones

It is very probable that all animals secrete substances into the external environment which influence the behaviour of other animals, particularly those of the same species. Many cases have been investigated and it is found that such substances, called **pheromones,** act as hormones in having a co-ordinating effect on animals of a species, particularly in regard to breeding. A few examples are given here.

The most outstanding cases of co-ordination and control by pheromones are found in the social insects, which form what seem to be well-organized communities, e.g. bees, ants, wasps and termites. The queen bee in a colony secretes a substance called **oxo-decenoic acid,** and spreads it over her body. The worker bees, which are sterile females, lick this substance and it inhibits the development of ova in their bodies; the same substance also attracts the males (drones) to the queen. The bees recognize members of their own colony by perception of this substance, and they will prevent any other bees from entering the hive or nest. If the workers are denied access to this substance, which can be achieved by screening the queen from them, the colony lapses into a state of unco-ordinated confusion. If the meshes of the screen are large enough to allow the workers to touch the queen, all is well. Similar secretions control and co-ordinate other insect colonies.

The male musk deer of Tibet mark their territories with a strong-smelling oil secreted from their abdominal glands; this substance warns males of the same species against intrusion, and also acts as a sexual attractant in the breeding season. There are many similar examples among other mammals. The lemurs mark trees with an odorous secretion from glands at the sides of the face. Male dogs and monkeys, and many other mammalian groups, respond strongly during the "heat" periods to substances secreted from the genital apertures of the females; the same substances are also present in the urine of the females at this time. In several cases, these substances have been found to be oestrogens.

Many female insects secrete sex attractants which are effective over great distances. In some cases investigated in moths, the substance has attracted males from a distance of 10 km.

Many examples of similar phenomena are known; they range through all the great groups of animals from protozoa to mammals. The secretions vary widely in chemical nature and thus they cannot easily be classified. It is best to regard them as hormones secreted externally which promote behaviour conducive to the survival of the species.

SUGGESTED EXERCISES

1. Examine the location and shape of the pituitary gland on the preserved brain of a rabbit or rat.

2. Locate the thyroid gland in the neck of a dissected rabbit or rat.

3. Examine a prepared slide showing a section of the thyroid gland. Note the rounded vesicles, which contain the thyroxin when the animal is alive.

4. Examine a prepared slide showing a section of the pancreas. Look for the groups of small cells, which are the islands of Langerhans.

5. What symptoms have you noticed that indicate that adrenalin is circulating in your blood when you are in a state of fear, tension or apprehension? Some novelists have described these symptoms very adequately. Make a list of them.

6. Dissect the thyroid gland to expose its dorsal surface; locate the small brownish parathyroid glands. Use the rat or rabbit from Ex. 2.

7. Remove the adrenal glands from a dissected rabbit or rat and leave them to harden for a few days in 70 per cent alcohol. Examine thin sections, cut with a razor, and note the medulla and three zones of the cortex.

8. The effect of the thyroid gland in promoting metamorphosis can be observed on tadpoles about ten weeks old. Separate five or six tadpoles from the main stock, and feed them with small portions of thyroid gland from a freshly dissected rabbit or rat. Observe the effects and compare the rate of metamorphosis of the two stocks.

Chapter 16

HEREDITY

All living creatures are normally capable of producing offspring of the same species as the parents. This capacity may take two forms, sexual or asexual reproduction (*see* p. 219). Human beings, like all other vertebrate animals, can produce offspring only by the sexual method, which involves the union of the gametes of both sexes in the act of fertilization (*see* Chapter 12). Unless it suffers accidental death or serious disease, a baby will grow up to become another adult human being, and the sort of person it will become depends on two factors. In the first place, it will inherit certain characteristics from its parents, and in the second place, all the factors in its environment will affect it. The determination of the relative importance of the parts played by heredity and environment is one of the most difficult branches of biology, and in many cases the proportions of each cannot be assessed.

The basic principles of heredity were discovered by an Austrian monk named Mendel who published an account of his work in 1857. Unfortunately, his writings were neglected at the time and rediscovered only in 1901. Since that time, enormous progress has been made in the study of heredity. Some of Mendel's work is described on p. 311.

What Do We Inherit?

Firstly, we inherit the ability to live and to develop as members of our own species and not as horses, mice or frogs. Secondly, we inherit all our physical and mental characteristics, but the development of these depends, at least to some extent, on our environment. In recent years, much attention has been given to improving the environment in which human beings grow and develop. Nevertheless, however great our efforts, we cannot improve the environment to such an extent as to make all free from disease, of equal stature or of equal capability. There is a limit to any one person's achievement in any one aspect of his development and that limit is imposed by his inherited characteristics.

The Chromosomes

It has been mentioned (*see* pp. 3 and 220) that every cell contains a nucleus, and that each nucleus contains structures called chromosomes. These chromosomes are essentially long chains of complex chemical substances called **nucleic acids.** The actual sizes of the molecules of nucleic acid are not known but it has been shown that they consist of paired chains arranged in a helix and joined laterally by hydrogen atoms. The variable part of each chain consists of the nitrogenous substances **adenine, thymine, guanine** and **cytosine**; each of these is united to a sugar group ($C_5H_{10}O_4$) and adjacent sugar groups are united by a phosphate group, thus—

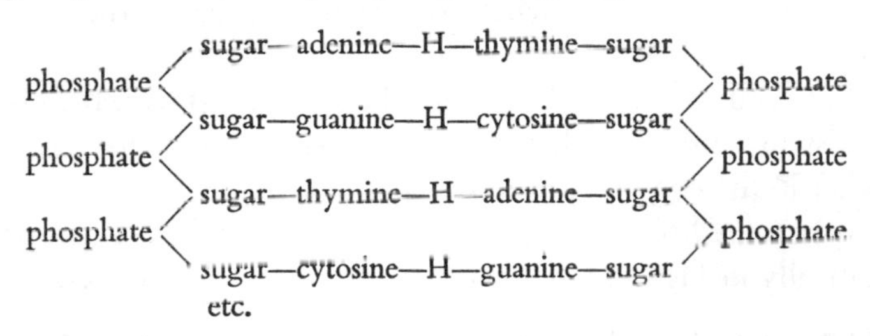

The sequence of adenine, guanine, cytosine, thymine along the chain varies, but the pairing between the chains is always the same, adenine pairing only with thymine and guanine with cytosine. Such chains, varying in length and each with a thread of protein, constitute the chromosomes.

The various species of animals and plants differ both in the numbers and lengths of their chromosomes and also in the sequences of adenine, guanine, cytosine and thymine along the lengths. Within a single species, the numbers and lengths of the chromosomes are identical, but the sequences may vary. Thus, all members of our own species, *Homo sapiens*, have forty-six chromosomes and the lengths of these chromosomes are identical throughout the whole population except for certain abnormalities which are discussed on p. 322. The exception, which may be noted here, is that one of the male sex chromosomes differs in form from the corresponding female chromosome.

It is now known that the sequences along the chromosomes determine the proteins that any cell can make, and, since all the enzymes are also proteins, we can say that these sequences determine all the chemical reactions that can take place in any cell.

Genes

Until the recent discovery of the chemical structure of chromosomes it was always stated that they consisted of linear series of **genes,** rather like beads on a string; each gene, or sometimes a group of genes, determined the presence or absence of a particular inherited characteristic. The genes are actually portions of the chains shown on p. 305, but we do not yet know what particular lengths of the chains are concerned.

In all the cells of the body, except the sperm and ova, the chromosomes exist in pairs, each having a maternal member inherited from the mother in the egg and a paternal member inherited from the father in the sperm. Chromosomes of a human female are shown diagrammatically in Fig. 174. Chromosomes of a male are identical with

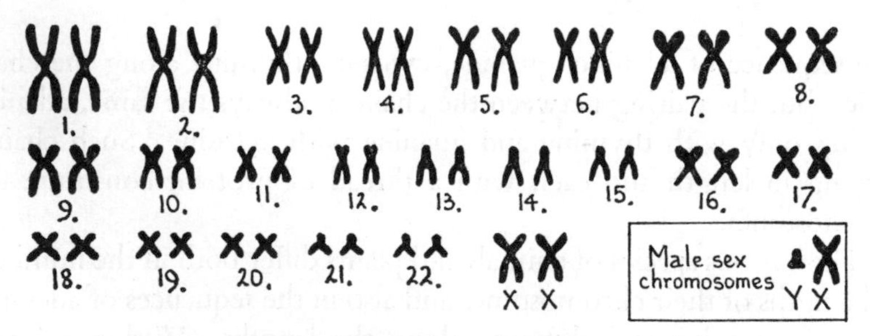

Fig. 174. The Human Chromosomes (Female), Arranged in their Pairs; Male Sex Chromosomes Shown Separately; × 2000

those of the female, except that, instead of having two equal sex chromosomes (*X* chromosomes), the male has one *X* and a shorter one known as *Y*.

Cell Division

In animals there are two kinds of cell division, known as **mitosis** and **meiosis.** Mitosis takes place in all cell divisions except those concerned in the formation of the sperm and ova. Meiosis takes place only in the two final divisions that give rise to these sperm and ova.

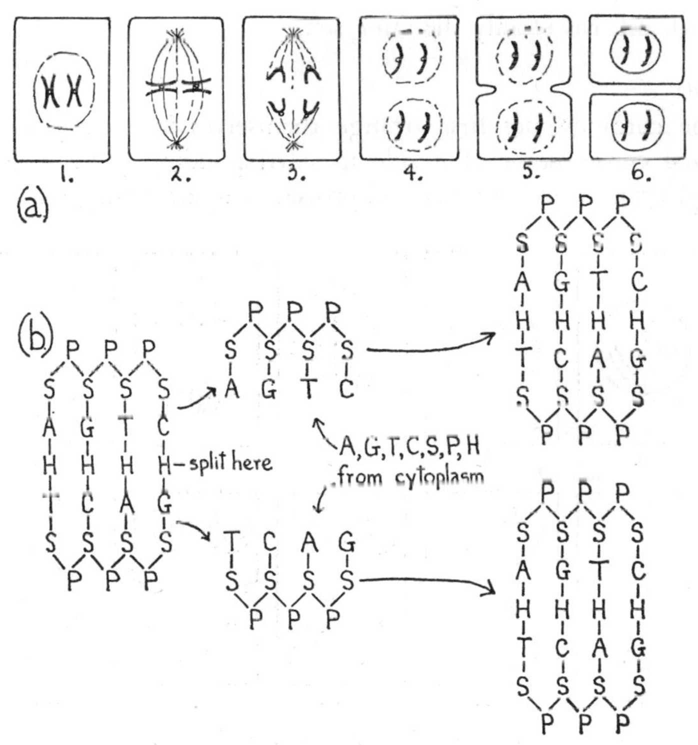

FIG. 175. Mitosis (*a*) Stages shown with one pair of Chromosomes (*b*) The Splitting and Duplication of a Part of a Single Chromosome

MITOSIS

The chromosomes each divide into two threads by splitting along the line of H atoms (*see* Fig. 175). Each half chromosome is then restored to a whole chromosome by the addition of the lost substances (adenine, guanine, cytosine, thymine, phosphate, sugar) from the surrounding cytoplasm. Thus two identical chromosomes are formed; they separate and move to opposite ends of the cell. This takes place for every chromosome and so the whole process results in two identical sets in the same cell. The cytoplasm is then divided and the cells separated by intercellular jelly or, in the case of plants, by a cell wall.

In our own case, every cell thus contains forty-six chromosomes and each cell has exactly the same set.

MEIOSIS

The chromosomes first arrange themselves in their pairs, each maternal chromosome alongside its corresponding paternal partner (*see* Fig. 176). After a complicated process of intertwining, when there

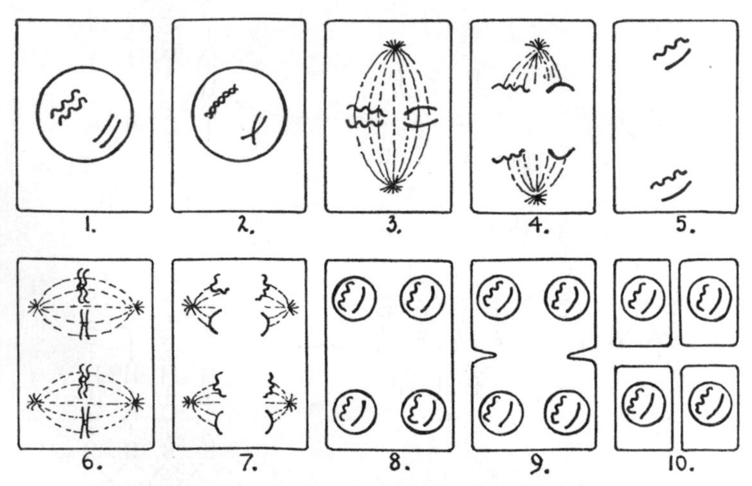

FIG. 176. Diagrams Illustrating Meiosis
Two pairs of chromosomes are shown: 1 to 5, the first division; 6 to 10, the second division

may be actual interchange of portions of the chromosomes, the members of each pair separate, the maternal and paternal partners moving to opposite ends of the cell. There is no standard pattern in this separation; all the maternal chromosomes do not move to one end and the paternal to the other, although there is a remote chance that this might happen.

Thus, in the case of the human being, there will be two nuclei in the cell, each with twenty-three chromosomes. This first division of meiosis is immediately followed by a second, when division of the chromosomes takes place in the manner described for mitosis, so that, from the original cell with forty-six chromosomes, there are now four with twenty-three chromosomes each.

Fertilization

The union of two gametes, a sperm and an ovum, each with twenty-three chromosomes, results in the formation of a zygote with forty-six chromosomes. From this zygote, by repeated mitotic division, the whole body is built up of cells each containing forty-six chromosomes identical with those in the zygote. But it must be noted carefully that these chromosomes will be unlike those of either parent, for two reasons.

1. In the first meiotic division there was haphazard distribution of maternal and paternal chromosomes to the two newly formed nuclei.

2. In the intertwining process the paired chromosomes may have interchanged portions of their chains. There may also be mutations (*see* p. 321).

Because of these two facts, and because each parent contributes twenty-three chromosomes to the zygote, it is impossible for any child to resemble either of its parents exactly and it is an extremely remote possibility that any two children in the same family will resemble each other exactly. (The case of identical twins is different, since they are developed from halves of the same zygote, not from two zygotes, and therefore they must have identical chromosomes.)

Inheritance of Characteristics

Each of us inherits twenty-three pairs of chromosomes, one set from the father and the other from the mother. From the moment of fertilization, each of us has a code that determines all the chemical reactions we can carry out. Provided that the environment is suitable, each will develop and show those characteristics for which the requisite code is present on the chromosomes.

The genes are portions of the lengths of the chromosomes; each gene has a specific length, as yet undetermined, and, since the chromosomes are in pairs, the genes are in pairs, each member of a genic pair

FIG. 177. Diagram Illustrating the Paired Nature of the Chromosomes and of the Genes

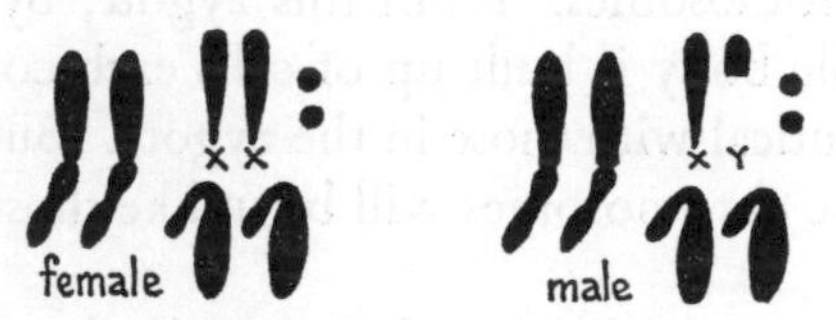

FIG. 178. Chromosomes of *Drosophila melanogaster*

being situated in identical positions on the two chromosomes. An example to illustrate this important idea of paired genes is shown in Fig. 177. Thus the gene marked A may affect eye colour and its partner on the other chromosome of the pair will also affect eye colour, but, perhaps, in a different manner. The small size and large number of human chromosomes have made it extremely difficult to locate particular genes, except for some on the sex chromosomes. But in some species of animals, particularly a little fruit fly called *Drosophila*, the actual locations of most of the genes have been mapped very accurately. *Drosophila* has only four pairs of chromosomes easily recognizable in shape (*see* Fig. 178), and, fortunately, the cells of the salivary glands have very large chromosomes.

Kinds of Genes

In their structure, all genes consist of lengths of the chain shown on p. 305, but each particular gene is different from all others in its length and in the sequences of adenine, guanine, cytosine and thymine along it. Some genes have the power of suppressing the action of their partner genes and thus they are called **dominant,** while the suppressed genes are known as **recessive.** For example, Mendel crossed pure-breeding tall pea plants with pure-breeding dwarf plants and found that the plants of the next generation were all tall, indicating that the dwarfness characteristic had been suppressed by the tallness characteristic. Therefore tallness was known as dominant and dwarfness recessive. It is customary to show the process that has occurred in the following form, where symbols are used for the sake of brevity. Thus $P =$ the original parent plants, a multiplication sign $\times$ = mating, F_1 = offspring of these parents, F_2 = offspring of the second generation with F_1 as parents, $G =$ gametes, $T =$ tallness (capital letter for the dominant gene), $t =$ dwarfness (small letter for the recessive gene). Thus in crossing tall and dwarf pea plants, we have—

P	$TT \times tt$
G	T and t
F_1	Tt (all tall since T suppresses t).

All these F_1 plants are **hybrids,** i.e. not pure-breeding, since they have one factor for tallness and one for dwarfness. Now if two of the F_1 plants are crossed, the result shows that the dwarfness characteristic has not disappeared but is suppressed when it is associated with the tallness characteristic, i.e.

P	Tt		$\times$		Tt		
G	T and t				T and t		
F_2	TT +	Tt +		Tt +		tt	in equal proportions
	pure-breeding tall	hybrid tall				pure-breeding dwarf	

The actual plants would show a ratio of 3 tall : 1 dwarf. It must be noted that, unless we use very large numbers of such matings, the ratio is not likely to be 3 : 1. This type of gene, which exists in two distinct forms, dominant and recessive, is very common in plants and in the lower animals.

There are some genes that do not exert their full effect unless present in the double dose. For example, if red antirrhinums are crossed with white antirrhinums, the offspring are all pink.

P	RR (red)	$\times$	rr (white)
G	R	and	r
F_1		Rr (all pink)	

But the whiteness characteristic has not disappeared because it emerges again in the F_2.

P	Rr	$\times$	Rr
G	R and r		R and r
F_2	RR +	$\underbrace{Rr + Rr}$ +	rr
	25% red	50% pink	25% white

Such examples are known as **blending inheritance.**

Very often the final appearance of a characteristic such as eye colour, skin colour and hair colour, is due to several pairs of genes. This **polygenic inheritance** is found in most human characteristics.

Some genes are **lethal** in their effect when present in the double dose; the possessor always dies. Thus pure-breeding yellow mice cannot exist; they always die before birth. Hence, all the yellow mice that exist must be hybrids. The mating of two such hybrid yellow mice is shown below—

P	Yy	$\times$	Yy
G	Y and y		Y and y
F_1	YY +	$\underbrace{Yy + Yy}$ +	yy
	dies	hybrid yellow	normal grey colour

A number of such lethal genes can occur in human beings.

Some Examples of Human Inheritance

It has already been pointed out that most human characteristics are due to polygenic inheritance, i.e. several pairs of genes affect a particular characteristic. There are, however, a few cases where inheritance is due to a single pair of genes.

Inheritance Due to a Single Pair of Genes

Some people can detect a bitter taste in the rind of a grapefruit, while to others it tastes like pure water. The substance causing the taste is **phenyl-thio-urea.** Tasting is dominant to lack of the ability to taste. Thus, in a mating between one parent who is a pure-bred taster, and one who is not, the following is the result.

P	*TT* × *tt*
G	*T* and *t*
F_1	*Tt* (all tasters)

Now, if two tasters of the F_1 type mate, the offspring on the average of a large number of cases will be—

P	*Tt*		×		*Tt*
G	*T* and *t*				*T* and *t*
F_2	*TT*	+	*Tt* + *Tt*	+	*tt*
	pure-breeding taster		hybrid tasters		non-taster
	25%		50%		25%

It must be clearly understood that the number of offspring in human families is too small for an average result to be obtained in any single family. The offspring in each case shown here indicate the ratios that would be obtained in a very large number of cases.

Some human beings can protrude the tongue and curl its sides upwards to form a kind of tube, while others cannot do this. Curling is dominant to non-curling, thus mating of a pure-bred curler and a non-curler would, on average, give the following result—

P	*CC*	×	*cc*
G	*C*		*c*
F_1	*Cc* (all hybrid curlers)		
P	*Cc*	×	*Cc*
G	*C* and *c*		*C* and *c*
F_2	*CC* +	*Cc* + *Cc* +	*cc*
	pure-breeding curler	hybrid curler	non-curler

Total colour-blindness in human beings is rare; people with this defect see every object as black, white or shades of grey. The defect is due to a single pair of genes in the recessive condition (*cc*).

P	*CC*	×	*cc*
G	*C*		*c*
F_1	*Cc* (no colour-blindness but all are hybrid)		
P	*Cc*	×	*Cc*
G	*C* and *c*		*C* and *c*
F_2	*CC* +	*Cc* + *Cc* +	*cc*
	pure-breeding with colour vision	hybrid with colour vision	colour-blind

Partial colour-blindness is sex-linked and is dealt with on p. 320.

Blood groups on the AB system are probably inherited through a single pair of genes. A group O individual carries *OO*, a group A person carries *AA* or *AO*, and a group B person carries *BB* or *BO*. The most interesting mating here is the following—

P	*AO*	×	*BO*
G	*A* or *O*		*B* or *O*
F_1	*AB* +	*AO* + *BO* +	*OO*

Thus, individuals of all four blood groups A, B, AB and O may be in the same family.

The inheritance of Rhesus blood groups, i.e. Rh positive and negative, has been shown to be due to three pairs of genes: *C*, *D*, *E*, all dominant, and *c*, *d*, *e*, all recessive. It seems fairly certain that Rhesus incompatibility between a mother and her child is due to the *D* factor. Those individuals possessing *D* (either *DD* or *Dd*) are Rhesus positive and those possessing *dd* are Rhesus negative. Thus we have the following possibility when a father is Rh positive and a mother Rh negative—

P	*DD*	×	*dd*
G	*D*		*d*
F_1		*Dd*	

This Rh positive child is carried in the uterus of a mother who is Rh negative, and thus in some cases, trouble occurs later in pregnancy (*see* p. 191).

Lethal Genes

One form of anaemia, which is commonest in countries bordering the eastern Mediterranean, is called **thalassemia.** It is apparently due to the presence of a factor *T* (dominant to the absence factor *t*). An individual *TT* invariably dies before maturity. Both the parents of such a child must, therefore, have been carrying *Tt*, thus—

P	*Tt*		×		*Tt*
G	*T* and *t*				*T* and *t*
F_1	*TT*	+	*Tt* + *Tt*	+	*tt*
	die with severe anaemia		suffer from mild anaemia		do not suffer from this type of anaemia

A dominant gene in the double condition causes the formation of dark and brittle skin on those parts of the body most exposed to light, i.e. the head and hands. Sooner or later, cancerous growths begin and death usually occurs before maturity. Therefore a child with this disease must have had parents hybrid for the condition, i.e. *Bb* and *Bb* (B = brittle). Thus—

P	*Bb*		×		*Bb*
G	*B* and *b*				*B* and *b*
F_1	*BB*	+	*Bb* + *Bb*	+	*bb*
	dies		hybrids show only traces		completely clear

A number of other conditions that have been lethal in the past are now brought under control. One is the condition where an individual lacks serum globulin and cannot produce antibodies against various diseases. Nowadays, such persons are kept alive with antibiotics and injected gamma globulin.

Polygenic Inheritance

Many human characteristics are due to the presence of several pairs of genes. Notable examples are skin colour, hair colour, eye colour and intelligence.

Skin colour is due to the interplay of a number of factors, some of which are inherited and some of which are due to effects of the environment. The inherited factors are the presence of one or both of two pigments: a brown substance, **melanin,** and a yellow substance, **carotene.** A pure Negro carries several pairs of genes that favour the production of larger quantities of melanin. An **albino** has no colour genes and therefore no ability to make melanin. We do not know how many pairs of genes are involved, but it is certainly more than two. If we take three pairs as a possibility, then the following would be the result of a mating between pure black (Negro) and pure white (albino).

P	*BB CC DD*	×	*bb cc dd*
G	*BCD*		*bcd*
F_1	*Bb Cc Dd* (all brown—mulattos)		

If two mulattos mate, then the children may have skins of any shade from black to white.

P	*Bb Cc Dd*	×	*Bb Cc Dd*
G	*BCD, BCd,*		*BCD, BCd*
	BcD, BcD		*BcD, BcD*
	bCD, bCd		*bCD, bCd*
	bcD, bcd		*bcD, bcd*

Any one type of gamete has an equal chance of fusing with any other and hence the children can vary between *BBCCDD* and *bbccdd*, i.e. six-sixths black and nought-sixths black (white).

Albinos are rare, and most so-called white people are one-sixth or two-sixths black. Persons who are one-sixth black could have any of the following genetic groups: *Bbccdd, bbCcdd, bbccDd*. The yellow colour of the Chinese and Japanese is due to a combination of genes for melanin production and genes for carotene production, while the bronze races, such as the Red Indians, also have both types of genes with relatively more favouring melanin production than in the Chinese.

Colour of hair is also due to several pairs of genes and may vary from black to white.

Similarly, eye-colour, which is really the colour of the iris, is due to several pairs of genes. The pigment may appear in the back layer of the iris, or in both back and front layers, and the deposition of the pigment may be heavy or light. Black eyes are due to heavy deposits of melanin in both layers; brown eyes have a heavy deposit at the back and a lesser deposit in front; blue eyes have a light deposit in the rear but none in front; grey eyes have a light deposit in the rear and a little scattered pigment in front; green eyes have a light deposit in the rear and scattered yellow pigment in front. True albinos have no pigment in the iris and the pink colour is due to the small blood-vessels. It seems certain that at least three pairs of genes are involved. In general, genes for dark eyes are dominant to those for lighter eyes, but it is possible that blue-eyed parents may carry genes for brown eyes, and if these genes are transmitted to a child by both parents, then the child may have light brown eyes.

The inheritance of height is also due to a number of pairs of genes;

among these, it seems that genes for shortness are dominant over genes for tallness. But it is more likely that two tall parents will produce tall children than that two short parents will produce short children. Tall people are not likely to possess any of the dominant shortness genes, but short people may possess recessive tallness genes. Thus, if we consider three pairs of height genes *A*, *B* and *C*, then a tall person is likely to have genes *aabbcc* but a short person would look the same with *AaBbCc* and *AABBCC*. In a mating between two short parents with genes *AaBbCc* an offspring could be either short or tall or somewhere between. But if both parents carry *AABBCC* then all their children will be short. Unfortunately, it is rarely possible to find out the actual types of genes, because we do not know how many pairs are involved, and also because human beings do not have enough offspring for a valid judgement to be made.

Much has been written about intelligence and about what fractions of it are due respectively to heredity and to environment. There is undoubtedly an inherited component but it is difficult, in most cases, to say whether the effect of the genes or the environment is more important. Careful measurements based on intelligence tests seem to indicate that more than half the variation is genetic in origin and the remainder environmental. If this is so, then the gradual equalization of environment and educational opportunity will gradually show greater variation among those tested. Certain famous families have often been quoted to show that "genius runs in families." The Bach family, in four generations, produced six notable composers, and thirty-one talented musicians; in the third generation, no less than seven were cathedral organists. The Darwin family, from the time of Erasmus Darwin, who was the grandfather of Charles Darwin, the biologist, produced, in five generations, ten Fellows of the Royal Society, the highest award for scientists in Great Britain (*see* Fig. 179). There was at least one Fellow in each generation. There are many other examples of such families, but one must be careful not to argue from this that music and science run in families. It is just as correct to argue that the children became musicians or scientists because they grew up in an atmosphere where music or science was the all-important consideration and the most frequent topic of discussion.

The general modern opinion is that intelligence and ability are at least fifty per cent inherited.

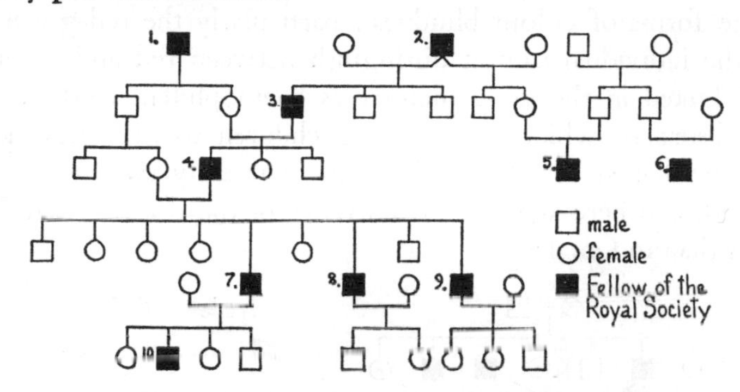

FIG. 179. Part of the Pedigree of the Darwin Family
1. Josiah Wedgwood, Potter 2. Erasmus Darwin, Doctor 3. Robert Darwin, Doctor 4. Charles Darwin, Biologist 5. Francis Galton, Geneticist 6. Douglas Galton, Engineer 7. George Darwin, Astroomer 8. Francis Darwin, Botanist 9. Horace Darwin, Engineer 10. Charles Darwin, Physicist

SEX-LINKED INHERITANCE

Characteristics that are described as sex-linked are those represented by genes on the sex chromosomes. It has been stated previously that females possess two *X* chromosomes, and males one *X* and one *Y*. On part of the *X* chromosome, a male may carry unmatched genes because of the shorter length of the *Y* chromosome

The inability to clot blood, causing **haemophilia,** is sex-linked, and is due to a recessive gene. Females with two *X* chromosomes may possess two dominants, *HH*, one dominant and one recessive, *Hh*, but they could hardly survive with two recessives, *hh*, since the first menstruation would probably be fatal. Males, on the other hand, can possess *H*– or *h*–, the latter suffering from haemophilia. Few affected males survive to maturity, but, fortunately, the condition is rare. Fig. 180 shows how the condition is transmitted from a carrier mother to her sons. There is a fifty per cent chance of each son being affected and of each daughter being a carrier. Queen Victoria was a carrier of haemophilia, and her daughters introduced the gene into the Spanish and Russian royal families. Her son, King Edward VII, did not inherit the gene and could not transmit it to his descendants.

Some forms of colour-blindness, particularly the red-green type, where the individual cannot distinguish between red and green, are also sex-linked in the same manner as haemophilia. Certain severe muscle disorders, which affect young children so that they quickly become permanent invalids and die in their early twenties, are also sex-linked, and here again the occurrence of the disease is far higher in males than in females.

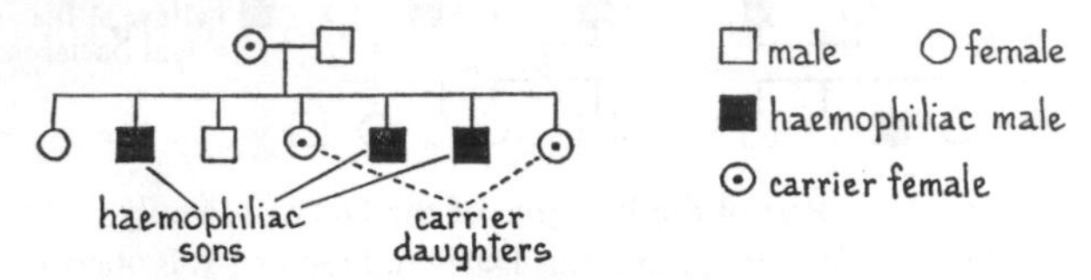

FIG. 180. Pedigree Showing Sex-linked Inheritance of Haemophilia

Evolution

A full account of evolution is not within the scope of this book. If the reader has grasped the essential facts about the structure of chromosomes, about cell division and about sexual reproduction, he should be able to understand something of the way in which evolution has occurred.

The essential idea of evolution is change. A boy may be interested in the evolution of the aeroplane, i.e. he wants to know how the earliest aeroplanes were gradually improved until the modern aeroplane was achieved. Similarly, we can speak of the evolution of language, of mathematics, or of any other study. In biology, evolution means the changes that have taken place in living creatures since life began; it is known as **organic evolution.**

We do not know how or where life began, but it seems certain that the first living creatures must have had the power of reproduction and hence of handing on their genes to their offspring. If these genes had remained unchanged, then all living creatures would still be exactly like the original forms, and there would be none of the millions of species that exist and have existed.

We now know that genes change and that chromosomes change: these changes are called **mutations.** There are many causes of mutation: among them are certain rays such as **cosmic rays** and **X-rays**; radioactive materials such as **radium, ultra-violet light,** and certain chemicals. But it is estimated that all these sources do not account for the actual rate of mutation that occurs, and it is suggested that there is a natural mutation rate for every gene and this is variously estimated as between one in ten thousand and one in a million occurrences of the gene.

A mutation may affect a single gene, a number of genes, a whole chromosome, or all the chromosomes. Unless, however, it occurs in the sex glands and affects some of the gametes, then it will not be inherited.

Some Examples of Mutation

1. There is a form of anaemia in which the red corpuscles, instead of being like biconcave discs, are shaped like small crescents; it is known as **sickle-cell anaemia.** The actual structure of the haemoglobin in these cells differs slightly from the normal, and it is considered to be due to a small change in one gene. It arises in persons who have no previous family history of this condition and is therefore a mutation that could arise in any person. The condition is usually fatal.

2. The antibiotic **penicillin** is extracted from a mould fungus called *Penicillium.* Numerous new strains of the fungus have been produced by subjecting the plants to X-rays. Some of these plants produce far more penicillin than the wild strain.

3. It has been shown that a severe muscle disorder, known as **muscular dystrophy,** is due in at least one-third of the cases to fresh mutations. One child in every twenty-five thousand is affected and death usually occurs in the early twenties.

4. The haemophilia transmitted by Queen Victoria probably

arose in her as the result of a mutation, since the condition could not be traced in the ancestry of either of her parents.

5. The condition of **mongolism,** which occurs in normal families with normal ancestry, is due to another type of mutation, the possession of an extra chromosome; there are three instead of two of chromosome 21 (*see* Fig. 174, p. 306). Mongol children (there is no association with the Mongolian race) constitute about two per thousand of the birth-rate in Western Europe. They are characterized by the small size of their facial features; the head, eyes, nose, mouth and ears are all small, the eyes are slanting, and the fingers are short. There is nearly always a high degree of mental deficiency. The risk of having a mongol child increases with the age of the mother. For a woman of twenty, the risk is one in three thousand, for a woman of forty-five, one in a hundred.

Very many examples of mutation are known and most are noticeable because they cause obvious and usually detrimental conditions. It is probably true that the majority of mutations are of no survival value. Nevertheless, there are numerous small mutations that have very slight effects, some of them beneficial.

Variation

Because of mutations, and because of recombinations in any interbreeding population, all the individuals vary, sometimes in small features, sometimes in large ones. If we can forget the artificial conditions under which human beings live, and think of natural populations, then it is obvious that there is a **struggle for existence.** Animals compete for food and for mates; plants compete for water and for light. Creatures with slight advantageous variations will tend to survive to maturity and produce offspring, which may inherit the advantageous variations. By accumulating advantageous mutations, a species gradually becomes changed; by accumulating disadvantageous mutations, it becomes extinct.

Earth movements and climatic changes in the past have caused isolation of groups of animals and, to a lesser extent, of plants, and, in isolation, they have evolved into various new species.

Charles Darwin propounded the idea of evolution by **natural selection,** which means that the creatures that survive do so because they are more fitted to their particular environment. They are better adapted than others of their species because they have advantageous variations; if these variations are to be inherited, they must be represented in the chromosomes.

Human Evolution

Human beings of our own species, *Homo sapiens*, have not existed for a very long period of time (*see* p. 26). We take a long time to reach the maturity necessary for reproduction, and the number of offspring is usually small. Therefore, human evolution is necessarily slow. Also, compared with wild animals in their natural state, we live a highly artificial existence, in highly artificial surroundings. We are in a stage that may be called **social evolution**; a struggle for existence is not the lot of all mankind, though it is for some. We strive to improve all aspects of the environment, but as yet we have explored only the fringes of the possibilities of improving human heredity. This is an important study, known as **Eugenics.** Certain aspects of this study are already applied in forbidding marriages between close relatives, who are likely to possess the same "bad genes"; in some parts of the world individuals who carry hereditary disease are undergoing voluntary sterilization; in many countries, such individuals are advised not to have children.

There is no doubt that all human beings would wish to see that all children are born free of serious hereditary physical or mental defects, but there is no way of ensuring this. There are several reasons why we may have serious doubts about any policy of **artificial selection.** In the first instance, it is difficult to feel that control of other people's lives is morally right. Secondly, in view of mutations and other factors, such as hidden recessive genes, it is doubtful whether any method of artificial selection would be really beneficial. Thirdly, we cannot be expected to know what qualities will be most beneficial to human beings in the remote future.

SUGGESTED EXERCISES

1. Examine on the high power of a microscope or microprojector prepared slides of longitudinal sections of the root-tip of a root of an onion or broad-bean plant. Various stages in mitosis can be seen.

2. Similarly examine prepared slides of transverse sections of the anthers (pollen sacs) of flowering plants, e.g. the lily. Various stages in meiosis can be seen.

3. Investigate the occurrence of tongue-curling in the members of your own class and of your own family. Try to draw up a family tree showing the inheritance of this characteristic.

4. Investigate the occurrence of the ability to taste phenyl-thio-urea. The substance can be purchased from biological dealers. The best strength of solution is forty milligrams per litre of distilled water.

5. Similarly investigate the ability to distinguish between saccharin and sugar. Use 300 cm^3 of water for each solution; in one dissolve two saccharin tablets, in the other, two lumps of sugar.

6. The occurrence of right-handedness and left-handedness, the colours of eyes, the variation in height at maturity, are interesting investigations in large families.

Do you consider that there are any particular inherited abilities in your own family?

7. Almost any species of animal or plant can be used to examine variation. Suitable examples are (*a*) the colour pattern on clover leaves, (*b*) human finger-prints, (*c*) the markings and colours on the shells of the common garden snail, *Helix aspersa*.

8. Most natural history museums show series of specimens that indicate the evolutionary changes that have occurred in groups of animals. Good examples are (*a*) the sea-urchins of the genus *Micraster*, (*b*) the horse, (*c*) the elephant.

CHAPTER 17

HUMAN DISEASE

BEFORE we can define disease, which means "not at ease," we must try to define health. If a person is happy, energetic, and suffering from no discomfort or disability, then we may say that he is healthy. Most people experience this desirable state at certain times, but it is rare to find abounding health persisting for long periods of time. At the least, the great majority of human beings suffer from minor disorders such as headaches, colds, or digestive troubles. We cannot say that the healthy state is the normal state, except possibly in the young, although even childhood has certain diseases associated with it, which seem to be almost inevitable—for instance: measles, scarlet fever, chicken pox, etc.

To most people, disease means drastic damage to some part of the body: damage that needs prompt medical attention. Everyone would call poliomyelitis a disease, but few would include the common cold in the same category, and yet it causes death in some parts of the world. Both, however, are caused by minute microbes called **viruses**. (*See Microbes* in this series.) In this chapter, the word "disease" will be used in the widest sense, to include any unusual or abnormal condition, even though this does not cause very apparent damage to the body.

Diseases may be classified in ten groups, which are briefly described below.

1. INHERITED ABNORMALITIES

These range from serious and often fatal diseases such as haemophilia to curiosities such as the possession of six fingers on each hand. A number of such inherited conditions were discussed in the last chapter and it was pointed out that in certain cases the condition can be considerably alleviated (*see* p. 316). Dwarfness, colour-blindness, mongolism and most other inherited abnormalities cannot be remedied; what can be done and is done in certain countries is to reduce the spread of such conditions among the population by advising persons who have the conditions, or those who may be carriers, not to have any children.

2. Infection or Injury Before or During Birth

Inherited abnormalities and conditions caused by some damage to the foetus during gestation are together known as **congenital diseases.** Some cases are clearly inherited, others are clearly due to damage occurring in the uterus or at birth, but some congenital conditions have not yet been assigned to one cause or the other. It is estimated that, of children dying in their first year, about one quarter die of a congenital abnormality, either inherited or acquired in the uterus. Of those who survive the first year, about one in every hundred is severely handicapped by congenital deformity. Some conditions that are clearly due to damage in the uterus are: "Rhesus babies," (*see* p. 190); congenital syphilis (*see* p. 230); cataract, deafness and deformed heart due to infection of the foetus with German measles from the mother; and the tragic cases of thalidomide babies. The drug **thalidomide,** taken by the mothers during pregnancy, in some cases had the extraordinary effect of preventing or restricting growth of the limbs. Some of the children, quite normal in other respects, have only remnants of limbs.

Other cases of congenital damage may arise at birth and may range from birthmarks to serious damage to the brain or limbs. If a foetus lies in a difficult position just before birth, then its emergence may necessitate assistance by strong pulling or by the use of forceps. In the past, cases have occurred where all the nerves of an arm have been damaged, leading later to complete inability to use the arm. Nowadays, a great deal is known about such damage and how it is caused, and correspondingly greater care is taken during the birth process.

A number of congenital abnormalities have not yet been definitely assigned to heredity or to intra-uterine damage, though they must be due either to one of these or to a combination of both. Such cases, which arise in families where there is no previous history of the abnormality, include mere oddities such as webbed toes or extra nipples on the breasts, and more serious conditions where the brain may be defective and mental development is considerably retarded, or where the heart is defective and the child dies sooner or later of cardiac failure. There is scope for a great deal of research into the conditions that affect the foetus while it is in the uterus.

3. Badly Adjusted Diet

It was pointed out in Chapter 7 that shortages of various essential substances in the diet lead to deficiency diseases (*see* p. 109). It was also pointed out that the diet needs to be balanced; all the essential food materials must be present in reasonably constant amounts.

Apart from diseases due to the lack of a single factor in the diet, lowering of the standard of nutrition leads to an increase in the incidence of certain other diseases, especially **tuberculosis**. During the two World Wars, the number of cases of this disease rose sharply in most of the participating nations, especially those subject to blockade.

Malnutrition is widespread in some parts of the world and is by no means unknown in some of the more enlightened nations. It is becoming increasingly obvious that even in prosperous communities there is need for universal education in simple **dietetics** (the study of diet).

4. Physical Injury

Physical injuries may range from simple cuts and bruises to sudden death. Forcible contact with blunt objects may produce little superficial damage but may have severe effects on internal organs. A blow on the head may not break the skull and yet may cause serious damage to the brain. The "kidney punch" is forbidden in boxing because of the very severe effect it may have on the kidneys.

Penetration into the body by a sharp instrument is serious if the penetration is deep. First, there is always the risk of introducing infection into the wound and, secondly, there is always the danger of excessive bleeding (**haemorrhage**). Also, in some cases, e.g. a bullet or stab wound, one cavity of the body may be put in communication with another. If the heart is perforated, blood will leak out into the chest cavity and the blood pressure will fall rapidly; if a lung is perforated, air passes out into the pleural cavity, the lung collapses, the patient is short of breath and soon loses consciousness from lack of oxygen.

Bone fractures may arise from forcible contact with blunt or penetrating objects; in order that a bone may heal in the correct position, some method of splinting is usually necessary.

A common cause of accident in older people is due to sudden strain on some part of the body. Most ruptures and "slipped discs" (*see* p. 62) are due to this cause. Those who play vigorous games will probably have experienced "pulled muscles," when fibres of the muscle are actually broken, and it is now realized that "warming up" is essential before a period of violent activity.

The dangers of destroying large areas of skin by burning or scalding were emphasized in Chapter 4. The skin performs such important functions that any large-scale loss is fatal, and further there is always serious danger of infection by microbes.

In this electrical age, most people are familiar with the use of many domestic appliances that derive their energy from electricity, but few realize the danger of shock from worn wires or from handling electrical apparatus with wet hands.

Since 1945, when the first atomic bombs were dropped on the Japanese cities of Hiroshima and Nagasaki, human beings have lived under a new threat—the threat of extermination by nuclear war. But it is not generally realized that we are all subject to natural radiations from **cosmic rays** and from **radioactive materials** present in the Earth's crust. To this natural radiation is added radiation from X-rays used in medical diagnosis and treatment, and from radioactive "dust" from every atomic explosion. Although the increase in radioactivity has not yet reached serious proportions it is now realized that any increase constitutes a further hazard to life.

In a time of rising population, coupled with increasing use of machinery in the home and in industry, we must expect a high accident rate, which can be reduced only by education in the causes of accidents and in safety measures that prevent them. Finally, everyone should have at least a basic knowledge of first aid.

5. Poisoning

In this account, the term "poison" is restricted to chemical substances, which, if taken into the body, have a damaging and sometimes a fatal effect. Many of the diseases caused by microbes are serious because of the effect of chemical substances produced by these microbes; such diseases are due initially to infection and are mentioned later in this chapter.

Apart from the use of chemicals for deliberate suicide or murder, poisons may be taken into the body by accident. For example, children may eat the poisonous black berries of the deadly nightshade, or the red berries of the wild arum (cuckoo pint). A poisonous fungus such as *Amanita phalloides* may be gathered with mushrooms and eaten. A person may be bitten by a venomous snake or by a poisonous spider. A bottle containing poison may be unlabelled or wrongly labelled and someone may drink it. Workers in many industries are subject to a gradual build-up of poisonous chemicals in their bodies. Some people become addicted to drugs such as opium, and eventually die of poisoning by them. Some substances, which are quite harmless to most people, have devastating effects on others. For example, **aspirin,** widely used to relieve headaches, causes immediate **asthma** in some people. We say that such a person is **allergic** to aspirin: his body cannot tolerate it. Many of these allergies are quite common; the most widely known are due to various kinds of pollen, which cause **hay fever.** Some people are allergic to tomatoes, to eggs, to bananas, to plums, to horsehair or to house-dust, and some are allergic to penicillin or to other drugs in common use.

In the First World War, the poisonous gases chlorine, phosgene and mustard gas were used. Even more lethal gases were held in readiness for the Second World War, but fortunately neither side used them. Carbon monoxide (CO) is a particularly dangerous poison because the first reaction to it is drowsiness, then sleep, and death occurs in the sleep. The gas combines with the haemoglobin of the red corpuscles, thus causing a gradual reduction of oxygen-carrying capacity. Escaping coal gas and the fumes of a car exhaust both contain carbon monoxide, and many accidental deaths have occurred in closed rooms or garages from this gas.

Overdoses of drugs, especially those used to induce sleep or relieve pain, such as **morphine, codeine** and **cocaine,** may cause death. More serious, however, is the fact that some people acquire the habit of using such drugs to reduce anxiety or tension, and gain feelings of confidence, happiness and peace. Though such drugs are not usually taken in fatal doses, the addicts begin to suffer from malnutrition, from loss of confidence, and eventually from inability to carry out a job.

The most serious risks of poisoning arise in various industrial processes where dangerous chemicals are manufactured or where less dangerous chemicals may be absorbed and may build up eventually into fatal quantities. Among such dangerous chemicals are compounds of **lead, mercury, arsenic, nickel, chromium** and **manganese.** These substances enter the body as dusts, which are inhaled through the mouth and nose. Some volatile chemicals, such as **benzene,** can cause fatal poisoning, and particularly dangerous are a number of derivatives of coal tar.

Inhalation of industrial dusts, especially very fine particles of **silica,** causes a variety of respiratory diseases, the worst of which is **silicosis.** Silica is the main constituent of sand, granite and quartz. Large particles do not penetrate further than the nasal cavities, where they are trapped, but very fine particles of $5\mu m$ in diameter or less penetrate into the lungs. They become enveloped in fibrous tissue and there is therefore a gradual reduction of the respiratory surface. Very often, tuberculosis arises as a secondary complication. Silicosis may occur in occupations where machining of hard rocks is necessary. Drilling in coal mines, blasting in quarries, machining granite blocks and sharpening tools on dry grindstones are all occupations where the disease has been prevalent. Nowadays, as far as possible, when these processes are performed, jets of water are arranged so that the stone surface is kept wet and very little dust rises. A great deal of study has been made of industrial disease and there is a long list of Government regulations about precautions that must be taken to lessen the effects of these hazards.

6. Degeneration with Age

Advancing age may bring its rewards in the shape of achievement, contentment and a more ready acceptance of the conditions of one's life, but it also brings its penalties in the gradual waning of physical and often of mental powers. Some of the degeneration is not serious; most men begin to become bald, the skin wrinkles and loses its elasticity, the sense of hearing becomes less sharp, the teeth become fewer and fewer in number, there are changes in the eyes and there is less ability and usually less desire to engage in active physical pursuits. Bones break more easily and are repaired with greater difficulty; wounds take longer to heal; illnesses require longer periods of convalescence. Mental processes gradually slow down; it is not so easy to grasp new ideas and it is difficult to remember new facts. There is great variation in the age at which degeneration becomes very obvious. Some people are old at forty while others are young at seventy. However carefully one lives, the onset of **senility** is inevitable.

The most serious consequences of ageing are those concerned with the blood vascular system and the nervous system. Degenerative changes occur in the blood-vessels, particularly in the muscular layer and the smooth inner lining. The elastic and muscular tissues are gradually replaced by fibrous tissue or are even finally calcified so that in extreme cases the vessels become rigid tubes; this condition is known as hardening of the arteries or **arterio-sclerosis.** The inner lining may become lumpy and distorted by deposition of fat, or may be torn and broken; either of these conditions is called **atheroma.** Fortunately, both arterio-sclerosis and atheroma have a sporadic distribution and are found in only a few places in the body. The most serious effects arise when either condition affects the smaller arteries, particularly the coronary, cerebral, retinal, renal and tibial. Some of these effects are listed below.

1. **Angina pectoris** (pain of the chest). This occurs when one or both of the coronary arteries become narrow due to atheroma. They may carry enough blood to the heart muscle to enable it to work efficiently while the body is at rest, but not enough blood to enable it to increase its output during effort. The pain that is experienced is very characteristic; it begins in the region of the sternum, spreads across the shoulder, down the inner side of the arm (usually the left) and continues into the fingers. This pain occurs every time the person exerts sufficient extra effort; it soon subsides with rest, and the affected person has to learn gradually that his activity must be reduced.

2. **Cardiac infarction,** generally known as heart failure, is the commonest cause of sudden death. A coronary artery, or one of its branches, becomes suddenly blocked, or nearly so, either by swelling of its inner lining or by the blood clotting at a torn portion. Blood stops flowing through the artery to part of the heart muscle and then venous blood flows back into this region so that it becomes **infarcted** (i.e. stuffed) with blood. If a main coronary artery is affected, death is almost instantaneous, but if it is only a branch artery, there is slow recovery after considerable pain. Some of the heart muscle has, however, been destroyed and replaced by fibrous tissue and the patient will find that he easily gets short of breath and is more liable to another infarction and to angina. Therefore he must live much more carefully than he did before.

3. **Angina cruris** (pain in the leg). This is similar to angina pectoris but the atheroma or arteriosclerosis occurs in the tibial artery

and the pain is felt in the calf muscles. The artery may carry enough blood for slow movements but not for walking fast or climbing hills. The pain disappears quite quickly with rest, but recurs with further effort until the sufferer learns that he must move about in a lower gear. If this condition worsens, the blood supply to one or more of the toes may be completely cut off. The tissue becomes blue and painful and then slowly turns black, and eventually it will be cast off by the body. This ailment is known as **senile gangrene**; its most serious aspect is infection by microbes, which penetrate the dying tissue very easily; the condition may spread up the leg and necessitate **amputation** above the knee or even at the hip. It is vitally important that gangrenous tissue should be kept thoroughly clean and dry.

4. **Cerebral thrombosis** (blood clot in the brain). This is commonly called a "stroke" and is caused by the blood clotting suddenly in one of the branches of the cerebral artery. The patient loses consciousness rapidly and, though partial recovery is usual, some permanent damage has been done and there is partial or complete paralysis down one side of the body; very often, speech is seriously affected. Sooner or later, there will be another stroke or a succession of them, and one will be fatal.

5. **Cerebral haemorrhage** (bleeding in the brain). This is due to the bursting of an arteriosclerotic branch of the cerebral artery; blood leaks out and gradually destroys the brain cells. It can be quickly fatal but normally the patient recovers and is left with some paralysis, usually less serious than in the case of cerebral thrombosis. It is strange that such spontaneous bleeding very rarely occurs in any other part of the body.

All these disorders of the blood vascular system are far more common among men than among women. Every man must learn to live at a pace that suits his physical condition, and must remember that he is as "old as his arteries."

Two common effects of ageing on the nervous system are **senile dementia** and **Parkinson's disease.** Again, both are far more common in men. Senile dementia, or madness of the aged, is associated with diminishing blood supply to the frontal lobes of the cerebrum. Although the physical condition is unaffected, the powers of reasoning and judgement gradually decline and, in addition, there is poor memory of recent events, though memory of earlier times remains surprisingly clear. King Lear, in Shakespeare's play of that name, provides a fine study of senile dementia.

Parkinson's disease affects the basal ganglia at the base of the mid-brain. There is increasing muscular stiffness, tremor of the hands and feet, and a set facial expression. Indeed, a mask-like face, lacking expression, is very characteristic. There is also difficulty in speech and lack of control of salivation. A number of causes of this disease have been advanced: in early adult life, it may be due to degeneration of the basal ganglia from no known cause; some cases are associated with manganese poisoning, and some with arteriosclerosis of the blood-vessels supplying the basal ganglia.

7. Mental Illness

The mind is not an anatomical part of the body; we cannot locate it or describe its position, its size or its nature. It has been described as "the brain in action"; expressions such as "feeble-minded" and "strong-minded," are in common use. We can be certain that there are three components at least that determine the quality of the mind; they are: the inherited condition of the brain and its physical development; the chemical environment supplied to the brain by the body fluids; all the conscious experiences that impress themselves to a greater or lesser extent on the brain. Response to environment constitutes behaviour and it is by observing behaviour that we are able to discern minds of differing qualities.

Again, if we take the average condition as being normal, it is very obvious that many persons are superior and many persons inferior to the normal. The latter show some type of mental disorder.

Serious **mental deficiency,** when the affected person can never become normal, may be congenital or due to some environmental accident. For example, mongolism, congenital syphilis and rhesus incompatibility are congenital causes of mental deficiency. On the other hand, serious blows, wounds, or growths in the brain, are environmental causes. Nowadays, great progress is being made in the education of mentally defective children and there is increasing provision of special schools for them, but, in most cases, they can never become normal.

Apart from serious and permanent mental deficiency there is a great variety of mental illness ranging from the relatively trivial to the extremely serious. Broadly, mental illnesses can be divided into two groups: those in which there is definite damage to the brain and those in which no apparent damage can be detected.

In the former group, the most common illness is that known as a state of **confusion.** This is not normally longlasting and is generally due to some disturbance of the chemical environment of the brain. Many infectious diseases that lead to fever (typhoid fever, pneumonia), and excess of alcohol, or of drugs such as morphine or cocaine, may lead to a confused state of mind. The patient shows riotous imagination and poor memory, and is very restless and often puzzled by what is happening around him. Confused states are usually cured when the cause is removed.

Permanent insanity may be caused by damage to the parts of the brain on which the mind depends, mainly the frontal lobes. For example, serious shortage of the B vitamins, severe poisoning by alcohol, cocaine, morphine, etc., the last stages of syphilis, meningitis, tumours in the frontal lobes, may all cause increasing and permanent insanity.

In the latter group, where there is no apparent damage to the brain, there is a great variety of mental afflictions ranging from very mild to very serious. It is quite possible that all people have slight mental derangements during the course of their lives. It is a well-known saying that "there is only a fine line between sanity and madness." **Schizophrenia** is a condition in which a person becomes increasingly engrossed in himself. The type of person usually affected is shy and indecisive and finds it difficult to make friends. A major event in his life may start the deterioration: failure in an examination, the death of a parent, a physical accident or a bad illness, these, and a host of other factors, may start the decline. Gradually, the patient comes to live in a world of his own imagination and suffers from strange delusions. Nowadays, many schizophrenics are cured by **insulin shock treatment** or by **leucotomy,** but we do not know why these treatments effect a cure. The insulin treatment consists of giving the patient doses of insulin sufficient to make him unconscious for an hour or so. Leucotomy is an operation whereby some of the nervous connections between the right and left frontal lobes are severed.

A **psychopath** is a person who does not appear to possess any sense of responsibility and does not feel obliged to obey the same laws and rules as other people do. He is anti-social in the sense that he tends to destroy the social order and in most cases becomes a criminal. It is questionable whether punishment has any deterrent effect on psychopathic criminals.

Mania occurs when there is a rise in emotional level to a point when the patient is uncontrollable. Mania may take a variety of forms. It may be caused by the effects of anger, laughter, rhythmic dancing, revenge or other emotional occurrences on persons who are susceptible. Electric shock treatment, in which an electric current is applied to the brain, renders the patient unconscious immediately, and he recovers fully within a short time. Some maniacs are too violent to be at liberty and are confined in asylums.

Hysteria is manifested by certain types of people when they fail to get their own way or when they wish to escape certain obligations. It begins as a kind of martyrdom and later there is exaggeration of minor ailments such as headaches, colds, coughs, etc. In severe cases, there is imitation of the symptoms of more serious diseases.

Many people suffer from **psychoneurosis,** in which there is a failure to face up to the difficulties and problems of life. They become worried by responsibility and are especially disconcerted by failure. Such people often find that life is too much for them and they are potential suicides.

There is a series of mental illnesses known as **anxiety states.** The patient is torn between two opposing desires, for example, sexual licence and morality; sharing good fortune and keeping it secret; work and play. Under guidance from a psychiatrist, most patients gradually learn to make compromises and to seek advice about their problems. Some, however, cannot compromise and cannot repress emotion; they suffer from mental fatigue caused by worry, and this continues until some physical function begins to go wrong.

These are the commoner mental illnesses. There are others and often several of these conditions are present in one patient. A great deal remains to be discovered about mental illness, and it is probable that it causes more human unhappiness than does physical illness. We must realize in our dealings with other people that human beings are not all the same; they differ genetically as well as in their environments. We are doing more and more to improve the environment and gradually we are realizing that something can be done about the genetic contribution to human make-up.

8. Deficiency or Excess of Hormones

The effects of deficiency or excess of various hormones were discussed in Chapter 15. Nowadays, hormone therapy (treatment) is an important part of medical practice. Sugar diabetes can be held at bay by administration of insulin; thyroxin is administered to make up for natural deficiency of the hormone; milk production can be stimulated or stopped by administration of suitable hormones.

9. Infection

By infection, we mean that other living creatures obtain entry to the body and there cause damage. There are three main pathways of entry: through the skin, through the mouth with food and drink, or through the nose and mouth during inhalation. When the infecting organisms are very small, they are often known as **microbes.** This is not a name used in biological classification, but it indicates any small living creature, whether it lives a free life or exists as a parasite in other living creatures. The parasitic microbes include the viruses, some of the bacteria, certain fungi and certain small animals.

Viruses are the smallest microbes; we do not know whether to regard them as living or non-living. Each virus is composed of nucleic acids (*see* p. 305) and protein, and can multiply only in living cells, each kind of virus being capable of infecting one type of cell and generally only one kind of animal or plant. Viruses range in size from 6 nm to 400 nm, and they cause a great many diseases in plants and animals. Examples are the common cold, herpes (sores on the lips), yellow fever, smallpox, influenza, measles, German measles, mumps and poliomyelitis.

Bacteria are, in general, larger than viruses, ranging from 100 nm to 20 μm. The majority are free-living and are especially plentiful in soil and in water, but a number of kinds are parasites. Some human diseases caused by bacteria are tuberculosis, pneumonia, diphtheria, typhoid fever, cholera, boils, meningitis, gonorrhoea, scarlet fever, blood-poisoning, plague and leprosy.

Although the **fungi** cause a vast variety of diseases in plants, they do not affect human beings very much. Two diseases caused by fungi are ringworm and athlete's foot, in which the skin between the toes becomes very irritable and peels off.

A number of serious diseases are caused by small animals of the phylum **Protozoa.** Among the worst are amoebic dysentery, malaria and sleeping sickness. Most human beings, at least at some time during life, are infected with **round-worms** of the phylum Nematoda. Nearly all young children have threadworms in the large intestine, where they cause much discomfort and irritation. More serious diseases are caused by hook-worms, Guinea worms, and minute worms called **filariae**; all these are also nematodes. Some **flatworms,** of the phylum Platyhelminthes, also cause human disease, the most serious parasites being various species of tapeworm, and flukes such as the bladder fluke and Chinese fluke. The study of worm diseases is a vast subject in itself; it is known as **Helminthology.**

The list of diseases due to infection is very long, but, in many cases, they can be avoided if the correct precautions are taken. Here again, we must blame the general lack of biological education. Many bacterial and virus diseases can be prevented by immunization (*see* p. 184). Thorough cooking of meat will prevent infection by certain tapeworms and nematodes. Cleanliness of food, purity of drinking water, care in the handling of articles of diet, proper disposal of faeces and correct preservation of various foodstuffs, are all important. But more important than any of these is the biological knowledge of the life-histories of the parasites, so that we may know at what points in their life-cycles we may destroy them and what means we can take to prevent infection.

For example, the common human tapeworm, *Taenia solium*, which lives in its adult phase in the small intestine of man, is transferred to the pig by the tapeworm eggs in human faeces, which a pig will eat. In the pig, the young stages, called **bladder-worms,** are found in the muscles; man may be infected by eating undercooked or raw pork or sausages. Therefore, if thorough cooking of meat and suitable sewage disposal are practised, there is little likelihood of infection.

Plague, which was responsible for many deaths in the Middle Ages (the Black Death), is caused by bacteria carried by rat fleas. When an infected rat dies, the fleas migrate from the cooling body and may alight on human beings. When the flea pierces the skin to suck blood, it transmits the parasites into the body with its saliva. Thus, control of rats is an important factor in the prevention of plague.

House-flies may carry, on the feet and mouth parts, a number of parasites causing human disease; for example, typhoid fever, infant diarrhoea and possibly poliomyelitis. Therefore, it is important that human food should be protected from contamination by flies.

Infection with colds, influenza and sometimes tuberculosis, results from the inhalation of air-borne droplets containing the causative organisms. Hence a handkerchief should be used by anyone coughing or sneezing, and contact with infected persons should be avoided as far as possible.

10. NEW GROWTH

Normal growth is a harmonious process, with cells in all parts of the body dividing at various rates so that the mature structure is eventually complete. Thereafter, growth slows down and eventually ceases, except in specialized replacement areas such as the germinative layer of the skin and the marrow of the bones, which produces most of the corpuscles of the blood. But sometimes, and particularly in old age, a single cell or group of cells may suddenly begin and continue unrestricted division, forming a mass of cells far in excess of the normal for that particular part of the body. Such growths are of two kinds, **benign** and **malignant**; both are often called **tumours.** A benign growth (benign means kindly or well-disposed) seldom causes much trouble. It may occur in any tissue of the body, and the only really serious cases are those of tumours in the brain or spinal cord. Warts on skin, some lumps in the breasts, polyps in the rectum, fibroids in the uterus are all common examples. The tissue of a benign tumour retains its normal functions and is restricted eventually by the growth of a fibrous connective tissue capsule around it.

A malignant growth is much more serious; it is called a **cancer** or **carcinoma.** Cancer is the Latin word for crab and the growth is so-called because it sends out long processes like the limbs of a crab. The cells of a malignant growth do not retain the normal function of the tissue in which they arise, but become very simple or undifferentiated, like the cells of a very young embryo. Furthermore, they cannot be restricted by the growth of a connective tissue capsule. Thus the branching mass extends into various organs and interferes with their functions. Worst of all, small groups of cells break away and are carried in the blood or lymph to other parts of the body where they set up secondary cancers.

A great many factors are thought to be responsible for these malignant growths. There is, in many cases, a genetic predisposition to certain types of cancer. In some families, cancer appears in every generation. Age is certainly another factor, since the great majority of malignant tumours occur in people over the age of fifty. Severe and persistent irritation of the skin, of the buccal cavity and of various other parts may cause local cancers. Over-exposure to X-rays may cause skin cancers. Many chemical substances, for instance, certain constituents of soot, coal tar, exhaust fumes and cigarette smoke are known to be **carcinogens.** Viruses may be responsible in some cases.

It seems certain that whatever the actual causative agent, the genes of a cell are altered, so that instead of the cell performing its normal work, most of its energy is devoted to division. It follows that there is no single method of prevention. Also it is unfortunate that most malignant tumours cannot be detected in their very early stages. There are various methods of treatment, all of which depend on the shape and size of the growth. Surgical removal, where possible, is still the best method, though treatment with radio-active substances is often effective. In some cases X-rays or radio-active materials are applied externally, but there are methods whereby short-lived radio-active materials are "planted" internally round the growth area.

Cancer is the most dreaded of diseases in the civilized countries; it is becoming more common with the gradual increase in average life span. It is a problem disease, teeming with unanswered questions. Why is it that some men smoke heavily and live to a ripe old age? Why are some people subject to warts, while others never have them? There are so many factors involved that it is difficult to arrive at any clear answers. All that we can do is to avoid obvious sources of danger and perhaps the chief of these is smoking. It is an undoubted fact that cancer of the lungs is far more common in smokers than in non-smokers, but even in this instance there is a possibility of inherited pre-disposition.

The story of disease is never very pleasant; consolation may be derived from the fact that many diseases have been conquered, and that an immense amount of research work is being carried out into those that still cause unnecessary suffering and loss of life. For the layman, the important thing is that he should understand the requirements of health: a good diet, sufficient fresh air and exercise, avoidance of excess in all things, and the striving to achieve some worthy aims or ambitions.

Human Success and Problems

Homo sapiens has been a very successful species on three counts: increase in numbers (500 million in 1700, over 3500 million now); dominance over other living creatures; and very great power of control of different types of environment.

Human achievements of importance would constitute an impressive list. Not only have we a vast heritage of culture, a great knowledge through science of our environment, but we are also increasing length of life and gradually conquering disease.

Nevertheless man's successful exploitation of his environment has brought many serious problems. Among these are the question of the rapid increase of population (three-fifths of human beings are under-nourished); the thoughtless interference with populations of other living creatures; the problem of racial conflict; the possession of violent weapons of destruction; the use of leisure; and the difficulty of acquainting each new generation with the increasing mass of human knowledge.

We might, in conclusion, note the remarks of one of the most famous geneticists of our day—

> "Man is an animal. He is just as much a product of interaction of genes and environment as are fungi, plants, insects, etc. The main task of the next century is a biological one. Man must begin to know where he is going and what kind of life he is likely to have thrust upon him." (From *Genetics and Man*, by C. D. Darlington, by permission of Messrs. George Allen & Unwin Ltd.)

CHAPTER 18

HYGIENE AND PUBLIC HEALTH

HYGIENE may be defined as the **promotion and maintenance of health** as distinct from the practices of Medicine and Surgery, which are essentially devoted to the cure of disease. There is, however, a branch of medicine which is **preventive**; it includes such aspects of health as immunization and dietetics. The word "hygiene" is derived from the name of the Greek mythological goddess of health, **Hygeia**, the daughter of Aesculapius, the god of medicine.

Throughout this book, there have been numerous references to hygienic practices. Here, there is essentially a summary of these, together with some discussion of those aspects of hygiene which are outside the control of the individual. Hence, the subject is considered at three levels, the individual, the home and the community.

Individual or Personal Hygiene

In the early years of life, the personal hygiene of a child is the responsibility of the parents, and it is therefore essential that they should have some knowledge of the subject. As a child matures, it should be trained in hygienic habits, until it is old enough to assume responsibility for itself.

NUTRITION

The importance of a correct and balanced diet has been explained in Chapter 9. The food should provide not only the satisfactory **energy requirements** and the material necessary for **growth** and **replacement,** but also the **vitamins, salts** and **water,** without which deficiency diseases will manifest themselves. For correct mastication, care of the teeth is essential; the importance of brushing, inspection and treatment have been discussed on p. 132. The necessity for thorough cooking of foodstuffs such as meat, which might carry disease-causing microbes (**pathogens**), has previously been emphasized (*see* p. 339), and it should always be remembered that variety and aesthetic presentation add to the enjoyment of a meal.

The importance of **roughage** in the diet is not generally realized. To keep the food-stream moving along the alimentary canal by peristalsis a certain amount of bulk is necessary, and it is almost entirely provided by the cellulose in the diet; it is this substance which is lacking in many tasty and highly refined foodstuffs such as chocolate, ice-cream, sugar, pure starch, meat extracts and fruit-juices. None of these is at all harmful in itself, but greater emphasis should be placed on the inclusion of whole fruits and vegetables in the diet; these provide the bulky, indigestible cellulose which is so necessary for peristalsis and for satisfactory defaecation.

With regard to defaecation, the material expelled from the anus contains not only indigestible substances such as cellulose, but excess digestible food, countless numbers of bacteria and dead cells, and the excretory products derived from the breakdown of bile in the liver. Satisfactory clearance of this material is essential, but the modern rather rigid time-tabling of schools and places of employment leads to the formation of unnatural defaecating habits. Regularity of defaecation is not always easy to accomplish and hence there is excessive straining of the rectal muscles; in a great many people this leads to dilation and loosening of the rectal blood-vessels, often to the extent of their protrusion through the anus. Then, defaecation becomes painful and tends to be deferred and avoided, leading to great accumulation of faeces in the rectum and colon; this is **constipation**. The swollen veins, called **haemorrhoids** or **piles,** may become so painful as to necessitate surgical removal. Defaecation should be carried out when there is a natural impulse to do so; regrettably this would interfere to some extent with a time-tabled existence.

Posture and Exercise

There has already been some discussion of posture on pp. 77 and 89. Since the final shape of the skeleton depends largely on the muscular pull on the bones, it is important to ensure that every effort is made to encourage correct posture during the growth period. Good posture is essentially a matter of the correct positions of the bones in the vertebral column, and hence many preventable deformities usually show themselves in this region. The common postural defects of the vertebral column are listed here and illustrated in Fig. 181.

1. **Kyphosis** is exaggeration of the backward curve of the thoracic vertebrae; in extreme cases, it is called **hunchback.**

2. **Lordosis**, or hollow back, is extreme forward curvature in the lumbar region; it is often associated with kyphosis.

3. **Scoliosis** is lateral curvature of the spine in the upper or lower part, or more rarely in both.

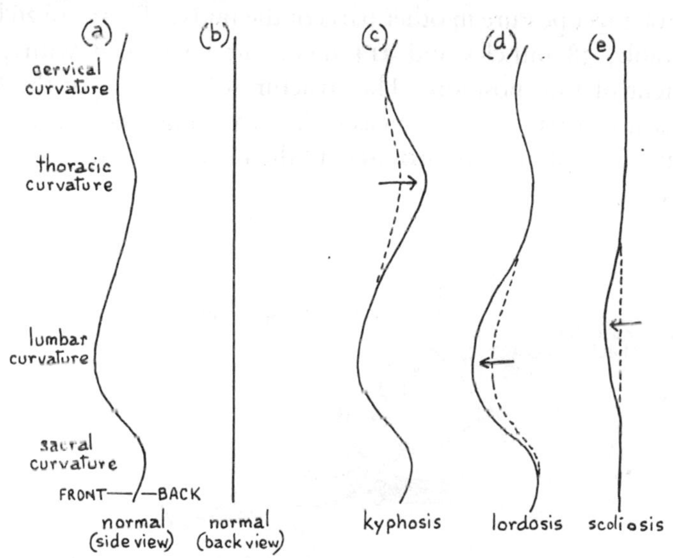

FIG. 181. Curvature of the Spine, (*a*) Normal, side view, (*b*) Normal, back view, (*c*) Kyphosis, side view, (*d*) Lordosis, side view, (*e*) Scoliosis, back view

These deformities are often due to unavoidable causes such as congenital abnormality, partial paralysis or injury, but they can be caused also by habitual bad posture. Practices which lead eventually to spinal distortion are: carrying weights always on the same side of the body, sitting with one leg doubled under the body, constantly standing with the weight of the body mainly on one foot, sitting obliquely in chairs with the base of the spine bearing much of the weight, and badly adjusted furniture, especially in schools and places of sedentary employment. It is absurd to expect one standard size of chair or desk to fit, for example, each of 30 nine-year-old children. Adjustable furniture, constant watchfulness by parents and others in charge of children, avoidance of strain caused by long periods in one particular posture, will all help in this connection.

Deformities of the feet are extremely common; they range from corns and callouses to fallen arches and misshapen joints; in their turn they must affect posture in other parts of the body. There are 26 bones, all movable, 38 muscles and 214 ligaments concerned with perfect adjustment of foot position. The structure of the arch of the foot is such that it is elastic in its action as well as carrying the main weight of the body. The **plantar ligaments** are the most essential structures for maintaining this arch (*see* Fig. 182).

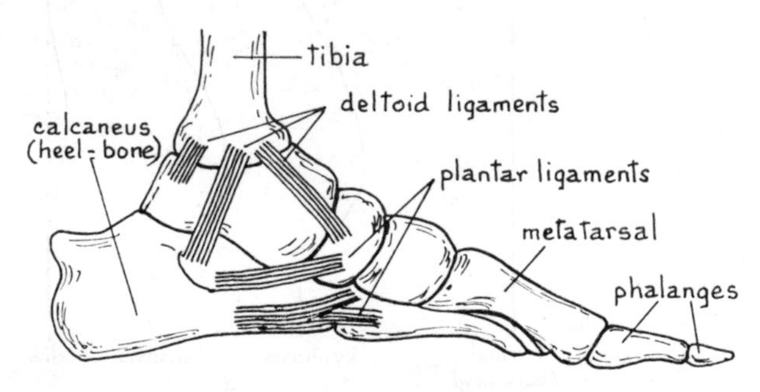

FIG. 182. Main Ligaments of the Ankle and Foot. View from the Inner Aspect of the Left Foot

Weak or fallen arches may be due to certain preventable conditions such as ill-fitting shoes, very high heels, unusual foot strain, or incorrect use of the muscles causing the weight to be thrown more on the inside than the outside of the foot. Any of these conditions may also cause lesser deformities.

Apart from injuries and congenital deformities, good posture is essentially a matter of close observation and correction by parents and teachers throughout the growth period. Specially devised appliances and exercises, as used by **physiotherapists**, can often correct defects. Other causes of bad posture are bad light for reading, undetected weak eyesight, weak muscles due to lack of exercise, poor diet, and excessive strain.

Exercise is a necessity, not only for the muscular system, but it plays a vital part in the maintenance of a healthy condition of the whole body. The most beneficial forms of exercise are those which increase activity of all parts of the body; common examples are walking, running, swimming, cycling, jumping; a whole range of games has been devised with the object of increasing interest and effort. Exercise is especially important during the growth period, because of its ultimate effect on skeletal shape and development of a high degree of co-ordination between the nervous and muscular systems. During exercise, the rate and depth of breathing are increased; the heart and general circulation are stimulated, and this occurs especially in veins where the squeezing effect of contracting muscles helps to force the blood along. The rate of heat production increases, and this leads to more sweating. In general, the effect is to tone up the whole body which is more ready to face rapid action. Also, and perhaps not the least important, graceful carriage and skilful movements eventually develop.

For all ages, a number of simple precautions are necessary. Violent effort should not be exerted immediately from a resting condition but there should be a short period of gradual warming-up; this tends to avoid muscular cramp or more serious injuries such as slipped disc, which can occur when a sudden powerful lifting effort is exerted. Such warming-up can be seen in all skilled and experienced sportsmen. It is essential that the nature and duration of the exercise should be suitable for the persons concerned. One cannot expect a man of sixty years old to play through a ninety-minute football match, or a child of five to run ten miles. Excessive exercise can lead to total collapse, in which case, the exercise has done far more harm than good. Considerable sweating entails loss of salt (sodium chloride); this must not continue for long periods, otherwise muscular cramp is inevitable. Many sportsmen take salt tablets during the actual performance of a long period of strenuous exercise such as a marathon race or a five-set tennis match.

Clothing should be suitable for the type of activity and the period of exercise; if long enough to develop profuse sweating then a bath and change of clothing should follow. The greatest benefit will be derived from exercise in the open air, or, if that is not possible, in well-ventilated accommodation.

Without exercise, the muscles will actually waste, and in some diseased conditions where immobility is enforced this does happen. The increasing knowledge embodied in the practice of **physiotherapy** has done a great deal to help such cases; the practitioners are fully trained to devise and teach suitable exercises to patients.

As human beings enjoy overcoming a physical challenge, so do they enjoy a **mental challenge.** A large portion of the art of teaching consists of devising, for each age-group, suitable mental exercise to stimulate that group; physical and mental stimulation are integral parts of harmonious development. In our modern technological age, work is often repetitive and boring and hence there is a vital need for **recreation,** both physical and mental. The background for this recreation depends largely on the stimulation and facilities provided by both the home and the school. This background rests on a wide provision of hobbies and interests which encourage both individual and group effort.

Fatigue, Rest and Sleep

It is usually considered that there are three distinct kinds of fatigue, though in many cases, two or even all three may be part of the total condition. **Muscular fatigue** is essentially due to accumulation of lactic acid in the muscles which are being used; to remedy this, the muscles concerned, and often the whole body, must have a period of rest until the lactic acid has been removed (*see* p. 171). **Mental fatigue,** due to long-continued activity of nerve-cells, especially when this activity is localized in certain special groups, is more rapidly attained than is muscular fatigue. The actual physical cause of fatigue in nerve cells is not known; a recent suggestion is the gradual cessation of production of acetylcholine at the synapses. In the case of neurons which supply muscles, they will cease activity while the muscles are still capable of further work; this will not occur in the absence of nervous stimulation. Hence, total collapse after excessive muscular effort is due to the breakdown of nervous stimulation. **Psychological fatigue** is due to boredom and repetition and is characterized by yawning, restlessness and discontent. It can occur in conjunction with both physical and mental fatigue. Many occupations in mass-production factories involve long periods of repetition of a relatively simple task. Various devices have been used to counteract this; such are: more frequent breaks and the use of broadcast or recorded music. The same type of fatigue can occur in members of an audience at a boring lecture, lesson or sermon. There, the only remedy is to enliven the performance and to allow for audience participation. However lively the performance, one of the other two forms of fatigue will eventually appear if the lecture, etc., is too lengthy.

All forms of fatigue eventually demand **rest**, which may not necessarily mean sleep. A change of activity often constitutes a rest; for example, after a vigorous game, the muscles need rest; this can be provided by sitting still and reading, or listening, or watching a television programme. In the same way, mental fatigue, after a concentrated period of study, is almost always relieved by some form of physical activity, such as a game or a brisk walk. Scholastic time-tables are, or should be, devised with all three forms of fatigue in view; there is also a further consideration—the younger the children, the shorter are the periods of concentration on one particular form of activity.

Eventually, in the course of a day, changes of occupation cease to provide rest, and **sleep** becomes essential. In the past, there has been a great deal of mystery about sleep, and most people have been influenced by tradition and "old wives' tales." Recent research has revealed a number of interesting facts. On average, human beings spend one-third of their lives in sleep, and although infants need more sleep than older persons, there is no truth in the saying that old people do not need as much as the average eight hours. Average figures are often misleading; the average requirement for babies in their first year is 16 hours, but a recent study showed that there is enormous variation in the individual amounts required; some very young babies were awake for only one hour out of 24, and, at the other end of the scale, some were awake for 16 hours out of 24. A large number of healthy, hard-working adults can manage adequately on 4 or 5 hours sleep in a day; others, equally capable, need 9 or 10 hours. In any measurement of this kind, the actual sleeping time must not be confused with the total time in bed. A person may go to bed at 22·00 hours and read or think for an hour; he may wake at 7.00 but not get up till 8.00. He has spent 10 hours in bed, but his sleeping-time is 8 hours.

Sleep is undoubtedly the great restorer, but we do not know why. Certainly most of the voluntary muscles are resting, the heart rate is a little slower and there is less production of urine. Evidently there is a great degree of physical rest, but the same degree can also be achieved by rest without sleeping. It seems that the critical factor is **rest for the brain.** Observation of people kept awake for two or three days has shown that physical performance is not seriously affected by lack of sleep, but there is always some degree of mental derangement in the forms of hallucination, suspicion, fear and heightened nervous tension. A young American, in New York, recently kept himself awake for more than 200 hours (over 8 days); the only serious symptoms of his long wakefulness were mental. When he eventually slept, it was for 13 hours, and he awoke in a perfectly normal condition. Reports from prisoner-of-war camps have stated that enforced wakefulness for long periods have resulted in death, even when all the other necessities of life have been provided.

It was discovered by Dr Hans Berger (Germany) in 1929 that there were various patterns of rhythmical electrical activity in the brain. He devised an instrument called the **electroencephalogram** whereby these patterns could be detected by two electrodes placed on the head.

The signals detected are magnified and a graphical recording can be produced; it is called an **electroencephalograph.** The device has been used to investigate electrical activity of the brain during sleep; the graph shows a type of cycle of alternating electrical activity. At first there is **orthodox sleep,** with little dreaming; if there is any, it is associated with recapitulation of very recent events. Throughout the whole period of sleep, this alternates with **paradoxical sleep,** characterized by random movements of the eyes beneath the closed lids and continuous dreaming, essentially of a dramatic nature. The dreams are often unreal, usually frightening but sometimes pleasant. Throughout the whole sleep, there are five or six periods of orthodox sleep together making 80 per cent of the total, alternating with five or six periods of paradoxical sleep together constituting 20 per cent of the total. It seems that both kinds of sleep are essential. The only dreams usually remembered are associated with the last paradoxical period.

There is no dreamless sleep (we do not know about very young babies), and no person "sleeps like a log." Apart from very young babies, the position of the body changes, on average, forty times a night. People who wake during an orthodox period usually think they have slept peacefully and dreamlessly; those who wake during a paradoxical period think they have been dreaming all night.

It should now be evident that sleep is essential; without sleep people would die much more quickly than they would die without food. The amount of sleep which is required varies considerably; nowadays most people are awakened by an alarm clock or by someone calling them; they have not finished their sleep and will tend to doze off during the day. Sleep cannot be enforced without the use of drugs, but regular use of sleeping pills, without medical prescription, is unwise. Sleep will come if the brain has reached a certain degree of exhaustion but uncomfortable conditions will lead to deferment. As a matter of hygienic practice, the best possible conditions should be provided—a comfortable bed, a well-ventilated room, a suitable temperature, and an absence of loud and unusual noise. Many mothers have become desperate about their children not going to sleep at the right time; there is no right time.

If conditions are satisfactory, a child will sleep when he is tired; if the child is put to bed before this time, he will not sleep, but he will rest if he is allowed to read, to listen to quiet music or to engage in some other restful pursuit.

Cleanliness and Clothing

The necessity for cleanliness should be obvious to any student who has read thus far, though it is difficult to convince young children on this point. Because of the natural function of sweating, the skin gradually becomes covered with a layer of dried perspiration consisting mainly of salts and urea; mingled with this layer is the greasy material secreted by the sebaceous glands and dead cells freed from the outer surface. Various microbes can live and multiply in this material; some of them cause disease and together they produce the unpleasant smell known as **body odour.** The material and odour permeate the underclothes which therefore require frequent washing. The more viscous type of sweat produced in the armpits and groins, etc. (*see* p. 54), also stains the clothing; such stains are difficult to remove, especially if they are of long standing. In most human beings, sweating is very profuse in the feet, giving rise to aching feet, increased friction with socks and shoes and hence to a whole range of foot troubles.

Deodorants, so extensively advertised in these days, disguise the odour; they do not remove the cause of it. Frequent washing and changes of clothing do much to mitigate body odour. Substances which are claimed to prevent sweating should not be used; they are preventing the body, or part of it, from carrying out its natural function.

There are numerous diseases essentially due to dirty skin; the commoner ones are summarized here.

1. Two types of **coccus bacteria** (spherical) are plentiful on dirty skin; they are various species of *Staphylococcus* (clusters) and *Streptococcus* (chains). If the skin surface is damaged, these bacteria can cause **necrosis** (death) of the surrounding skin cells. If the bacteria penetrate blood vessels, **septicaemia** (blood-poisoning) may result. This was commonly fatal until the discovery of antibiotics such as penicillin;

even so, prompt action is usually necessary. **Boils** arise from infection of hair follicles by *Staphylococcus aureus*; the skin damage is usually caused by friction of tight clothing on the skin, hence they are common around the neck because of tight collars.

2. **Blackheads** on the face cause a great deal of disfigurement and concern, especially to adolescents; they are caused by blockage of the pores but can be less prominent with frequent and careful washing. It must, however, be pointed out that during the adolescent stage, they may be present in the cleanest of persons; there is no obvious explanation of this.

3. **Pimples,** which may develop into small boils, constitute the condition of **acne,** due initially to blockage of sebaceous ducts, followed by staphylococcus infection.

4. **Ringworm** and **athlete's foot** are caused by fungi on the skin. The former is a ring-like loss of hair on the head due to the fungus, *Microsporon audouini*. The latter is due to infection by spores of *Tinea pedis*, and it is often conveyed from one person to another in communal baths or on towels. The common name for the condition arose because the disease was, and possibly is, common among sportsmen who shared a communal bath after games. It is characterized by great irritation, and possibly loss of skin, between the toes.

5. **Scabies** is caused by a tiny mite, *Sarcoptes scabei*, which belongs to the same class as spiders, **Arachnida.** The mite bores into the skin causing great irritation and often sores, which may be secondarily infected by coccus bacteria to cause **impetigo** or **eczema.** Scabies usually occurs on the wrists and hands; it is estimated that 2 per cent of the population of Great Britain are afflicted with it.

6. The **body louse,** *Pediculus humanus corporis*, and the **head louse,** *Pediculus humanus capitis*, are not particularly common in Great Britain nowadays, though the former was once a great scourge since it carried the microbes responsible for **typhus (gaol) fever** and **relapsing fever.** The body louse feeds by piercing the skin and then sucking the blood; clotting is prevented by a substance present in the insect's saliva. As the saliva is pumped into the wound, the disease-causing organisms,

present in the saliva, are able to obtain entry into the body of the person attacked. The pathogens causing **trench-fever,** which was so common in the Great War of 1914–18, are also transmitted by the body louse. When they are not feeding the lice live in the clothing and especially in the seams. The best way to eliminate them is by thorough cleanliness of skin and clothing.

Head lice feed on dead cells of the skin of the scalp; they cause a great deal of irritation and subsequent scratching may break the skin and allow secondary infection by more dangerous pathogens. These lice have in the past been so plentiful in some people, that the whole mass of hair appears to be moving. In the first 20–30 years of this century, head lice were so common on children, that there were regular inspections by school nurses; also, in those days, the "small-toothed comb" was an essential toilet article. The incidence of head lice decreased considerably up to 1955; since then, it is undoubtedly on the increase; the reason for this is not obvious in view of improved standards of cleanliness.

The value of frequent and thorough washing, in warm water and using a good soap, should now be abundantly clear. Apart from disease and body odour, dirty people are shunned by their fellow-men, and then they tend to develop anti-social attitudes. Hands should always be washed before meals, and certainly after using the toilet. Children should be consistently discouraged from sucking their thumbs or fingers and biting their nails for the simple reason that by so doing, they may risk infection with a whole range of dangerous organisms. Such, for instance, are the eggs of **tapeworms** and **threadworms,** and the bacteria causing **enteritis** and **dysentery**. The common human threadworm, *Oxyuris vermicularis*, is a nematode about 1 to 2 cm long and as slender as a fine grade of cotton. It lives in the large intestine, causing irritation, pain, and often inflammation. Mature females migrate down to the anus, especially in warm conditions at night, and lay their eggs in the skin around the anus. The puncturing causes intense irritation which is not relieved by scratching; it is thus that reinfection occurs, by carrying the eggs to the mouth on the fingers and especially under the nails. The eggs are swallowed; they hatch in the intestine and the cycle begains again. It is important to notice that in this, as in many other cases of gut pathogens, infection occurs by sucking the fingers.

Nails should always be kept short and with smooth edges; there is then less danger of scratching oneself, and less space for accumulating dirt and pathogenic organisms.

Finally, in connection with washing the body, part of the value lies in thorough drying and massaging; these stimulate the circulation in the dermal blood-vessels. In dry, cold weather, thorough drying is even more essential, to prevent or reduce the painful cracks in the skin, known as **chapping**; chapped skin often provides means of entry for pathogenic organisms.

Clothing

Apart from prevailing fashions or the desire to present an attractive appearance, there are certain basic principles to be observed in choosing clothes. In such a highly variable climate as that of the British Isles, one has to be prepared for all emergencies. Some kind of mackintosh or similar waterproof is essential, so that the rest of the clothing is not saturated during rain; if it is, the body loses a great deal of heat and is much less resistant to various types of infection such as **colds, coughs, influenza** (virus), **pneumonia** (bacteria) and **pneumonitis** (virus).

For cold weather, the best types of clothing are those which allow trapping of air, which forms an excellent insulating layer. Such are woollen materials and some of the synthetic cellular fabrics. The underclothes should be absorbent enough to take up the moisture of perspiration. Wool is almost certainly the best material for both purposes, though some synthetic fabrics may have almost identical properties.

For warm weather, the main concern is to lose heat by evaporation of sweat and hence the materials used should afford maximum ventilation of the skin; such are the lighter materials such as cotton. Choosing suitable clothing is largely common sense; if it causes discomfort for any reason, then it is unsuitable.

Tight, and hence restrictive, clothing can interfere with blood circulation, respiratory movements and muscular action, and it can cause soreness of the skin by undue friction. The same uncomfortable condition occurs frequently in babies, from wearing damp or coarse napkins. It is important, especially in young children that the main weight of the clothing should be borne by the shoulders. Young children have no well-marked waist region, and it causes constriction if lower garments are fixed in position with tight elastic bands or belts.

Perhaps more than any other articles of clothing, shoes and boots must fit well; the evils of friction due to tight or ill-fitting shoes have been pointed out previously. During the growing period, there is the necessity for frequent (and expensive) purchase of larger sizes, for most cases of ugly and deformed feet are due to the wearing of shoes which are too small. The shape of footwear is important, not only to preserve the correct shape of the arches, but to accommodate all the ranges from very wide or very long, to very narrow or very short. Shoes with very high heels will always constrict the toes and lead to unnatural curvature of the arches. There is almost as much foot trouble due to shoes which are too large, as for shoes which are too small. In the former case, the toes and the arches have to be held in unnatural positions especially if the shoes are not laced or otherwise fastened.

In general, there is a very simple rule; any footwear which causes discomfort is unsuitable. Regular washing of the feet and cutting of the toe-nails will do much to maintain a healthy condition. Suitable shoes and care of the feet will reduce the incidence of pedal ailments such as blisters, hard skin, corns, callouses, verrucas (warts), enlarged joints and fallen arches. Any persistent or painful ailments should receive the skilled attention of a chiropodist.

Body Apertures

The normal functions of the various body apertures have been discussed earlier in the book. The mouth is the aperture for the entry of food and sometimes for the entry and exit of air; it also serves for the expulsion of vomit, when conditions in the stomach are deranged. The nostrils are the normal apertures for the entry of air and the exit for exhaled gases. The anus serves for the exit of faeces. In males, the penis is the exit for urine and seminal fluid. In females, the vulva serves a greater variety of purposes. It is the exit for urine, for blood and discarded tissue at menstruation, and through it, the new-born baby passes to reach the outside world. It is also the aperture for the entry of the penis during copulation. No material substances pass in or out of the external auditory canals. Apart from their normal functions, the apertures provide avenues of entry for pathogens; many cases have been described in this book. Obviously they can all provide sources of infection for other persons, and thus special measures must be taken to ensure cleanliness of these apertures.

Cleanliness and care of the mouth by cleaning the teeth regularly and rinsing with clean water has already been mentioned. It is very unwise to develop the habit of mouth breathing, because, as compared with the nostrils, there is no filtration mechanism. The nostrils, by their important filtering action, eventually tend to become partially blocked by the accumulation of dust particles embedded in mucus; this is very obvious in districts where there is a great deal of pollution in the atmosphere. If the nostrils become completely blocked, then mouth breathing must take place, and with no filtering mechanism, there is obviously a greater risk of infection. Blowing the nose to clear the accumulated material should not be accompanied with great force; there is a real risk of bursting small blood-vessels. One nostril should be closed by lateral pressure of a finger, and the other cleared by gentle blowing into a handkerchief or tissue; the same method is utilized for the other nostril. It is always wise to use disposable tissues which can be burnt; this applies particularly when there is any respiratory disease such as a cold, or catarrh, or even tuberculosis. Burning the tissues will ensure the destruction of all the microbes, whereas with an ordinary handkerchief they remain as a source of infection until the handkerchief is washed, and that may not destroy all the pathogens. If the accumulation in the nostrils has hardened, it should be softened by bathing in warm water, and then blowing.

Irritation, infection or accumulation of mucus in any part of the respiratory passage may lead to sneezing, coughing or spitting. The material thus expelled may contain pathogenic organisms such as those which cause the common cold, catarrh, and possibly pneumonia and tuberculosis. Such **droplet infection** causes the spread of many diseases, including the influenza epidemics which take a heavy toll of human life. Therefore, out of consideration for other people, one should always use a destructible tissue for sneezing, coughing and spitting; it should then be burnt, immediately if possible, otherwise it should be kept in a paper or polythene bag and burnt when the opportunity arises.

Thorough and regular washing of the urinogenital openings is very important; there are natural discharges in both sexes; these materials accumulate and form ideal breeding-places for pathogens. In females, absolute cleanliness during the menstrual periods is essential. Most important of all is the necessity for careful washing of the vulva with suitable disinfectants for several days after the birth of a baby. Until 1940, death from **puerperal fever** was a common occurrence, especially in hospitals; the disease was called **the scourge of motherhood.** It is known to be due to infection of the damaged tissues by bacteria of the species *Streptococcus pyogenes*; almost invariably septicaemia ensued, usually followed by death. Modern care after birth is based on absolute cleanliness together with use of antiseptics and aseptic precautions. This has been outstandingly effective and the disease is now rare in civilized countries, though it is still common among the backward nations.

Faeces of infected persons almost always contain pathogenic organisms and the eggs of various parasitic worms. Traces of faeces are always present around the anus, and hence there should be very frequent washing of this region. This will help to prevent re-infection of oneself such as occurs with threadworms and with the pathogens of **dysentery.**

The external auditory canals are closed by the tympanic membranes and hence they are not used for the entry or exit of any materials. Nevertheless, they accumulate dirt and wax (*see* p. 262) and require careful washing. In some cases there may be a solid plug of wax against the membrane, which is then unable to carry out its normal vibrations, thus leading to partial deafness. The canals should be gently washed with warm water and carefully dried. Above all, hard or pointed instruments should not be used to remove a wax plug because piercing of the ear-drum will lead to almost total deafness in that particular ear. A little warm olive oil will soften the wax which can then be removed by a soft cloth or a roll of cotton wool. More drastic treatment, such as syringing, requires expert medical attention.

Cosmetics

Cosmetics are intended to adorn and beautify the body, and provided they do not cause any damage to the skin, they are normally quite harmless. However, healthy skin cannot be achieved by their use; there is no suitable substitute for thorough washing with a good soap and gentle massaging with a towel. Many people have skin blemishes which are unsightly and naturally they wish to conceal them. Most cosmetics are harmless to most people, though occasional allergies do occur (*see* p. 329); in these cases, the use of the substance must be discontinued. Decoration of the body with gaudy colours is an ancient and primitive practice; it has been, and still is, common to both sexes in many parts of the world. In civilized communities, it is almost entirely confined to females.

People should beware of substances which are stated to be skin foods or nourishing creams; the only food or nourishment available for the skin comes from what we eat. Some of the cheap soaps are very alkaline, and although they remove dirt more easily, they often cause irritation and dryness of the skin. During shaving with a sharp blade, thin layers of dead cells are usually removed; the use of aftershave lotions based on alcohol will tend to kill bacteria and prevent infection of such unnaturally thinned skin; before their advent, rashes on the face and neck were very common.

Healing creams used on damaged or diseased skin are based on **lanolin,** a grease obtained from sheep's wool, or petroleum jelly, a by-product of petroleum refining. Both may contain antiseptics; such creams are beneficial, if used on clean skin, since they provide a sterile barrier against infection and allow the damaged tissues to heal.

Lipstick, eye-shadow and talcum powders are generally harmless; if one considers them to be decorative, there is no danger in using them. Excessive use of powders tends to block the skin pores; all powder should be thoroughly removed by washing, especially before going to bed. They should never be used merely to conceal dirty skin. Perfumes give a pleasant smell to the body and they disguise body odour. Again, they should not be used as a substitute for washing.

Recreation, Fresh Air and Sunlight

Recreation is essentially the re-direction of energy and attention to some occupation distinct from one's daily work, an occupation in which one has both interest and pleasure. We re-create ourselves by re-directing our energy into some activity we desire to carry out. Hence one attacks what may be an arduous task with zest. The pastime may require mainly physical work, or mainly mental work; it will usually have a modicum of both.

Recreation embraces all those activities called hobbies, interests or pastimes, and it covers all possible fields of action. It may take place indoors or out-of-doors and may range from highly-organized group projects such as team games, expeditions, orchestral or choral music, etc., to individual activities such as gardening, painting or collecting stamps, reading or examining microscopic life. Many of us are gregarious and thus prefer group activity; many of us are highly individual and thus prefer solitary pursuits; many partake of both.

The foundations of recreation are laid down in youth and adolescence; parents try to devise activities in which their children will show interest; schools and youth clubs cater for a wide variety of tastes; these activities and others are extended by adult organizations. With proper early training, providing above all a wide range of activity, no person should ever be in a position where he cannot find anything to occupy his time.

People employed in sedentary indoor occupations should spend a good deal of their leisure time in outdoor activities where they can get the benefits of fresh air and sunshine. There is no source of absolutely "fresh" air; the same layer of atmosphere surrounds and circulates around our planet, and we cannot add to it from any source. What we mean by "fresh" air is less polluted air. After the dust, smoke and gases of some of our towns, it is a great pleasure and a tonic to breathe the comparatively clean air over mountains, open stretches of country or over the sea.

The light rays emitted by the sun are of no direct value to us except that they enable us to see. They are, of course, essential to us indirectly because they supply the energy which is trapped by green plants during photosynthesis. This process eventually provides all our food and all our oxygen. The more directly beneficial rays are the infra-red and ultra-violet. The infra-red rays transmit radiant heat from the sun; without them our planet would soon be a cold dead world. The ultra-violet rays benefit us directly in two ways; firstly, they convert certain sterols in the fat of the dermis into vitamin D, and secondly, they are lethal to most of the organisms which live on our skin. These are the only advantages of sun-bathing; it is unwise to indulge in it for long periods, because the ultra-violet rays damage the skin unless there is enough pigment in the skin to absorb them. Deliberate tanning of the skin has no real value; it is the body's attempt to react to the ultra-violet rays by manufacturing more pigment cells. Sun-tan develops in persons who work regularly out-of-doors in full sunlight; they must develop it to protect themselves from ultra-violet light. It is not permanent but will gradually be lost when exposure is lessened. The only truly dark skins are due to inheritance (*see* p. 316). Albinos can never acquire sun-tan; they will only become red and blistered. Other fair-skinned persons should only expose themselves to full summer sunlight by slow degrees.

REPRODUCTION

The personal and social consequences of indiscriminate sexual relationships are essential parts of the study of hygiene. In Chapter 12 a full account has been given of the reproductive system of both sexes, the manner in which fertilization is brought about, and the birth and development of a baby. The book *Microbes* in this series gives a full account of the symptoms and drastic consequences of the venereal diseases, **syphilis** and **gonorrhoea.** Both are easily avoided but this avoidance demands strength of character among both sexes.

In the great majority of countries, society is founded on the family, consisting of the father, mother and children, and in some cases, grandparents as well. The bond between mother and father is consolidated not only by mutual affection but by some type of marriage ceremony by which they show their desire to be recognized as a married couple. Thus family units are bound together to perpetuate the name, the family traditions and the good characteristics of which every family is proud. Children born into such closely bound units have every chance of receiving the love, care and encouragement which they need.

Nowadays, in western civilization there is a much greater increase in the amount of freedom of action attained by young people. This is quite admirable in many ways; it develops self-reliance, encourages initiative, and enables the youngsters to take part in "progressive" movements much earlier than would have been possible fifty years ago. It is unfortunate that this hard-won freedom has in many cases become licence, and thus has developed what is often called the **permissive society**; at its worst, it is a society in which there are no rules and no responsibilities. Social organization has never been perfect, and never will be; human beings (except identical twins) are not born equal and never can be. We should all strive to improve and equate environmental conditions, so that we can all make the most of our inherited potential.

Extreme permissiveness in sexual relationships is undesirable on several grounds. It can lead to the spread of venereal disease, the incidence of which fell until about 1955, but doubled between 1955 and

1964 and is still rising. Much of the increase is in the age range 18–25 years, and can perhaps be attributed to the evident decline in moral standards.

Other undesirable consequences are the production of illegitimate offspring who may bear throughout life the stigma of being born a bastard; the difficulties which beset unmarried mothers; and widespread recourse to abortions, legal and illegal, and even infanticide.

Among parents, there is often a natural reticence about communicating sexual knowledge to their children. Some parents are ignorant of all but the most obvious facts. It seems patent that the task of acquainting children with all the facts must be a part of their general education; there are excellent films and television programmes about these matters. Alone, they are not enough; teachers and parents must be able and willing to have full and frank discussion with the children in their care. After all, we are discussing the most important function of living creatures.

Habit Formation

Habits are actions, simple or complex, which at first require thought and attention, but after constant practice, they become almost reflex. For example, a man can traverse a daily complicated journey to his place of employment and arrive without ever having consciously thought about his route. People differ enormously in the way they perform regular routines such as bathing, dressing, eating, driving a car, etc. Some ways are probably better than others and they may be called good habits.

Children learn their early habits from the members of their own family; once these habits are thoroughly established, they are difficult to change. In the earliest stages, children tend to mimic their mother's way of doing things; from example, constant correction and insistence, they form habitual methods of washing, eating, dressing, etc. Many habits thus acquired concern personal hygiene.

Good habits help to maintain good health; they make for sequences which save time and energy; in some cases they help to prevent disease and accidents. It is probably true to say that in most civilized countries, the majority of people acquire and maintain good hygienic habits. Many of these have been mentioned in other parts of this book; some of the more important are summarized here.

1. All the practices which ensure cleanliness, and the regularity of these practices.
2. Habitual good posture.
3. Plenty of exercise.
4. Good habits of eating.
5. Timely defaecation and urination.
6. Consideration for others when coughing, sneezing, etc.
7. Road safety in all its aspects.
8. Tidiness, care of clothing, punctuality and many other habits which promote harmony in the family, the group and the community.

Bad habits, the reverse of all these, can very easily develop. For example, **smoking** can soon become addictive; there is no doubt at all that lung cancer, bronchitis and other respiratory diseases are more common among smokers than among non-smokers. **Narcotic drugs,** ranging from cannabis to heroin are even more habit-forming than the use of tobacco; even pep-pills are dangerous to anyone who becomes addicted. Access to these drugs eventually becomes almost the sole aim of the unfortunate addicts, and all other considerations are neglected.

Alcohol, while it is relatively harmless in small quantities, can also become addictive, and eventually lead to the chronic condition of **alcoholism,** where again the craving outweighs everything else. Moderation in this, as in all things, is the standard to aim at. In general, any practice which is liable to cause permanent damage, physically or mentally, should be discontinued, or better still, never begun.

Hygiene in the Home

It is the responsibility of parents, or of those acting in lieu of parents, to ensure that adequate hygienic standards are maintained in the home. With regard to the actual building and its adequacy, recommended standards were first laid down in 1946; the local authorities were held responsible for maintaining these standards which applied to any building started after the date of the recommendation. The main provisions are summarized below.

1. Dry in all respects, and in a good state of repair.
2. Each room adequately lighted and heated.
3. A sufficient supply of wholesome water for all purposes, inside the building.
4. Efficient means of supplying hot water for domestic purposes.
5. Internal, or readily accessible, water closet.
6. Fixed bath, preferably in a separate room.
7. A proper drainage system for all water used in the house and for running off rainwater.
8. Satisfactory facilities for cooking.
9. A well-ventilated larder or food store.
10. Suitable provision for storage of fuel.
11. A satisfactory path to outbuildings.
12. Convenient access from the street or road to a back door.

Such minimum standards are reasonable enough and one must remember that it was the first attempt to establish national standards. They are open to criticism and abuse on many points. For example, the bath should "preferably" be in a separate room; even today, probably the majority of new houses in the lower price ranges, have the bath and the only water closet in the same room. "Adequate facilities for heating" has often been interpreted as a single power point in most rooms, even though electricity is probably the most expensive form of heating. Today, practically all new houses have some form of central heating, though it is noteworthy that the cheapest to install are often the most expensive in running costs.

A serious problem is the difficulty, often impossibility, of bringing old houses up to the required standard, even though part of the capital cost of renovation projects is borne by the local authority; often, the householder cannot find his share of the cost, and all local authority projects are partly controlled by the amount of money available. Hence, appalling conditions are still to be found in city slums and in some country districts. The process of establishing suitable housing conditions for everyone is still very slow.

Fitness for Habitation

A house may become unfit for habitation if it is allowed to become so overcrowded as to endanger the health of the inhabitants. The exact meaning of the phrase "endanger the health" is uncertain; each local authority has its own powers of decision. There are obvious cases of considerable overcrowding, but whether health is definitely endangered is not so easy to prove. Such cases occur, for instance, where unscrupulous landlords overfill rented accommodation, or where immigrants without much capital, crowd together both for mutual support, and from inability to afford more spacious accommodation.

If a house is allowed to degenerate so that it becomes very damp or unsafe, insanitary or infested with pests and vermin, it can also be declared unfit for habitation. Under the Public Health Act of 1946, every local authority is obliged to employ a sufficient staff of qualified Public Health Inspectors (formerly called Sanitary Inspectors). One of the duties of such staff is to ensure that houses are fit for habitation; they must investigate complaints and, if necessary, declare houses unfit. In many cases, however, though unfitness may be declared, there is little action that can be taken; the shortage of housing in many areas is so chronic that really deplorable conditions have to be present, before any action, usually temporary, can be taken—the local authority then has the burden of providing suitable accommodation.

The Housing Act of 1936 contained the first attempts to define overcrowding; minimum standards were laid down; they are summarized here. **A living room** should have a minimum floor space of 50 square feet ($4 \cdot 6\ m^2$); a bathroom, scullery or kitchen is not counted as a living room in this context. The family is counted in units with every person over ten years old having one unit, children from one to ten, half a unit, and babies under one year old are not taken into account. On this basis, the following standards were stipulated:

1. A two-roomed house—three units.

2. A three-roomed house—five units
3. A four-roomed house—seven units, etc.

These standards are obviously minimal, but recent statistics show that a surprisingly high percentage of dwellings are below these standards.

HEATING

To prevent undue loss of heat from the body to its surroundings, especially in the colder parts of the year, it is desirable that all rooms in a home should be maintained at a suitable temperature; a living-room should be between 15°C and 18°C, while bedrooms and passages should be 10°C to 13°C. With old-fashioned fireplaces with open grates, it is impossible to maintain these temperatures throughout the house except at a great cost. It is still a common source of complaint that only one room in a house is warm. It is estimated that 70 per cent to 75 per cent of the heat escapes up the chimney with such fireplaces, also they cannot easily burn smokeless coal products, and hence they increase pollution of the atmosphere. Fortunately, there are many good designs of modern grates available which will burn smokeless fuels. Many houses of older types can be adapted to use slow combustion stoves with back boilers; much of the heat which formerly escaped is used to provide domestic hot water and possibly a few radiators.

The great majority of new houses built during the last ten years have some form of **central heating.** The power source depends partly on the local available supplies, partly on the cost, and partly on the preferences of the builder (or other person who is financing the project). In many rural areas there is no gas supply, but there is usually electricity. Coal and oil are universally available. In towns and most villages, all forms of power are usually supplied. Each form—solid fuel, oil, gas, electricity, has advantages and disadvantages; if used sensibly, all can provide adequate space and water heating. It is the concern of the individual householder to decide on such questions as initial cost of installation, annual cost of fuel, amount of maintenance and attention needed, etc.

Open fires or heaters, irrespective of their source of heat, solid fuel, electricity, gas or oil, can constitute a serious hazard, especially if there are young children in the home. They must be provided with adequate **guards,** of a type not easily removed by children. Gas taps should be out of the reach of the young, or of a safety type not easily manipulated.

Considerable loss of heat takes place through the glass of windows. **Double glazing** reduces this loss to a great extent and it also provides good **sound insulation.** To reduce loss of heat through ceilings into a loft, **lagging** with cheap and non-inflammable materials such as glass wool will prove effective.

Ventilation

With all systems of heating, ventilation is an important factor. The composition of the air in an inhabited room changes; the rate of change depends on the number of people, the temperature and the ventilation. If a large number of people are present in a warm room with inadequate ventilation, the air soon becomes saturated with water vapour, depleted of oxygen, while the temperature and the amount of carbon dioxide will rise. Vapour-saturated air prevents further loss of water vapour from the lungs, and to cope with the situation there is increased perspiration which cannot evaporate under the prevailing conditions. Discomfort soon becomes apparent; people become hot and drowsy and fainting may occur. Normally, doors and window-frames are never absolutely draught-proof and sufficient air may enter through narrow apertures; it may be necessary to have perforated air bricks or metal grilles. Many excellent, but expensive, heating systems provide for constant circulation of "fresh air" which is filtered and heated before distribution. Many old buildings, including some schools (condemned long ago but not yet replaced), lack proper heating and ventilation. Finally, every room in a house should be thoroughly ventilated for some part of the day, by opening windows widely.

Lighting

To ensure adequate lighting of a room in full daylight, it is officially recommended that the **window area** should be not less than one-third of the total wall area, and that windows should be present on at least two sides of the room. Quick and approximate measurement of most living rooms will convince anyone that this recommendation is more honoured in the breach than the observance. In many of the picturesque cottage-type dwellings, now so popular, the window area is less than one-tenth of the wall area.

Artificial lighting should be adequate for the purposes for which a room is to be used. Ideally, a living room should have several grades of lighting; a main light and smaller shelf or desk or wall-lights. A room 4 m × 3 m requires a single 150 watt bulb for adequate overall lighting, whereas a movable 60-watt lamp will be adequate for most close work. To use a main light when watching television, or when sitting and chatting, is obviously wasteful. It is perhaps instructive to notice that needlework with dark materials needs twice as much light as reading; a watch-repairer would require twice as much again.

Adequate lighting for all purposes is important; there is less likelihood of headache, eye-strain and bad posture.

CLEANLINESS

The person or persons in charge of the home should endeavour to ensure that there are adequate facilities for washing, bathing and cleaning teeth, and that careful attention is given to washing and airing clothes, particularly underclothes. With modern detergents, washing machines and launderettes, there is no excuse for dirty clothing; laundrywork is by no means the wearisome chore it was in the first quarter of this century and before. Crockery and cutlery need special attention because of the possibility of unremoved food particles carrying dangerous organisms.

Food particles such as crumbs of bread, on floors or furniture, and exposed food on accessible shelves or cupboards, inevitably lead to infestation by house-flies, blow-flies, cockroaches, mice and rats. All of these can carry pathogenic microbes which they may transfer to the food, and hence to human beings.

Water-closets, sinks and drains need special attention; persistent putrid smells always indicate clearly that bacterial decomposition of organic material is taking place. Hence, it is important that a high standard of cleanliness is maintained by using good disinfectants frequently and by cleaning out drain-traps and sink-traps. With regard to W.C.s, it is unfortunate that quite often the only W.C. in the house is in the bathroom. Thus, emergencies with respect to defaecation and urination can arise. However, the site in the bathroom does at least allow of washing of the hands after defaecation. Ideally, a W.C. should be in a separate room, with a small wash-basin and disposable paper towels also at hand.

Cleanliness of windows will allow penetration of more light and the cleaning process will remove excreta of flies and other insects; excreta (faeces + urine in insects) is always a source of infection. For the same reason, cleanliness of all internal surfaces is desirable.

Food Hygiene

This is a very important aspect of health in the home and it is unfortunate that there is a great deal of ignorance about it. There are several aspects to consider: the purchase of food; its treatment, if any, before eating; the short-term preservation of food not entirely consumed at meal-times, and the disposal of food remains.

In highly developed countries, food on sale is, in most cases, quite safe; public health inspectors make regular routine checks, especially of meat, fish, cream and milk. Any products which fail to reach the required standard must be compulsorily withdrawn from shops. Articles which are prepacked and sealed in glass jars, cans, plastic or polythene containers are, in almost all cases, perfectly safe with regard to their microbial content. Greater caution is required in purchasing meat and fish which have been exposed to the air for long periods, though this is rare nowadays. Any foodstuff which has a putrid smell, however faint, is a source of potential danger.

Many meat and fish products such as the various potted pastes, pies, fish fingers and fish cakes normally require no cooking, but some have to be warmed. They should be eaten soon after the packages are opened; recent laws compel manufacturers to state, on food containers, the date of packaging. Fresh meat and fish require thorough cooking **Boiling** will kill most bacteria but not many types of bacterial spores, some of which can remain viable after many hours of boiling at 100°C. **Baking, frying** and **roasting** need temperatures above 140°C, which is high enough to kill all forms of microbes.

Trouble in the form of food-poisoning stems largely from remnants of some food materials left after meals and exposed to the air. Such materials can be infected by microbial spores in the air or by those carried by various insects. Keeping such left-over foodstuffs in a warm place will greatly encourage the multiplication of bacteria. For example, exposed meat is not really fit to eat after three or four days in winter; the safe period is considerably less in summer. **Refrigeration** or even **freezing** will merely inhibit the multiplication of microbes; neither will kill them. The best practice with cooked meat or fish, and even with fresh meat or fish, which are to be kept for a while, is to wrap the foodstuff in greaseproof paper or a sheet of polythene, immediately the required amount has been consumed. In simple terms, carve the meat which is required, and then cover it; if no more is required, wrap it carefully and place it in a refrigerator or other cold place.

Fuller accounts of food poisoning and food preservation are given in *Microbes* in this series. Food poisoning may range from comparatively mild attacks of vomiting to severe abdominal pain, violent vomiting and diarrhoea, and in extreme cases to death. The causative agents are bacteria of the genera *Staphylococcus, Clostridium Salmonella* and *Escherichia.* In any particular case, the microbes may have been present on the foodstuffs at the time of purchase; this is rather unlikely nowadays. More commonly the organisms have come from the air, from the hands, clothing, hair, etc. of persons handling the food, or from insects such as flies. Hence, it is of the utmost importance that high standards of cleanliness should be maintained in all those persons who handle food. The most serious, and often fatal, attacks of food-poisoning are caused by bacteria of the species *Clostridium botulinum*; they produce highly poisonous by-products. More uncommon is food-poisoning due to chemical contamination of food, as, for example, remnants of poisonous sprays on fresh fruits and vegetables. All of these should be washed thoroughly before they are eaten. Another possible source of food-poisoning is excess of chemical preservatives.

All food remains are suitable breeding places for bacteria. Indeed, any food substance buried in the garden will eventually decay completely due to the attacks of various micro-organisms. Such remains can, of course, be fed as scraps to domestic animals, or put outside for the birds, provided that they will be completely consumed. If this is not possible, the materials should be wrapped in paper and placed in a dustbin with a secure lid; these precautions will prevent further contamination. If none of these is possible, then burning is the best method of getting rid of such potential sources of disease.

Food Preservation

The problems of preserving food in times of plenty, to last through times of scarcity, has exercised the minds of men since prehistoric times. By slow degrees, such methods as drying, salting and smoking came into existence, coupled with the preservation of fruit and vegetable juices in many forms of wine-making. It was not until the work of **Louis Pasteur** (1822–1895), which made the causes of decay evident, that any real progress, beyond trial and error, was possible. Decay is caused by microbes, i.e., very small living creatures ranging from viruses, through bacteria and fungi, to very small animals of the phylum Protozoa. Hence, any process which will kill microbes or render them inactive, is suitable for food preservation, provided the food is not so damaged in the process as to become inedible.

The principal methods of preserving food are summarized below; some are industrial processes, but a number of these have their counterpart in the home.

1. **Drying.** Microbes, like all living creatures, cannot grow or multiply if there is insufficient water present, hence drying of suitable foodstuffs has been employed for a very long time. Even today, in parts of Africa, strips of lean meat are dried in the sun; locally the material is

called **biltong.** It can be softened by soaking in water and cooked normally. Raisins, currants and sultanas will keep indefinitely after proper drying. In the 1939–1945 war, dried egg powder was used in this country in vast quantities, and nowadays we have dehydrated milk and potato and several other comestibles.

2. **Sterilization.** Various forms of treatment will kill all microbes; provided the foodstuffs thus treated are packed so that no type of microbe can penetrate, they will remain sterile. This prevention of contact with microbes is known as **aseptic storage.** The most common form of sterilization is by using **heat**; a common application is in the **canning** industry where the food material in tinned-steel cans is heated to the required temperature, sealed, and then allowed to cool slowly. Sterilization by **radiation,** using ultra-violet or X-rays, is simpler because the food material is not cooked; it is also very rapid, and is probably destined to become of great importance in the future.

3. **Smoking** is essentially heat treatment at a lower temperature. Microbes on the outside of the material are killed by poisonous substances such as phenols present in the smoke. The food material will keep well, provided the outer surface is not damaged afterwards. Such are kippers, smoked haddock and smoked bacon.

4. **Refrigeration** preserves foodstuffs because the microbes are incapable of multiplying at low temperatures; it is important to note that it does not kill the microbes—hence frozen meat and fish may already contain bacteria and soon after they are thawed, multiplication of the bacteria begins, hence cooking soon after thawing is essential. Practically all our meat and fish are frozen, e.g., New Zealand lamb, Argentine beef, Arctic cod. An important adjunct to every butcher's shop is the cold store. Nowadays there are techniques for freezing a vast range of foodstuffs apart from meat and fish, e.g., fruits and vegetables, all kinds of pastry products, and even complete ready-cooked meals. A **freezer** will eventually become an indispensable article of household equipment.

5. **Chemical preservatives** are still used on a large scale. All pickled products are preserved in **vinegar,** usually with the addition of

various spices; they include onions, cabbage, walnuts, cauliflower, gherkins. All wines are essentially fruit or vegetable juices preserved in **ethyl alcohol**; the juices give the taste and the alcohol the "strength." Further addition of **sulphurous acid** preserves those drinks such as light wines and beers, where the alcohol content is too low. Other substances used, especially for meat or meat products, are **benzoic acid, boric acid** or **borates,** and **salicylic acid**; these are added in such small quantities that they do not affect the taste, but inhibit the growth of bacteria.

6. **Salting,** essentially a chemical method of preservation, is ancient enough to deserve a separate mention. Common salt, which is sodium chloride, if in sufficient strength, will prevent any bacterial activity. Hence, for many centuries salt fish, salt pork and beef have been common articles of diet. Salted butter will keep better than unsalted.

7. **Conversion** of important constituents of a natural food is also of long standing. Milk is notoriously difficult to keep under normal temperatures and hence arose the practices of butter and cheese-making. They will both keep much longer than milk, but not indefinitely.

Accidents in the Home

Public health authorities are becoming very concerned about the increase in numbers of avoidable accidents which occur in people's homes. Several reasons are advanced to account for the increase. Firstly, a great many mothers with young children go out to work, so that children, often too young for such responsibility, are left unsupervised at home. Secondly, there has been a large increase in the numbers and kinds of domestic appliances, some of which are potential sources of danger.

The commoner kinds of accidents range from simple cuts and bruises, to scalding and burning, often to a serious extent, and even to fractures by slipping on highly polished floors. One of the most serious dangers nowadays is that of electrocution.

The authors realize, from experience, that it is very easy to give advice, but it is not always easy to carry out all the recommendations that can be given. Accidents often happen quickly, "while one's back is turned."

First of all, young children should not be left unsupervised in the home. The perils may be numerous; exposed fires, insecure fire-guards, gas and electric cookers, the temptation to boil a kettle or to cook something, switches and power plugs; all are possible sources of accidents. The remedies are obvious. Faulty switches, plugs and sockets are particularly dangerous; they should be repaired or renewed without delay. Gas taps have a peculiar fascination for children; the taps should be of the safety type or the gas turned off at the meter and the meter cupboard locked. Polished floors may look very nice, but they are sources of danger to all who walk on them. Medicines of all kinds should be securely locked away.

Parents and children old enough to be responsible should learn the rudiments of First Aid, at least enough to be able to deal competently with the commoner types of accident, e.g., cuts, bruises, scalds and burns. Above all, everyone in the house should know how the doctor, police, and ambulance service in the locality, can be contacted quickly.

THE HAPPY HOME

Most children love their home; it is, or should be, a happy place where they find security, affection, and often a refuge from the in-hospitable world outside. In nearly all cases, similar conditions exist at their school, their church or club. The creation of these conditions is the function of the adults in control; a little caution here will not be amiss.

Though all the physical conditions of good living may be provided in the home, one must remember that a child should develop as a complete all-round person; he (or she) should be encouraged to do things, not always to watch or listen to other people doing them. Excessive use of radio and television is stultifying, especially if there is no special selection of suitable programmes. We all need to be quiet sometimes to do our own thinking and to pursue our own interests and hobbies; the facilities necessary for this should be provided at home.

Public Health Services

The **National Health Service Act** of 1946 had the express purpose of providing **"a comprehensive medical service to secure improvement in the physical and mental health of the people, and the prevention, diagnosis and treatment of illness."** Since the Act became operative (July 1948) the controlling body, formerly the Ministry of Health, has now extended its functions and is called the **Department of Health and Social Security.** It works in close co-operation with the Department of Education and Science, and the Department of the Environment. Each of the three departments is under the overall control of a Minister, who is responsible to Parliament. Governments, policies and ministers change, but the continuation of the work of each department is ensured by a permanent and very experienced staff of civil servants.

The Department of Health and Social Security

Here, we are concerned essentially with the functions of the department with regard to health. These functions are clearly divided into two groups: **Personal Health Services,** concerned with the health of individuals, are controlled directly by the department; **Environmental Health Services** are controlled by Local Authorities which may be County Councils or County Borough Councils (in some large towns or cities). The County Councils delegate some aspects of control to Borough Councils, Urban District Councils or Rural District Councils. Local Authorities have power to levy **rates** which, with capital grants from the Department, provide the money necessary to finance their services. A simplified scheme of the essential organization of the Department is shown on facing page.

The whole system is very well organized, and now with over twenty years' experience, it does a magnificent job, compared with the facilities which existed before 1948. At present, the scope and efficiency of its work is limited, for two reasons. One is the insufficiency of staff at all levels, and the other is shortage of finance. We are particularly short of doctors, nurses and ambulance staff.

The **Personal Health Services** are essentially concerned with the individuals and their ailments. The administration is divided into two sections: **Regional Hospital Boards** and **General Practitioner Services** (medical, dental, etc.).

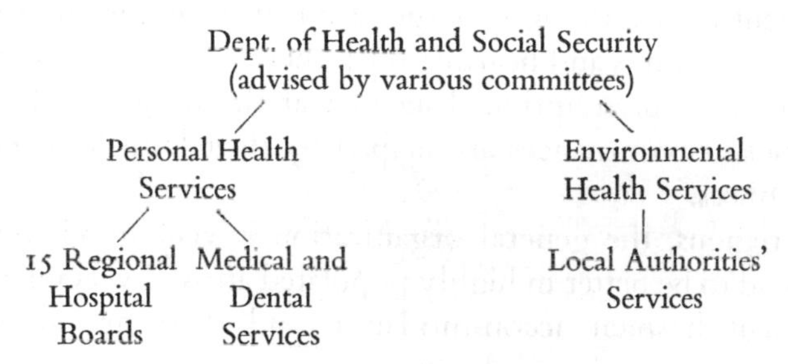

REGIONAL HOSPITAL BOARDS

All the hospitals in the country, with a few exceptions such as those with medical schools (the teaching hospitals), are controlled by regional boards, and for this purpose, England and Wales are divided into 15 regions. The regional boards control facilities throughout their respective areas through **Hospital Group Management Committees,** each of which supervises the day-to-day administration of a small group of hospitals. Each committee delegates the various aspects of its work to sub-committees, each with a limited area of responsibility (*see* Table below).

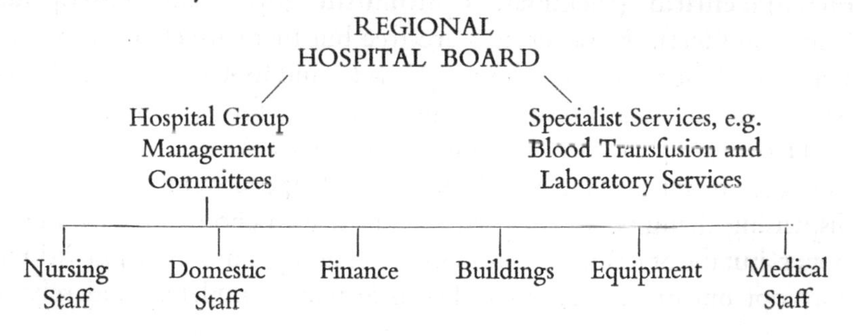

Each regional board is responsible for ensuring that there are sufficient specialist services in its area, e.g., blood transfusion service, X-ray departments, laboratory services, etc. Some of the individual hospitals are too small to provide all services.

Each regional board can provide comprehensive facilities for in-patients and out-patients, sufficient to meet almost any emergency. All treatment is free, though charges are levied for various appliances such as artificial limbs and hearing aids; there is also, at the moment, a charge for every prescription dispensed at the hospitals. Patients in poor financial circumstances are helped by their local Supplementary Benefits Offices.

In all regions, the general organization is very good, but actual facilities tend to be better in highly populated areas. Everywhere, there is insufficient hospital accommodation and there are always staff shortages. Patients whose ailments are not serious enough to cause immediate concern, often have to wait long periods before a hospital bed is available. The position is further aggravated by private patients, who are prepared to pay the whole cost of all facilities and treatment provided. There have been frequent complaints that such patients have priority.

General Practitioner Services

Throughout the country there are 138 **Executive Councils,** each responsible for ensuring that there are sufficient practitioners to supply the following services in their respective areas: **medical, dental, pharmaceutical** (chemists), **ophthalmic** (eyes) and **chiropodal** (hands and feet). Most services are free but there are charges for prescriptions, false teeth and fillings, spectacles and foot appliances. Again, persons unable to afford such charges can apply for assistance.

Here again, the organization is essentially sound, but there are shortages of all general practitioners, with the possible exception of dispensing chemists. Every person is entitled to choose his own practitioner, but the practitioner is not bound to accept him. Any practitioner may opt out of the National Health Service completely and rely on private patients for his livelihood. Most practitioners deal with both private and "National Health" patients; the former may have some priority. For each patient on their panels, the practitioners are paid a standard fee by the local executive council; the chemists collect a standard charge for every prescription and the extra fee is paid by the executive council (some drugs are very expensive).

The authorities are becoming highly perturbed about the disinclination of young people to enter these specialized services. Certainly, the long and arduous training deters many; in some cases the hours of work are long. Also, for those who are shrewd enough, the fields of business and finance offer greater rewards.

Environmental Health Services

These are delegated to the local authorities and are directed largely towards the prevention of disease rather than cure. In many aspects of their work, the local authorities must have close liaison with the hospital boards and the practitioner services. Each local authority is obliged to establish a **Health Committee** consisting of members of the council and co-opted experts; this committee deals with the following aspects of public health:

1. **Ambulance service.**
2. **Vaccination and immunization.**
3. **Health centres**
4. **Maternity and child welfare clinics.**
5. **Health visitor service.**
6. **Home nursing service.**
7. **Midwifery service.**
8. **Domestic home help service.**
9. **Mental health service.**
10. **Prevention of illness in general**; this is mainly the work of the Public Health Inspectors.

The local authorities are responsible for maintaining and where necessary improving standards of hygiene in all aspects of the environment. Their duties and powers are clearly defined in the following Acts of Parliament:

1. **Public Health Act** 1936; this refers to sewage and refuse disposal, water supplies, offensive trades and public nuisances, and the spread of infection.

2. **Housing and Slum Clearance Acts** of 1936 and 1946; these refer to housing standards, overcrowding, and fitness for habitation (*see* p. 365).

3. **Food and Drugs Act** 1955; this gave local authorities powers to examine and test all foodstuffs and drugs, and if necessary to order their withdrawal from shops.

4. **Clean Air Act** 1956, concerned with smoke abatement and smokeless zones.

Sewage Disposal

Sewage consists of all the waste material from houses, schools, factories, etc., which is carried away by water in a system of pipes called sewers. Nowadays the word is used also for any form of disposal of these waste materials whether waterborne or not. Thus sewage may be collected in a **cesspool** or even in **buckets**; the local authority is responsible for collecting the material. In general, sewage consists largely of **human excreta** (faeces and urine), toilet paper, detergents, all water used for domestic purposes and industrial wastes which are waterborne.

The ultimate safe disposal of sewage is one of the great problems of our time, especially in view of the increasing population and of the high degree of pollution caused. Human organic waste material can be effectively decomposed by various microbes; they use parts of it to provide their growth and energy requirements, and transform the remainder into simple chemical substances which can enter natural cycles such as those operative for Nitrogen, Carbon, Sulphur and Phosphorus. Hence we say that excreta is **biodegradable**; in fact, all the materials in sewage are eventually biodegradable except for a few detergents and certain industrial wastes such as cyanides and phenols which are toxic to all living creatures. In very dilute solutions, even these poisons can be rendered harmless by the action of certain microbes. Trouble arises when huge quantities of sewage are deposited in localized areas; they upset the balance of nature and interfere drastically with natural cycles; this is **pollution**.

Disposal of sewage is a big problem for two reasons. Firstly, there has been a vast increase in population, and it is still increasing. Secondly, since the Industrial Revolution, there has been, and still is, a large-scale movement of people from the country districts to the large towns. At one time, say 400 years ago, the majority of houses had a **privy** in the garden, supplied with a seat and a bucket; a heap of soil and a small shovel were conveniently placed, so that the excreta could be covered, thus preventing noxious smells. The material would periodically be buried in the garden where it would decay and add its manurial value to the soil. This is still the practice in many parts of the world and possibly in a few remote households in this country. However, as the population increased rapidly in the towns, houses often had no gardens; in London excreta was simply thrown out into the streets where it stank and bred huge numbers of flies, and hence diseases were rife, especially epidemics of typhoid and cholera. When the rains came the excreta was washed into the Thames which thus became almost an open sewer.

By the end of the seventeenth century, sanitary conditions in towns became intolerable, but by 1860 most towns were supplied with a piped system which discharged the sewage into the sea or into the nearest river. Sea-water kills practically all fresh-water microbes, so that if the sewage is discharged far enough from the shore, the method is perfectly satisfactory. Marine organisms will cause the decay of the sewage into harmless and useful materials. Often, however, it is discharged too near the shore where it causes unpleasant pollution of beaches and disrupts the natural living communities between high and low tide. The legal limit is still just beyond the lowest ebb tide.

Discharge into rivers is a different matter; it depends on the quantity of sewage and the speed of flow of the rivers. If slow-moving, then a great deal of the solid excreta will merely sink and form a layer of evil-smelling ooze which decays very slowly. The **sulphate-reducing bacteria** which are the most active of the microbes under such conditions produce the evil-smelling and poisonous gas, Hydrogen sulphide; **methane bacteria** produce methane (marsh gas). Between them, they can kill almost all forms of plant and animal life, and it is thus that many of our rivers are really nothing but open, black, malodourous sewers. The black mud and water, and even black sand, are caused by the formation of iron sulphide by the action of hydrogen sulphide on the yellowish or reddish iron salts which colour sands. Poisonous industrial wastes increase the pollution. Now, at last, we are becoming concerned about the state of our rivers, and surveys are being made to find the degree of pollution. The next step is to remedy it. Already there are signs that fish, not seen at Tower Bridge for 50 years or more, are beginning to appear.

The best methods of disposal are far out at sea, or by means of **sewage processing works**; the use of rivers on any large scale must gradually be discontinued. A fuller account of sewage processing is given in *Microbes* in this series; here a brief outline is given.

1. The material is passed through **screens** (large filters) which remove the larger solid objects, which are then buried, burnt, or chopped into small pieces which can then pass on to the next process.

2. **Sedimentation Tanks** allow most of the solid matter to settle, forming a **sludge**; the process can be accelerated by the use of several harmless chemicals. The water above the sludge contains very fine suspended sediment; it is filtered through beds of sand or stones at a slow rate controlled by rotating sprayers (these can often be seen in action at sewage works). The beds contain various groups of microbes which can remove all traces of organic material. The water which emerges from these filter beds is almost as "pure" as our drinking water, and it can be discharged into streams or rivers without causing any pollution.

3. The sludge remaining in the tanks is removed to large containers which are heated to 35°C, which is a favourable temperature for microbial action; the material decays with the production of methane,

carbon dioxide and traces of other gases. The methane can be piped off and burnt at jets to provide for the heating process. Finally the material is dried over beds of sand or stones; the final product has no objectionable smell and is useful as a fertilizer. In fact, a great deal is sold for this purpose. Thus the human excreta is finally deposited in the soil, where it should be.

It is unfortunate that these processes are very costly, hence they are mainly used in some highly populated areas. A fine example of such a sewage plant is to be seen at Mogden in West Middlesex. A fully modern plant can salvage much useful material from the process:

1. **Methane,** used for the heating process and any excess can be supplied to the Gas Board (coal gas is 25–35 per cent methane).
2. **Fertilizer.**
3. **Vitamin B_{12}** (*see* p. 110); sewage is the richest source of this valuable vitamin.
4. **Sulphur,** which is becoming a rather scarce raw material.

Such future sewage plants will kill two birds with one stone; they will not add to existing pollution and they will contribute materially towards conservation (*see* p. 396).

Refuse Disposal

Refuse is the assortment of solid materials which are normally deposited in dustbins and collected by local authority refuse vans. It consists of paper, cardboard, glass, tins, plastics, polythene, metallic objects, rags, remnants of food and all sorts of other odds and ends. Some authorities salvage clean paper and cardboard; this is sold to scrap merchants and eventually made into wood pulp which is used to make more paper and cardboard; this makes a small contribution towards the cost of collection. In some large institutions, food scraps are collected in pig bins; the material is collected and used by pig-keepers, after boiling, to fatten their animals. By law, local authorities have to find permissible ways for disposal of refuse; a number of such methods are in use.

1. **Controlled tipping.** The material is deposited on waste land,

marshland, excavations such as old gravel and chalk pits; some authorities use it to raise the level of low-lying land near the seashore. Such tips can become highly insanitary and often have most offensive smells; all the trouble comes from food scraps. Enlightened authorities cover the refuse with soil, as it is tipped; some practice burning as well. In many places such tips can eventually be used for agriculture or sports fields etc; there will certainly be very good drainage.

2. **Burning.** In some parts of the country, all the material is burnt in furnaces at high temperatures with forced air draughts. This has the objection that noxious gases are liberated, though the clinker left after the burning process is useful for road-making.

3. **Separation and Salvage plants.** This is a complex and expensive system, but in view of pollution and conservation it might very well be the system ultimately used in all areas. Firstly, all dust is removed in draught chambers; this is spread on the land. Secondly, there is removal of metals by magnetism since most metal remnants will be of iron or steel; the metals are sold to scrap-merchants and eventually used again in industry. Further sorting processes can extract paper for salvage; bones, used for making glue or bone-meal fertilizer; glass is pulverized to fine powder and spread on the land; the remainder, mainly food scraps, rags etc. are burnt and tipped or spread.

There are two widespread and highly undesirable practices to be noted here. One is **illegal dumping of poisonous chemicals** which might easily percolate into the water supply. In recent years, there has been great concern about **cyanides.** The other is the dumping by individuals of awkward objects such as old cars, prams, bicycles, mattresses, and of smaller objects carelessly and thoughtlessly thrown anywhere, paper, polythene, glass, tins, etc.

WATER SUPPLY

The Department of the Environment is responsible for national water supplies and any large-scale new project must be approved by them. In general, the responsibility for maintenance of supply is delegated to local authorities and in many cases they have their own Water Boards, or they receive water from Water Companies.

For drinking purposes, each person needs an average of 1·3 litres a day, but far greater amounts are used for cooking, washing, flushing lavatories and commonly for watering gardens and washing cars. The total daily average national consumption per head is about 180 litres (40 gals)—this includes all mains water used for industrial purposes. Many industrial concerns based near rivers pump in river water and discharge it into the river after use, often highly polluted.

It is obviously a large undertaking to supply the whole population with sufficient and suitable water (note that pure water is obtainable only by distillation; this is unpalatable for drinking). When one considers that a small hole 1 mm in diameter in a water pipe will cause the loss of more than 1000 l in a day, it is obvious that we should all be careful about dripping taps and all other ways of wasting water.

Sources of water supply are **rivers and streams,** natural or artificial **lakes,** underground water from **wells** and, in rare cases, the **sea.** In general, most homes are nowadays fed by a water supply travelling in pipes, such being known as **mains water supply.** All mains water is subjected to various kinds of treatment, before it is fit for drinking.

1. **Natural purification.** The water from the source is collected by streams or pumped into **primary reservoirs,** or more rarely, large tanks. These reservoirs eventually form large and beautiful lakes. The water is subjected to a **resting period** when solids settle, and exposure to wind, sun, and air, also affects it beneficially. Plant and animal life is encouraged, and much unwanted organic matter is decayed. Fishing is allowed in most reservoirs. Water is pumped out as required to maintain the levels in **secondary reservoirs.** Outgoing water from primary reservoirs is filtered through sand beds; this holds back all but the smallest forms of life; also the filtering beds soon develop a layer of microbes which cause decay of any small portions of organic material. All outgoing water is pumped through a **metering house** which measures the quantity.

2. A pumping station next drives the water through further **filters** and **settling tanks.**

3. The water next passes through a **sterilizing station** for destruction of remaining pathogens (normally **chlorination** is used). Water finally emerges suitable for drinking—it is rigorously tested for its bacterial content.

4. The water is then pumped into the **secondary reservoirs** each of which supplies a particular area. Such reservoirs may be high enough to supply all houses in the area by **gravity feed**; usually the water is pumped to the houses, both to overcome gravity and to keep the tap supply at a fairly constant pressure.

It is certain that water thus supplied to houses is absolutely safe for drinking, unless secondary pollution has occurred after the sterilization process. Careful precautions are taken to guard against this.

It is rarely safe, in this country, to drink natural water from streams or lakes; one must remember that even rain-water is polluted. Small streams in remote mountainous areas may contain water fit to drink, but there is always the possibility of a dead animal or excreta in the water upstream. The safest "natural" water is that from a mountain spring where it emerges on the surface.

Water extracted from the sea by various desalination processes is in use in some parts of the world. One or two experimental schemes have been operated in this country; they have been abandoned temporarily because of the cost of the process.

Water supply authorities have freedom of action to consult local opinion about the advisability of adding carefully measured quantities of **fluorides** to the water. It is claimed, and probably rightly, that much of our water (and diet) lacks fluoride; this might be a contributory cause of bad teeth (*see* p. 132).

Supply authorities are constantly searching for new sources of water; certainly the bulk of the rain-water eventually flows away into the sea. The great problem is providing sites for the large primary reservoirs. No-one wishes to flood agricultural land or destroy a beautiful valley for the sake of a reservoir. Every person immediately affected wants to have it somewhere else, and yet the authorities are bound to ensure sufficient supplies. The final answer may be the use of sea-water and, if we believe in conservation, we must be prepared to pay more water-rate.

Offensive Trades and Public Nuisances

In many trades there are by-products which may be poisons, smoke, dusts, etc. which incite a great deal of protest from local residents. Nowadays such trades or industries would not be allowed to start in residential areas, at least, not without a very exhaustive public inquiry. Sometimes the trouble is with excessive noise such as may occur at an airport; a new airport may be necessary but few people would want it in their own particular area. The same reasoning applies to atomic power stations.

Most of the trouble arises, however, from old-established industries, which were there before houses aggregated around them. For example, a coal mining site would be discovered and then rows of miners' cottages were built close to it. This had to be, because there were no reasonable means of transport to and from work. Similarly, gasworks, iron-works, breweries, vinegar factories, soap factories and tanyards, etc. are now incongruously surrounded by the dwelling-houses. No committee or any other power can order the demolition of such a factory, without providing for a suitable site and compensating the aggrieved owner. However, all complaints are investigated, and if substantiated, various measures are suggested: complete removal, or precautions to lessen the nuisance. For example, smoke from a factory can be almost entirely eliminated by modern heating using smokeless coal, or by using gas or electricity. Thus, in many industrial cities, smokeless zones have been created.

On a smaller scale, a fish-and-chip shop may be a great convenience to some but a great nuisance to those who have to tolerate perpetual smoky and grease-laden pollution of the local air.

"Public nuisance" is a term which can have very wide implications but it does not include offensive trades.

Some examples are given here.

1. Excessive and irritating noise from a dwelling-house or meeting hall.

2. Persistent disturbance of private property by irritating practices such as throwing stones; by neighbours' dogs or other animals; by diverting surface drainage water into a neighbour's garden, etc.

3. Persistent production of bad smells in a public or private place, e.g., keeping pigs or poultry too close to dense housing.

4. Burning rubbish in undue quantities and at inconvenient times.

5. Urination and defaecation in unsuitable places.

6. Growing hedges, trees or bushes which interfere with a neighbour's light.

7. Any wilful damage to public property.

8. Continuous use of unsuppressed electrical appliances which interfere with radio and television reception.

9. Keeping dangerous animals ranging from dogs to lions.

All the above examples are based on actual cases brought before magistrates' courts.

If any person considers any practice is a "public nuisance" he can report it to the Local Authority, who will investigate it, and if necessary take measures to prevent or alleviate the trouble.

Spread of Infection

Infectious diseases are all caused by pathogens which enter the body through one of its apertures or through the skin. Each local authority has a duty to do all in its power to prevent the spread of infection particularly of certain dangerous diseases. This duty is carried out by the Medical Officer and his staff, and by the Public Health Inspectors. Every person, every family and the community in general, all have a moral duty to help in this work; they are frequently reminded of precautions by notices, pamphlets, lectures and posters. It is wise to remember that various hygienic measures at all levels are designed to help in preventing the spread of infection. A complete list of infectious diseases would be long indeed; some of the most important, with the method of spread, are given below.

1. **Sewage leakage** or unsatisfactory sewage disposal may cause pathogens to get into drinking water and into some foods. Typhoid, paratyphoid, dysentery, cholera, gastro-enteritis and summer diarrhoea may all be spread widely from this cause.

2. **Insects,** especially flies, can carry the pathogens of all the above diseases, particularly if they breed and feed on discarded food or on faeces. They may also carry the poliomyelitis virus. Lice can spread typhus fever.

3. **Contaminated food.** Milk can spread bovine tuberculosis, typhoid, diphtheria, cholera and dysentery. Since milk is mainly distributed from large centres, it is very difficult to locate which particular batch is contaminated, and whether the contamination occurred at the farm or subsequently. Infected **meat** can spread food poisoning, dysentery, and various worm diseases. Other forms of food rarely carry or spread infective organisms. One interesting exception was a typhoid outbreak due to a particular bed of watercress; really the water was contaminated.

4. **Human contact.** Boils, chicken-pox, smallpox, German measles, measles, syphilis, gonorrhoea, impetigo, mumps, pneumonia, ringworm, scabies, scarlet fever, tuberculosis, warts, whooping-cough, colds, influenza, etc. Some of these are transmitted by **skin contact** and some by **droplet infection.**

5. **Rats** and other vermin may carry bacteria causing food poisoning and deposit them on human food. Rat fleas carry the pathogens causing **plague** (the last British occurrence was in Suffolk in 1918).

6. **Soil, dust** and **dirt** can carry bacterial spores causing anthrax, tetanus and gangrene; they normally enter through skin wounds.

Obviously to limit the spread of infection is a huge task involving the co-operation of various specialists. First of all the disease has to be identified; if it belongs to the group of **notifiable diseases,** it must be reported to the Medical Officer of Health for the locality. The main notifiable diseases are: poliomyelitis, meningitis, cholera, diphtheria, dysentery, erysipelas (skin infection), sleeping sickness, malaria, paratyphoid, plague, pneumonia, scarlet fever, smallpox and tuberculosis. These are particularly important since their rate of spread is rapid and the diseases are serious, and hence the patients must be isolated, and questioned, and the source of infection found as quickly as possible.

A great deal of good work to prevent the spread of infection is carried out by those responsible for testing water supplies, those in control of sewage disposal and by the public health inspectors who regularly check many foodstuffs.

All families and individuals can help in these matters by attention to all the details of personal and family hygiene.

Food and Drug Inspection

All foodstuffs, all drinks and all drugs supplied to the public are required to reach standards which will make them fit for human consumption, and to prevent cheating of the public. The dangers associated with milk and all meat products have been emphasized previously. Packaged foodstuffs can be tested for weight, composition, and examined for date of packaging. All alcoholic drinks are manufactured to contain alcohol of certain minimum and maximum concentrations; the vendors must comply with those limits.

All medicines and drugs (pills, tablets, capsules, ointments) must conform to British Pharmacopoeia standards. As with all foods and drinks, samples are withdrawn from shops by public health inspectors and their contents examined by public analysts. If they do not come up to the required standards, the vendors can be prosecuted.

Clean Air

Pollution of the atmosphere really began when primitive men first discovered the use of fire. Since that remote time, the discovery of coal and oil as fuels, together with the increase of population and the rapid rise of industry, have led to another serious problem of our time—the damage caused by polluted air.

There are three main, and several subsidiary, causes of the pollution.

1. The incomplete combustion of coal in industry and in households.
2. The combustion of petrol, diesel oil, gas and other similar materials in cars and lorries, in factories and houses.
3. The production of industrial dusts in many industries.
4. Natural decay of organic material—normally only causes objectionable smells.

The burning of coal in industry produces large quantities of **dust**, especially soot (carbon) particles, **poisonous gases** such as **sulphur dioxide** and **carbon monoxide**, and **tars** of varying composition. The use of oils as fuels is somewhat less dangerous; petrol used in various types of engines releases through the exhaust systems considerable amounts of carbon monoxide, which is highly poisonous because when it is inhaled and absorbed into the blood, it combines with haemoglobin to form **carboxyhaemoglobin,** and the capacity to absorb oxygen is reduced. Deaths from carbon monoxide poisoning can occur quite easily in cars if the exhaust system leaks into the body of the car, and in closed garages where the engine is kept working. Having no smell, the gas is not easily detected, and since its first effects are simply yawning and drowsiness, fatal accidents can easily occur. The victim falls asleep and dies peacefully.

Sulphur dioxide is also poisonous; when liberated after burning coal or diesel oil, it dissolves in the droplets of moisture in the atmosphere forming **sulphurous acid.** This acid

$$H_2O + SO_2 = H_2SO_3$$

is not very stable and it is oxidized in air to form the dangerous **sulphuric acid**;

$$2H_2SO_3 + O_2 = 2H_2SO_4$$

this not only corrodes stone and ironwork, but also attacks the lungs.

Soot and other dusts blacken buildings, roads and almost everything in the industrial landscape. Some industrial dusts are lethal; such are silica (from coal processing plants and stone-crushing), and asbestos, largely used as a heat insulator. The former causes **silicosis,** the latter **asbestos poisoning.**

These and other air pollutants have, in the past, made industrial areas lethal to plants and animals and have shortened many human lives. When atmospheric conditions are correct, i.e., the atmosphere still and cold, with a large amount of water vapour, all these gases and dusts remain in the atmosphere, colouring the normal white fog from yellow to almost black. Such were the great **smogs** in London (1952 and 1956); the former is said to have caused 4000 deaths of elderly people and of patients with respiratory diseases. These lethal smogs led to great public pressure on the government, and the **Clean Air Act** was passed in 1956.

This act made a number of recommendations, and local authorities were given powers to impose local by-laws which would ensure cleaner air in their own districts. The main recommendations of the act were:

1. The use of smokeless fuels.
2. Increased use of gas and electricity in industry.
3. Power stations burning coal to be gradually converted to use oil; also increased development of nuclear power stations.
4. Electrification of the railway (some of the large termini, e.g., King's Cross and Paddington, were described as "smoky, sulphurous caverns").
5. The establishment of smokeless zones, e.g., London, 1955, Manchester 1952, and others.
6. Extraction of grit, dust and soot from furnace smokes.
7. Government grants to enable these provisions to be implemented.

There is no doubt that atmospheric pollution has been considerably reduced by application of these measures but there is still a great deal to be done. The vast increase in numbers of motor vehicles may cause, in the end, greater pollution than all industrial smokes. There is no need for this; if sufficient pressure were exerted, the car industry could soon find measures of reducing pollution from exhaust gases. Cars would cost more, but if we really want clean air, and reduction of respiratory diseases, then we must pay for it. A great part of the extra cost would be saved in less illness, less cleaning of large buildings, and though less substantial, there would be much more sunlight penetrating the cleaner atmosphere.

The field of **public health** and **hygiene** is enormous. It is highly organized and competently administered; the avowed object is to improve health and decrease the burden of disease. Every individual has a part to play in this field, not only in his own hygienic practices, but in his role as a member of a community. The greatest single factor in the increase of population in civilized countries has undoubtedly been the steady improvement in public health services and hygienic practices. The average expectancy of life (at birth) has risen from 48 years in 1900 to almost 70 years now in Great Britain.

Chapter 19

POPULATION, POLLUTION AND CONSERVATION

For the last twenty years or so there has been increasing concern among scientists and others about the future of the human race. This concern has arisen because we are aware of three factors on which human survival depends. They are:

1. Increase of population.
2. Increasing pollution of our planet.
3. Necessity for conservation.

Population

At the time of the birth of Christ, we estimate that the total human population was about 250 million, mainly living in the East. By the year A.D. 1600 the total was approximately 500 million, i.e., it had doubled in 1600 years. In 1850, the figure had doubled again to 1000 million; in this case the doubling had taken 250 years. The next doubling took about 85 years bringing the population to 2000 million in 1935. Now, in 1972, there are more than 3500 million people alive, and at the present rate of growth, by 1980 it will have reached 4000 million. In the year 2000, the human species will reach the staggering total of 6000 million.

Our own population in Great Britain was 55 million in 1971 and it is still increasing. At the present rate of increase, in the year 2000, there will be 66 million. Experts maintain that for the maintenance of our high standard of living, the number ought to be reduced to 30 million.

It should be quite obvious that this rate of increase cannot continue; if it does, then by the most optimistic calculations, the total resources of our planet will be consumed in less than one thousand years. If one can imagine such an absurdity, we could arrive at a state where the whole land surface will be covered with people standing tightly packed. At present, at least 60 per cent of the population are undernourished, particularly in protein. This fact has been known for many years and tremendous efforts have been made by governments, by charitable organizations and, most of all, by the United Nations Food and Agriculture Organization, to remedy this sad state of affairs. All these efforts have made very little impact on the problem. Vast sums of money have been spent on improving agriculture in backward countries, and on better distribution of food supplies from the wealthy nations, but progress in overcoming undernourishment is intensely disappointing. Even if all possible agricultural land were fully utilized, and this would cost fantastic amounts of money, the present population could not be adequately nourished.

When we inquire into the reasons for the "population explosion" we find that they are comparatively simple. The birth-rate (i.e., the number of births per 1000 of the population) is very variable, but **infant mortality is decreasing,** and **people are living longer.** In simpler terms, it is not that more are being born, but that fewer are dying. To put the position more dramatically, on a world scale, **every second, four are born and two die.** If we compare the figures for infant mortality (i.e., before the age of five), we can see a very significant difference between 1900 and 1950.

Country	Deaths per 1000 babies born	
	1900	1950
India	232	116
Great Britain	154	30
U.S.A.	122	29

At the other end of the scale, we have increasing average length of life (*see* p. 40). In 1961, there were 5 750 000 people over the age of 65 in England and Wales; by 1971, there were about 6 500 000. At the same rate of increase there will be 7 500 000 in 1981.

It is a strange contradiction that some of our greatest achievements, in preserving life, in lessening disease, and in establishing better conditions of living, are the very factors which have brought us to our present serious problem of over-population.

The solution to this problem is exercising the experts in all governments, and most of all, in the United Nations Organization. There is obviously no easy solution which would be universally acceptable. The longer we defer the problem, the worse it will become. The size of a family is in many cases a matter of tradition, and sometimes of religion. Some parents want a large family; in some parts of the world, fathers pray to their gods for "many sons." **If we could persuade people that two children per family is ideal, the problem would be solved.** Even if we educate people about the urgency of the problem, we still have to find measures for limiting the family to two children. There are many methods of contraception in use nowadays but they do not appear to have made a great deal of difference. There are strong movements which advocate free, legalized abortion; religious and moral considerations often forbid this. Sterilization of fathers is widely practised, but the percentage willing to undergo this operation is extremely small. The opposite proposition, sterilization of all women after their second child, would solve the problem; there are again very strong arguments against this. Perhaps cessation of payments, such as family allowances, and of income tax rebates, would discourage some parents but certainly not all and, in any case, poorer parents seem to have the most children. Possibly governments could pay people annual allowances for not having more than two children!

The problem is one for clear-headed discussion at all levels, from the family, to the government and to the whole world. There are arguments for and against every suggestion that could be made. One thing should be made quite clear to everyone; something must be done about over-population, and it must be done soon.

Pollution and Conservation

There is no possible doubt **that we are now actively destroying the environment in which we have evolved** by pollution and by lack of conservation. A great deal of the pollution process, by sewage and refuse, has been described in Chapter 18. When any material is discarded in a position where it interferes with natural living populations, it constitutes pollution of the environment. Thus rivers can often become open sewers, and by a continuation of the same process, eventually all the seas will become the same. Only certain microbial forms of life would be able to survive; indeed that is already the case in some of our rivers. To aggravate this poisoning process, we practise large-scale arable cultivation, we return no humus to the soil and hence have to rely on artificial fertilizers to promote growth, and poisonous sprays to reduce insect and fungus pests. These fertilizers and poisons largely find their way eventually into rivers and hence to the sea, again aggravating pollution. These practices on the soil lead to its erosion, so that it becomes structureless dust which is blown or washed away. This **soil erosion** is again a world-wide problem.

Undoubtedly the main cause of pollution is a financial one; it is cheaper to dump our rubbish and poisons somewhere else. By that means we have got rid of our problem and we are not worried that soon it is someone else's problem. All petrols have **lead tetraethyl** added; this undoubtedly increases the efficiency of the engines, but eventually the combustion process liberates lead compounds into the atmosphere; winds spread the poison all over the world; it can even be detected in the snows of the Arctic. Eventually the level will become lethal to human beings. To do without the compound would mean increased cost of motoring. There are scores of similar examples that could be cited.

The opposite of pollution is **conservation** which means preservation from destructive influences. There is an important movement afoot to conserve the natural environment as far as possible. Hence we have areas for nature conservancy set aside so that natural populations of animals and plants can be maintained. There is also a growing movement for the restoration of areas which have been laid waste by industry. These movements do not merely promote conservation so that future generations shall be able to see wild animals and plants, though that in itself is important. The fact is that when we interfere with natural groupings of living creatures, we cause unforeseen consequences. The study of **ecology** concerns living creatures in their natural environments. When we study any ecological grouping in even a small area, we find that it is very complex. Each species, plant or animal, forms part of an interconnected network, each species relying on some other species for food. When we change the conditions in such an area, we change the grouping, often with grave consequences. Thus, if we destroy the large carnivores in Africa, the herbivores will flourish exceedingly for a while, and eventually could destroy all the herbage and thus exterminate themselves.

Conservation in its entirety involves a knowledge of world-wide ecology so that we can ascertain the effects of man and his pollution on the whole range of living creatures. We are far from that complete knowledge yet, but piece by piece, it is accumulating.

Population, pollution and conservation are the concern of everyone. We should all be educated in the essentials of these three topics which relate to our own survival. It is becoming evident that we shall have to rethink our ways of living so that ultimately the human race may continue as a smaller but happier population which will nurture, not destroy its ecological environment. We must consider carefully the ultimate value of some of our great scientific achievements, with especial reference to radioactive fall-out from atomic explosions and from nuclear power stations. The problems are for the experts to solve; this they can do if there is enough public pressure to demand these solutions.

Appendix

SLIDE PREPARATION

Instructions for simple preparations are given here; lists of chemicals and equipment follow.

(*a*) Smear preparation of cells from the inside of the human cheek.

(i) Scrape cells from the cheek, with some saliva, and smear **thinly** in the middle of the slide.

(ii) Dry in air and as soon as the surface appears to be dry, add 70 per cent alcohol, to kill and fix the cells. (2 minutes)

(iii) Drain off the alcohol and cover the area with **haematoxylin.** Agitate occasionally. (5 minutes)

(iv) Wash off excess stain with 70 per cent alcohol and blue in tap water or alkaline alcohol.

(v) Examine under low power of microscope. The ideal condition is to have bright blue nuclei and practically unstained cytoplasm.

(vi) If too dark, stain can be removed with acid alcohol. This is rapid and should be watched under the microscope. Wash off the acid alcohol with 70 per cent alcohol when the desired effect is obtained. If acid alcohol has been used, blue again in alkaline alcohol and wash off with 70 per cent alcohol.

(vii) Add 90 per cent alcohol. (2 minutes)

(viii) Counterstain red, if desired, in **eosin.** 90% alcoholic. (3 minutes)

(ix) Wash fairly quickly in absolute alcohol.

(x) Add xylol (xylene); if milky in appearance there is still water present; dehydrate again in absolute alcohol. The xylol clears the preparation as well as testing for the presence of water and is a solvent for balsam.

(xi) Mount in Canada balsam; do not use until quite dry, except for a cursory examination of the final effort.

(xii) Label.

(*b*) Smear preparations of blood

Use blood from a freshly-killed rat, or prick the thumb between the base of the nail and the first joint. Sterilize a needle in a flame, and the skin by wiping with alcohol.

(i) Place a drop of blood near the middle of the slide and smear by drawing the edge of another slide over it.

(ii) Dry over an electric lamp.

(iii) Stain for 5 minutes in Leishman's or Wright's blood stain.

(iv) Wash in water; shake off excess water.

(v) Dry over an electric lamp.

(vi) Mount in Canada balsam.

(*c*) Teased preparation of striped muscle.

Cut a small piece of muscle from the leg or arm of a freshly-killed rat. Size of piece should be about 1 cm long and one-quarter of the thickness of a match-stick. Tease as fine as possible on a slide in 70 per cent alcohol, using a pair of sharp needles (*see* Fig. 183). It is best to leave one end intact for handling; this can be cut off when

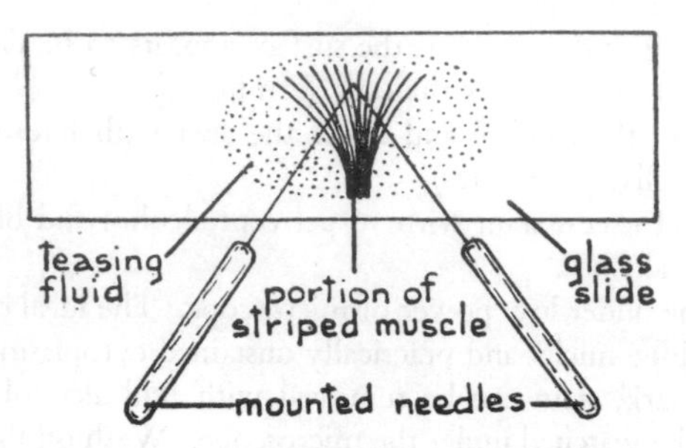

FIG. 183. Teasing Striated Muscle

clearing in xylol. Other technique as in (*a*). The same method can be used for teased preparation of nerve, using brachial or sciatic nerve.

(*d*) Stretch preparation of mesentery.

Use mesentery of a freshly-killed rat. Flatten a cover-slip against a stretched portion of mesentery and with the fingers press the mesentery to the cover-slip so that it adheres. Cut round the cover-slip with scissors.

Stain and carry out all operations as in (*a*), using the whole cover-slip.

For connective tissue, cut out a portion when skinning a freshly-killed rat. Smooth and stretch it on a slide, using the blunt sides of two scalpels.

(*e*) Sections of cartilage.

Using a section razor or safety-razor blade in a holder, cut sections of cartilage from the sternal ribs or xiphisternum of a freshly-killed rat. Examine in 70 per cent alcohol and proceed as in (*a*), using two or three of the thinnest sections.

Draining Liquids from a Slide

Always drain off to the left into a container, e.g. a beaker, and do not flood the slide with any of the reagents used.

Mounting

Try to estimate the amount of balsam needed, i.e. one, two, or three drops. Good estimation will come with practice. Slide the cover-slip so that its lower edge makes contact with the balsam, then lower the cover-slip slowly with a needle. Do not press it on the specimen but let it settle under its own weight. (*See* Fig. 184.)

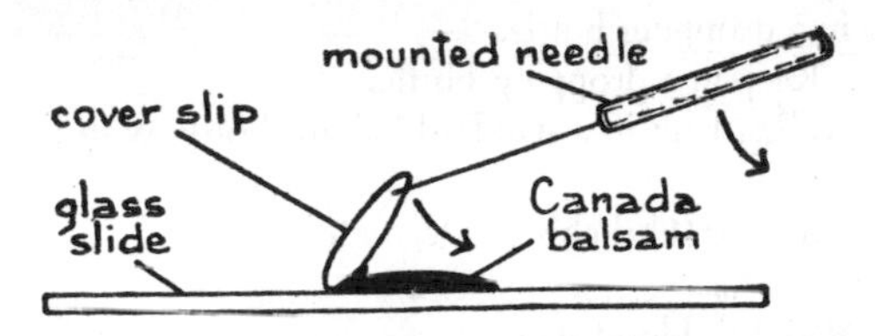

FIG. 184. Lowering the Cover-slip on the Balsam

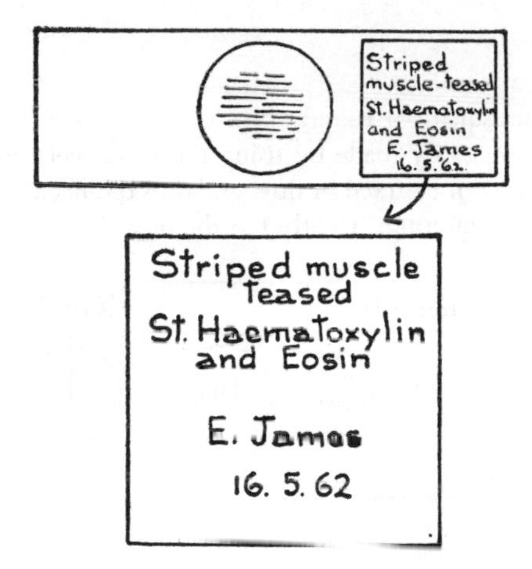

FIG. 185. Labelling of a Slide

Labelling

Stick label on firmly and pencil faintly the initials of the owner. When thoroughly dry, write in ink the name of the specimen, the stains used, the name of the owner, and the date. (*See* Fig. 185.)

Chemicals

Formalin 4 per cent solution. (Dilute commercial formalin twenty-four times with water.) Useful for storage of specimens or organs.

Alcohol. 70 per cent, 90 per cent, and absolute. It is best kept in dropping bottles so that only small quantities are used.

Acid alcohol: 100 cm^3 of 70 per cent alcohol + 0·5 cm^3 of concentrated hydrochloric acid. Keep in a dropping bottle.

Alkaline alcohol: 99·5 cm^3 of 90 per cent alcohol + 0·5 cm^3 of 880 ammonium hydroxide. Keep in a dropping bottle.

Xylene (xylol). Keep in a dropping bottle.

Canada balsam is best kept in a standard balsam bottle with glass cover and glass rod.

Haematoxylin, aqueous (Ehrlich's or Delafield's).

Eosin, alcoholic. 90%.

Leishman's or Wright's blood stain.

Keep all stains in dropping bottles for use.

Equipment

Slides, 75 mm × 25 mm

Cover-slips, 20 mm square or round. No. 1.

Mounted needles are easily made by using 6 mm dowel rod, sawn into 75 mm lengths. Sewing needles can be fixed in thus. Clamp the needle in a vice with about 13 mm of the eye end projecting. Gently tap the dowel rod on to the needle.

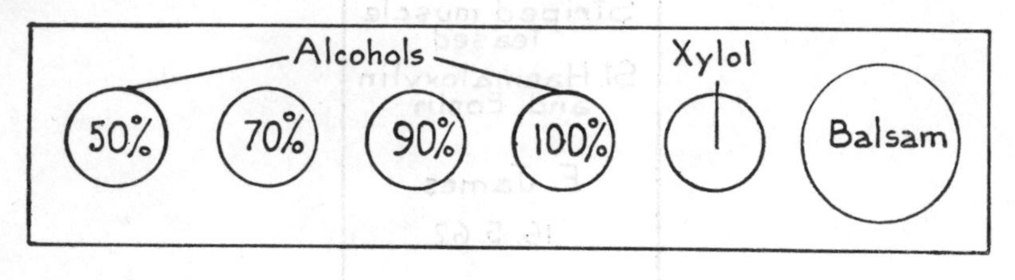

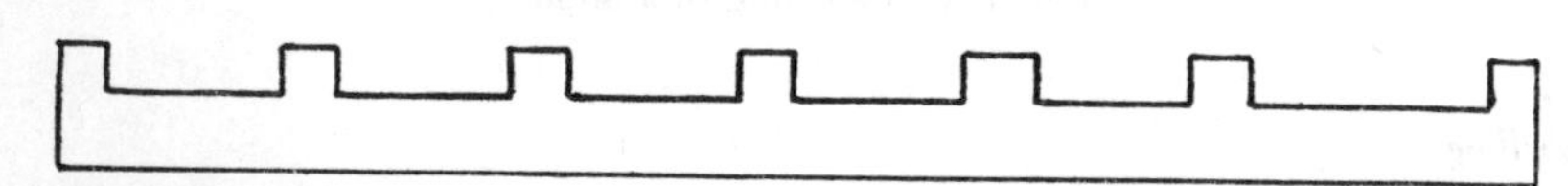

FIG. 186. Bottle Rack in Plan and Section

Section razor or razor-blade holder and blades.

Camel hair paint-brush, small, for lifting small pieces of tissue.

Slide labels, square, plain.

Racks for dropping bottles. In a 75 mm × 25 mm strip of the length required, drill 13 mm deep holes with brace and bit, to fit dropping bottles. (*See* Fig. 186.)

Recommended for information on solutions, reagents, and techniques: *Elementary Microtechnique*, by H. Alan Peacock, published by Edward Arnold and Co.

INDEX

ABDOMEN, 44–5, **46**
Abnormalities, inherited, 325–6
Abscess, 184
Absorption of food, 151–2
Accidents, 374
Accommodation, eye, 258–9
Acetabulum, **75**, 76
Acetylcholine, 269, 301
Acid, deoxyribonucleic, 101, 305, 307
  fatty, 104
  lactic, 154, 171
  nucleic, 101, 305, 307
  pyruvic, 171
  stomach, 149
Acidity, 147
Action, reflex, 277, **284**, 285
Adaptation, dark, 257
Addison's disease, 296
Adenine, 305, 307
Adenosine triphosphate (ATP), 99–100
Adiposity, 111
Adrenalin, 291
Afterbirth, 243, 248
Air, complemental, 166
  exhaled, 172
  inhaled, 172
  pollution, 390
  reserve, 166
  residual, 166
  sac, 163, 168
  tidal, 166
Alanine, 107
Albino, 32, 316–17, 361
Alcohol, 364, 374
Alimentary canal, 127–51, 200
Alkalinity, 147
Allantois, 239–40
Allergy, 232, 329
Alveoli, **163**, **168**
Amino-acids, 106–7, 149, **150**, 152–3
Amnion, 239, **240**, **241**, **242**
Amoeba, **5**, **219**
Amylopsin, 149
Anaemia, 109–10, 181, 315, 321
Androgens, 295, 299
Angina, cruris, 332
  pectoris, 332
Anthropoidea, 19
Antibodies, 178, 184–5, 296
Antigen, 184–5
Antitoxins, 150
Anxiety state, 337
Aorta, **197**, **199**, **201**, **204**
Apertures, body, 256–9
Appendicitis, 143
Appendix, **46**, 142
Area/Volume ratio, 28–9
Arteries, 192, **193**, **194**
Arteriosclerosis, 332
Astigmatism 260–1
Atheroma, 332
ATP, 99–100, 108, 153–4, 157, 170–1
Auricles, 196, **197**, 198, **199**, 200, **201**
Australiform race, **33**, **35**, **36**
Australopithecus, **22**, 23
Axis cylinder, **269**, 270, **271**
Axon, 267, **268**, 269

BACH family, 318
Bacillus, tubercle, 122
Bacteria, 132, 183, 338, 352, 354, 382
Behaviour, 283–7
Beri-beri, 110
Bile, 149, 217
Birth, 245, **246**, **247**
Bladder, 45, **46**, **212**, 216–17
Blastocyst, 233, **234**, **235**, **236**
Blister, 52
Blood, 178–92, **180**
  count, **181**, 182–3
  functions, 191–2
  glucose, 179, 296–7
  groups, 188–9, **190**
  ions, 179
  platelets, **180**, 186
  serum, 185–6, **187**
  smear, preparation, 327
  transfusion, 188–91
  velocity, 195
  vessels, 192–6, **193**, **204**, **205**
Body, ciliary, 252
Boil, 184
Bone, **58**, **59**
Bones, middle ear, **262**, **263**
  sesamoid, 65
Brain, **43**, **272**, **273**, **274**, **275**, **276**, **277**, **278**
  cavities, **274**

Brains, vertebrates, **276**
volume, 29–30
Breathing, **94, 163, 164, 165**
capacity, 166, 168
rate, 165
Bronchus, 44, **45, 162**

CAECUM, **142**, 143
Calcium, 58, 108, 132, 294
Camera and eye, 256, **257**
Canal, alimentary **43, 45, 46**, 127–46, **139, 142**
Haversian, **58**, 59–60
semicircular, **262**, 263, **264**, 265
Cancer, 340–1
Capacity, vital, 166
Capillaries, **168, 169, 170, 183**, 194–5, **206**
Capsule, Bowman's, 213, **214, 215**
Capsule, synovial, **63**
Carbohydrates, 100–3, 116
Carbon dioxide, in air, 172
blood, 179
Carbonic anhydrase, 179
Carcinogens, 340
Carcinoma, 340
Cardiac infarction, 332
Carnivora, 17, **130**, 131
Cartilage, **57**, 58, **59**, 160, **161, 162, 163**
bones, 59, **60**
Catalysts, 146
Cataract, 260–1
Cavity, buccal, **43**, 127, **128**
pericardial, **198**
pulp, **129**
Cell, 3, **4**
division, 307
Cellulose, 102, 344
Cerebellum, **273, 274**, 275, **276, 278**
Cerebrum, **273, 274, 275, 276**
Cetaceans, 17
Chain, autonomic, **281, 283**
Cheilosis, 110
Chiroptera, 17
Chlorophyll, 2
Chordata, 11, 14, **15**
Chorion, **240, 241, 242**
Choroid, **252, 256**
Chromosomes, 220–1, 305–10
Chyle, 206
Chyme, 140
Cilia, 158, **159**, 161–2
Circulation, general, 198, **199**, 200, **201, 204, 205**
lymphatic, 152, **170, 206, 207, 208**
pulmonary, **199**, 200, **201**
systemic, **199**, 200, **201**
Classification, 11, 14, 26
Cleanliness, 352–5, 369
Cleavage, 233, **234**
Clefts, visceral, 14–15
Clitoris, **224**, 225
Clothing, 355
Clotting, 178, 186, **187**
Coccyx, **70**
Cochlea, **262, 264**, 265, **266**
Coelenterate, **7, 219**
Coelom, **237, 238, 240, 241**
Coenocyte, 4, **5**, 79
Colon, **142**, 143
Colostrum, 249
Column, vertebral, 67, **70, 71**, 72
Columns, spongy, **229**
Condyles, occipital, 67, **69**
Cones, retina, 256, **257**
Confusion, 325
Conjunctiva, **252**, 253
Conjunctivitis, 253
Conservation, 397
Constipation, 217, 344
Contraction, muscle, 80, 82–3
Cooking, 120, 339, 371
Cord, spermatic, 227
spinal, 278–9, **280, 281**
umbilical, **241, 242**, 243, **246**, 247
Cornea, **252**
Corpus albicans, **223**, 226
luteum, **223**, 226, 293
Corpuscles, platelets, **180**, 186–7
red, 168–70, **180**, 181–2
white, **180, 183**, 184
Cortex, kidney, **213, 214**
suprarenal, 294, **295**
Corticosteroids, 295
Cosmetics, 359
Coughing, 175
Creatures, non-cellular, 5
Cretin, 109, 293
Cro-Magnon man, 25
Crying, 175
Crypts of Lieberkuhn, 140, **141**
Cycle, Krebs', 171
menstrual, 225–6
Cytoplasm, 3
Cytosine, 305

DARWIN family, 318–19
Decay, dental, 109
Defaecation, 146, 344
Deficiency, mental 334
Dementia, senile, 333
Dendrites, 267, **268**

Dentine, **129**
Dermis, 49, **50**, **51**, 52, **53**
Diabetes, 291, 297
Diaphragm, 44, **45**, **46**, **162**, 163–4, **165**
Diaphysis, **59**
Diet, 97, 111–21
Digestion, 144–51
  carbohydrate, 148–50
  chemical, 146–51
  fat, 149
  physical, 144–6
  protein, 149–50
Dipeptide, 107
Disaccharides, 101
Discs, intervertebral, **62**
  slipped, 62, 328, 347
Disease, bacterial, 338
  congenital, 326
  deficiency, 109–10, 327
  fungus, 338
  growth, 340
  notifiable, 389
  Parkinson's, 333
  venereal, 230–2, 326, 362
  virus, 338
  worm, 338
Drugs, 330, 364
Duct, bile, **46**, **140**, 141
  ejaculatory, **227**, 288–9
  endolymphatic, **264**, 265
  pancreatic, **46**, **140**, **141**
Duodenum, 140, **141**
Dystrophy, muscular, 321

EAR, 159, **160**, 261, **262**, **263**, **264**, 265, **266**
Ectoderm formation, 235, **236**, **237**, **238**
Edentates, 17
Effectors, 251
Embryo, development, 233–48
  growth, 47–8, 244–5
  monthly, 244–5
Emulsification, 145
Endoderm formation, 235, **236**, **237**, **238**
Endolymph, 265, **266**
Energy, 157
  of foodstuffs, 103
  relationships, 99
  requirements, 112–14
Enterokinase, 148
Enuresis, 217
Environment, external, 250
  Dept. of, 379–93
  internal, 109, 250
  intracellular, 250
Enzymes, 146–51, 179–87
Enzymes, (*contd.*)—
  activation, 148, 301
  properties, 146–7
  secretion, 148, 300
Epidermis, **49**, **50**, **51**
Epididymis, **227**, **228**
Epiglottis, 58, **136**
Epimysium 79, **80**
Epiphysis, **59**, **60**, 61
Epithelium, germinal, ovary, 221–2, **223**
  testis, 227, **228**
Eugenics, 323
Europiform race, **33**, **34**, 35
Evolution 9, 320–3
  human, 37, 323
  social, 37, 323
Evolutionary tree, 8, 9
Excretion, gut, 217
  kidney, **214**, **215**, 216
  lung, 168, **169**
  skin, 54
Exercise, 78, 347
Exhalation, 164
Eye, 251–61
  ball, **252**, 253
  brows, 253, **254**
  colour, 317, 320
  defects, 260, **261**
  lashes, 253, **254**
  muscles, 254, **255**

FACTORS, releasing, 291
Family, 11
Fat, 103–5
  storage, 52, 152
Fatigue, 95, 154, 171, 348
Fertilization, 221–3, 230, 309
Fibrin, 187
Filaments, muscle, 79, **80**, **81**
Filtrate, kidney, 214
Flavours, 159
Fluid, cerebrospinal, **272**, **274**, 280, 282
  coelomic, 164
  seminal, 229–30
  synovial, **63**
Fluorides, 58, 132, 386
Foetus, **241**, **242**, **243**, 244–5, **246**, **247**
Folds, vocal, 160, **161**
Follicle cells, 222, **223**
  Graafian, 222, **223**
  hair, **50**, 51
Food, 100–11
  absorption, 121, 144, **151**, 152
  composition, 123
  digestibility, 121

Food (*contd.*)—
digestion, 144–51
energy value, 7–14
hygiene, 343, 370
inspection, 290
packaging, 371
preservation, 370–4
psychology, 122
variety, 121
Foot, man and ape, 27
Foramen, 67
nagnum, 67, **69**
obturator, **75**
Foreskin, **229**, 230
Formulae, dental, 131–2
Fructose, 101

GALACTOSE, 101, 149
Gametes, 220, 222, **223**, **228**, 298, 306, 309
Gamma globulin, 178, 184
Gangrene, senile, 333
Gases, exchange, 168, **169**, 172
poisonous, 330
Gastric juice, 149, **154**
pits, **139**
Gastrin, 300
Genes, 220–1, 306–21
Genus, 11
Germ, 183
layers, 235, **236**, **237**, **238**
Girdle, pectoral, **74**, 75
pelvic, 75–6
Glands, 289
Brunner's, 140, **141**
Cowper's, **227**, 229
endocrine, 289–300
lachrymal (tear), 253, **254**
lymphatic, **207**, **208**, 209
parathyroid, 294
pineal, 290, 297
pituitary, 277, **278**, 289–93
prostate, **227**, 229
salivary, **43**, 134, **135**
sebaceous, **50**, **51**
suprarenal, **290**, 294–6
sweat, **50**, **51**
thymus, 44, **45**, 208, 296
thyroid, **43**, 44, 109, **290**, 293
Glenoid, 74
Glomerulus, **214**, **215**
Glottis, **136**
Glucose, 99–101, 296–7
Glycerol, 104
Glycine, 107
Glycogen, 102, 153
Glycolysis, 171
Goitre, 109, 293
Gonorrhoea, 231, 326, 362
Grasping, 92, **93**
Growth, benign, 340
malignant, 340
skulls, 61
Guanine, 305, 307

HABITATION, fitness for, 306–7
Habits, 286, 363
Haemocytometer, **181**, 182
Haemoglobin, 169, 180–1
Haemophilia, 187, 319, 321
Haemorrhage, 328, 333
Haemorrhoids, 344
Hair, **50**, **51**, **52**, 53
straight, **32**
wavy, **32**
woolly, 31, **32**
Hay fever, 329
Head, **43**, 44
Health, 343
public, 376–93
Heart, 196, **197**, **198**, **199**, 200, **201**, **202**, 203
circulation, 198, **199**, **200**, **201**
control, 201, **202**, 203
failure, 332
pacemaker, 83, **202**
Heating, 367
Height, 39–40
inheritance, 317–18
Hemispheres, cerebral, **273**, **274**, **275**
Hiccoughing, 175
Hilum (kidney), 212
Histamine, 184, 301
Hominidae, 21
Homo, 24
races, 33–6
sapiens, 26, **27**, **28**, 29–30, 33–6
Hormones, 289–303
alimentary, 300
female, 226, 293, 295, 298
male, 228, 293, 295, 299
unbalance, 337
Housing Acts, 364–7
Humour, aqueous, **252**, 253
vitreous, **252**, 253
Hybrid, 311
Hydra, **7**, **219**
Hymen, 225
Hypermetropia, 260, **261**
Hysteria, 336

ILEUM, 140, **141**

Ilium, **75**, 76
Image formation, 255–6, **257**
Immunity, 184–5
Implantation, embryo, 233, **235**
Impulse, nerve, 270, **271**
Index, cephalic, 31, **32**
Indole, 142–3
Infection, 337–9, 388
Inflammation, 184
Inhalation, 163
Inheritance, blending, 312
  blood groups, 314–15
  colour-blindness, 314, 320
  height in peas, 311–12
  human, 313–20
  intelligence, 318–19
  polygenic, 312, 316–19
  Rhesus factor, 315
  sex-linked, 319–20
  skin colour, 316–17
  tongue, curling, 313–14
Injury, physical 327
Innervation, reciprocal, 85, 195
Insanity, 335
Insectivora, 17
Insight, 286
Instinct, 286
Insulin, 107, 296
Intelligence, 318–19
Intercourse, sexual, 230
Intestine, large, 140, **142**, 143
  small, 140, **141**
Invertase, 149
Iodine, 109, 293
Ionization, 108
Iris, **252**, 253, **254**, 317
Ischium, **75**

JEJUNUM, 140
Joints, 61–5

KERATIN, 49
Ketonuria, 115, 297
Ketosis, 115, 297
Kidney, **45**, 46, 211–16, 299
Knobs, synaptic, **268**, 269
Knot, embryonic, 233, **234**
Krebs' cycle, 171

LABYRINTH, bony, **262**, 263, **264**, 265
  membranous, **262**, 263, **264**, 265
Lacteal, **141**, 152
Lactose, 101–2, 149
Lamprey, 15, **16**
Lancelets, 15, **16**
Langerhans, islands, **290**, 296
Laryngopharynx, **136**
Larynx, **44**, **136**, 160, **161**, **162**
Laughing, 174
Leg, **65**, **76**
  bird, **96**
Lens and camera, 256, **257**
Leucopenia, 186
Leucotoxin, 184
Leukaemia, 186
Leukocytosis, 186
Lever, 88
Ligament, capsular, **63**
  falciform, 143
  suspensory, **252**, 253, 258–9
Lighting, 368
Limbs, 47, **76**, 77
  pentadactyl, 76–7
Lipase, 149
Liver, 45, **46**, 148, **151**, 152–4, 203
Longevity, 40, **41**
Loop of Henlé, 213, **214**, 215
Lymph, 170, 179, 195
Lymphocytes, **180**, 183, **208**, 209, 296

MALE/FEMALE ratio, 41
Maltase, 149
Maltose, 101–2, 149
Mammals, 11, 16
Mania, 336
Marrow, 59, 180
Marsupials, 17
Mastication, 144
Mater, dura, 271, **272**
  pia, **272**, 280, 282
Matrix, bone, 58
  cartilage, **57**
Matter, grey, 271, **272**, 279, **280**
  white, 271, **272**, 279, **280**
Meconium, 248
Mediastinum, **165**
Medulla, brain, **273**, **274**, **276**, 277, **278**
  suprarenal, **290**, 294–5
Meiosis, 300, 307, **308**
Melanin, 50, 316
Membrane, arachnoid, **272**, **282**
  extraembryonic, 239, **240**, **241**, **242**
  foetal, 239, **240**, **241**, **242**, **246**, **247**
  nictitating, 253
  plasma, 3, **4**
  tympanic, **262**, **263**
Mendel, 304, 311
Meninges, 271, **272**, 280, 282
Meningitis, 271
Menopause, 230

Mesentery, 143, **144**
Mesocephalic, 31, **32**
Mesoderm formation, **236, 237, 238**
Metabolism, 211–12
Microbes, 337–9
Micromeasurement, 3
Mid-brain, **273, 274,** 277
Migration, human, 36
Milk, 122–5
  bacteria, 122
  composition, 124
  digestion, 149
  pasteurized, 124
  proteins, 122
  salts, 122
  sterilized, 124
  T.T., 124
  uterine, 233
Mitosis, 307–8
Mongoliform, 33, **34,** 36
Mongolism, 316
Monosaccharides, 100–1, 149, 151
Monotremes, 16
Mucin, 158
Mucus, 158
Mulatto, 310–11
Mumps, 135
Muscle, antagonistic, **86, 87**
  biceps, 85, **86, 87**
  cardiac, 83
  extensor, 85
  eyeball, 254, **255**
  facial, **20**
  fatigue, 95
  fibres, 79, **80, 81**
  fixation, 87
  flexor, 85
  group action, 87–8
  hair erector, **50,** 52
  nerve endings, **84**
  origin, 85, **86**
  prime mover, 27
  stomach, 138, **139**
  striated, 79, **80, 81**
  synergist, 88
  triceps, 85, **86, 87, 89**
  unstriated, **81, 82**
Mutation, 312–2
Myopia, 260, **261**
Myxoedema, 293

NASO-PHARYNX, 135, **136**
Navel, 247
Neanderthal, 24, **25**
Neck, **43,** 44
Negriform, **33,** 34
Neisseria gonorrhoeae, 231
Nephrons, 213, **214, 215**
Nerve, **269,** 270, **271**
  constrictor, 55, 195
  cranial, **278,** 279
  dilator, 55, 195
  endings, 51, **52, 84,** 85
  mixed, 270
  motor, 270
  parasympathetic, 84–5, 203, **281,** 282, **283**
  plexus, 280, **281, 283**
  sensory, 269–70
  spinal, **280, 281**
  sympathetic, 84–5, **203, 281,** 282, **283**
Neurons, 267, **268, 269**
Neurosecretion, 289, **292,** 300
Node, auriculo-ventricular, **202**
Nor-adrenalin, 294
Nose shape, **32**
Notochord, 11, 14, **15, 62, 237, 238**
Nucleus, 3, 299–300, 305–9
Nuisance, public, 387

OBESITY, 111
Odontoblasts, **129**
Odours, 158, 253
Oesophagus, **43,** 44, **128, 136,** 137, **138**
Oestrogens, 295, 298
Omentum, great, 143, **144**
  lesser, 143
Opalina, **5**
Orbit (eye socket), **68, 69,** 253
Order, **11**
Organizers, 301
Organs, **7**
  of Corti, 265, **266**
  sense, 251–67
Origin, human, 36
Oropharynx, **136**
Osmoregulation, 215, 291, 295, 300
Ossification, 59, **60,** 61
  at birth, **56**
  embryo, 244–5
Osteitis fibrosa, 294
Ova, 220, 222, **223**
Ovaries, 45, **46,** 221, **222, 223, 224, 290,** 298
Oviducts, 45, **46, 222,** 223, **224**
Ovulation, 222–4
Oxyhaemoglobin, 169, 180–1

PAIN, 52, **53,** 301
  labour, 247
Palate, **128**
Pancreas, 45, **46,** 140, **141,** 149, 296

Passage, respiratory, **43**, 158, **159**, 160, **161**, **162**, **163**
Patella, **65**, 66
Pellagra, 110
Pelvis, **75**, **76**
  kidney, 212, **213**
Penicillin, 321
Penis, **227**, **229**, 230
Pepsin, 148–9, **150**
Peptide, 107
Peptones, 149, **150**
Perception, 250
Pericardium, **198**
Perichondrium, **57**
Perilymph, 265, **266**
Periosteum, **59**, **60**, **129**, 130
Peristalsis, 81–2, 137
Peritoneum, 138, **139**, **141**
Peritonitis, 143
Peyer's patches, 209
pH, 147
  blood, 178
  enzymes, 148–9
  urine, 215
Phagocytes, **180**, 183–4
Pharynx, **43**, 135, **136**, 137, **159**, **160**
Phenyl-thio-urea, 313
Pheromones, 302
Phylum, 11–12, 14–15
Pigments, bile, 149, 181, 217
  iris, 317
  retina, 256–8
  skin, 32, **50**, 316–17
Pinna, 261–2
Pitch, 265–7
Pituitary, **290**, 291, **292**
Placenta, 235, 241, **242**, **243**, 248, 299
Plague, 339
Plasma, 178–80
  membrane, 3
Pleistocene period, 22, **23**, 24
Pleura, **165**
Pleurisy, 165
Pleurococcus, **5**
Point, far, 259
  near, 259
Poisoning, 329–331
Pollution, 380, 390, 396
Polypeptide, 107, 149, **150**
Polysaccharides, 102–3
Pons, **273**, **274**, **275**, 276–7, **278**
Population, control, 396
  increase, 394
Porifera, **6**
Posture, 77–8, 89, **90**, 91, 344–7
Presbyopia, 259
Primates, 17, **18**, **19**
Problems, human, 342
Proboscidea, 17
Process, odontoid, **64**, 65, **71**, 72
  turbinate, 158, **159**
Profile, **19**, **32**
Progesterone, 298–9
Prolactin, 248, 293
Proportions, of body, **48**
  man/ape, **27**
Proteins, 105–8, 114–15, 149–50
  blood, 178
Pro-thrombin, 187
Protoplasm, 2, 105
Psychoneurosis, 336
Psychopath, 336
Ptyalin, 148, **150**
Puberty, 222, 227–8, 298–9
Pubis, **75**
Pulse, 193
Pupil, **252**, 254
Pus, 184–5
Pyramids, kidney, 212, **213**, **214**

QUICKENING, 244
Quotient, respiratory, 172

RACES, human, 31, **32**, **33**, 34–7
Radiations, 328
Rate, breathing, 165–6
  heart-beat, 196–7
Reaction, reversible, 100, 104, 107
Receptors, 251
  skin, **53**
Recreation, 360
Refraction, 255
Refuse, disposal, 383
Relaxin, 298
Rennin, 149
Reproduction, 219–20, 362–3
Respiration, artificial, 172–3, **174**
  cell, 157
  external, 157
  muscle, 154
  tissue, 170–1
Rete testis, **228**
Retina, **252**, 253, 255–6, **257**
Rhesus factor, 190–1, 315
Ribs, **73**, **74**
Rock formations, **10**
Rodents, 17
Rods, retina, 256, **257**, 258
Roughage, 102, 334

SAC, scrotal, **227, 228,** 245
  yolk, **236, 237, 238,** 239, **240**
Sacculus, 262, **264,** 265
Saline, physiological, 181
Saliva, 135, 148–9
Salts, 108–9, 118
Scab, 187
Scapula, **74,** 75
Scar, 187
Schizophrenia, 335
Sclerotic, **252,** 254
Scurvy, 110
Sea squirts, 15, **16**
Secretin, 300
Segmentation rhythmical, 140, **141**
Selection, natural, 323
Senility, 331
Senses, 251
Sewage, 380
Sex determination, 220
Sighing, 175
Sight testing, 259
Silica, 330
Silicosis, 330
Skatole, 142–3
Skeleton, **66,** 77
Skin, **49, 50, 51,** 52, **53,** 54–5
Skull, fish, 67, **68**
  human, 65, 67, **68, 69**
  man and ape, **28**
Sleep, 349–52
Slide, preparation, 399–402
Sneezing, 175
Sniffing, 176
Snoring, 176
Sobbing, 176
Sodium chloride, 109
Sound, frequency, 265–7
  intensity, 161, 267
  pitch, 161, 265–7
  quality (timbre), 161, 267
Species, 11
Sperm, 220–1, **223,** 227, **228,** 229, 306
Sphincters, gut, 127–8, 137, **139, 142,** 143
  precapillary, 195, **196**
Spirochaetes, 231
Spirometer, **167,** 168
Spleen, **46,** 208
Sponges, **6**
Spot, blind, **252,** 253
  yellow, **252,** 256
Stalk, body, 238, **240, 241**
Starch, 100, 102, 116–17, 148–9
Sterility, 110
Sternum, 72, **73, 74**
Stimulation, parasympathetic, 84–5, 282, 301
  sympathetic, 84–5, 282, 301
Stimulus, 250
Stomach, **46,** 138, **139,** 149
Stroke, 333
Struggle for existence, 322
Stye, 253
Subrace, Alpine, 34
  Mediterranean, 34
  Nordic, 34
Substance, intercellular, 4
Success, human, 342
Suckling, 248
Sucrose, 100–2, 149
Sugars, 100–2
Sulcus, **272,** 273
Sunlight, 361
Suture, **62**
Swallowing, 136, **145**
Sweat, cold, 55
  composition, 54
  gland, 50
Syphilis, 231, 326, 362
System, 9
  arterial, **204**
  autonomic, 83–4, **281,** 282, **283**
  lymphatic, 203, **206,** 207, **208,** 209
  urinary, 211, **212, 213, 214, 215,** 216–17
  venous, **205**

TAIL, human, **70**
  true, 15
Taste, sense, 133, **134**
Tastes, 134
Tears, 254
Teeth, **129, 130,** 131–2
  milk, **130,** 131
  permanent, **130,** 131
  preservation, 132–3
  V.S., **129**
Temperature regulation, 54–5, 294
Tendon, 85, **86,** 255
Tentacle, 7
Testes, **46,** 226, **227, 228**
Thalassemia, 315
Thalidomide, 326
Threads, protoplasmic, **4**
Thrombin, 186
Thrombocytes, **180,** 186–7
Thrombosis, 187, 333
Thrombus, 187
Thymine, 305, 307
Thyroxin, 109, 291
Tissue, 7
  connective, **50,** 51

Tissue (*contd.*)—
fluid, 203
lymphoid, **208**, 209
muscular, 79, **80**, **81**, **82**, 83
nervous, 267, **268**, **269**
Tongue, **43**, **128**, 133, **134**
Tonsils, 133, **134**, **136**, 208
Toxins, 183
Trachea, 44, **45**, **161**, **162**
Trades, offensive, 387
Transfusion, blood, 188–91
Tree of life, 275, **276**
Treponema pallidum, 231
Triolein, 104
Trophoblast, 233, **234**, **235**, **236**, **237**, **240**, **241**
Trypsin, 149–50
Tube, Eustachian, 58, **136**, **159**, **160**, **262**
Fallopian, **222**, 223
neural, **238**
Tuberculosis, 321
Tubules, renal, 213, **214**, **215**
testis, 227, **228**
Twins, identical, 309

UMBILICUS, 247
Ungulates, 17
Urea, 54, 152–3, 179, 215
Ureter, 45, **46**, **212**, **213**, 216
Urethra, 45, **46**, **212**, **224**, **227**
Urination, 217
Urine, 153, 215
Uterus, **46**, **224**, 233, **235**, **236**, 242, **243**, **246**, **247**, 248
Utilization, food, 152–4
Utriculus, **262**, **264**, 265
Uvula, **128**, 245

VACCINATION, 185–6
Vaccine, 185
Vagina, **46**, **224**, 225, 230, **246**, **247**
Valve, bicuspid, 198, **199**, 200, **201**
Eustachian, **160**, 262
Valve (*contd.*)—
ileo-caecal, 127–8, 140, **142**
sphincter, 127–8
tricuspid, 198, **199**, 200, **201**
Variation, 322, 324
Vas deferens, **227**, **228**, 229
Vein, **193**, 194
coronary, **197**, 203
hepatic portal, **151**, **205**
mesenteric, **151**
umbilical, 242, **243**
valves, **194**
varicose, 194
Venae cavae, 193, **197**, **199**, 200, **201**
Ventilation, 368
Ventricles, 196, **197**, 198, **199**, 200, **201**, **202**
Vertebrae, 67, **70**, **71**, 72
Vertebrata, 11, 15
Villus, 140, **141**, 151
trophoblastic, **235**, **236**, **240**, **241**, **242**
Virus, 135, 338
Viscera, 44, **45**, **46**
Vision, 255–6, **257**, **258**, 259–61
Vitamins, 98, 110–11, 117, 120
Vomiting, 137
Vulva, 45, **46**, **224**, 225, **247**

WATER, importance, 118–19
supply, 384
Waters, breaking of, 246
Waves, peristaltic, 81–2, 145–6
Weaning, 249
Weight, 39–40
body components, 42
Window, oval, 262, **263**, **264**, 265
round, 262, **263**, **264**, 265
Work, 99, 157
Worm, acorn, 15, **16**

YAWNING, 175

Zygote, 221, 233, 309